Advances in Neuroethics

Advances in neuroscience research are bringing to the forefront major benefits and ethical challenges for medicine and society. The ethical concerns related to patients with mental health and neurological conditions, as well as emerging social and philosophical problems created by advances in neuroscience, neurology and neurotechnology are addressed by a specialized and interdisciplinary field called neuroethics. As neuroscience rapidly evolves, there is a need to define how society ought to move forward with respect to an ever growing range of issues. The ethical, legal and social ramifications of neuroscience, neurotechnology and neurology for research, patient care, and public health are diverse and far-reaching — and are only beginning to be understood. In this context, the book series "Advances in Neuroethics" addresses how advances in brain sciences can be attended to for the benefit of patients and society at large. Members of the international editorial board for this series are:

More information about this series can be found at: http://www.springer.com/series/14360

Eric Racine • John Aspler
Editors

Debates About Neuroethics

Perspectives on Its Development, Focus, and Future

Springer

Editors
Eric Racine
IRCM, Université de Montréal,
and McGill University
Montréal, Québec
Canada

John Aspler
IRCM and McGill University
Montréal, Québec
Canada

Advances in Neuroethics
ISBN 978-3-319-54650-6 ISBN 978-3-319-54651-3 (eBook)
DOI 10.1007/978-3-319-54651-3

Library of Congress Control Number: 2017939127

Printed on acid-free paper

This Springer imprint is published by Springer Nature
The registered company is Springer International Publishing AG
The registered company address is: Gewerbestrasse 11, 6330 Cham, Switzerland

Acknowledgments

As editors, we are tremendously grateful to all our colleagues who generously accepted to take the time to write for this volume. We know how busy academics are nowadays, and we appreciate their willingness to engage in a series of debates which we hope will have a lasting impact on the field. We wish to thank the staff at Springer, notably Dr. Sylvana Freyberg and Ms. Madona Samuel, who have been highly supportive of this project since the beginning. Sincere thanks to Drs. Veljko Dubljević, Fabrice Jotterand, and Ralf Jox, co-editors of the series Advances in Neuroethics, for their confidence in this project. Finally, we wish to extend our thanks to all members of the Neuroethics Research Unit in Montreal for their constructive feedback on different aspects of this project. Special thanks to Ms. Dearbhail Bracken-Roche for her help and feedback, and to Sonja Chu for editorial assistance. Eric Racine acknowledges support from a career award from the Fonds de Recherche du Québec—Santé, and John Aspler receives support in the form of a doctoral award from the same funding agency.

Contents

Introduction

Twenty-first-century advances in neuroscience have generated novel insights into human behavior and cognition, which in turn raise profound ethical, societal, and philosophical questions. For example, neuroscience studies about decision-making and agency may contradict common paradigms through which moral phenomena are typically understood (Decety 2015; Levy 2007). Some neuroscientists have even claimed that society needs to radically overhaul its conception of "free will," going so far as to claim that free will is an illusion or nonexistent (Greene and Cohen 2004; Haggard 2011; Harris 2012; Wegner 2002). Not without controversy and serious methodological questions (Mele 2014), much of this discussion was prompted by studies suggesting that the actual level of voluntary control human beings have over their actions may be more limited than previously imagined (Libet et al. 1982, 1983) or that implicit cognitive processes may thwart conscious and deliberate choice (Bargh and Chartrand 1999).

The development and use of novel neurotechnologies based on new insights from neuroscience (or generated by sheer serendipity) also call for serious ethical analysis. Here, the list of important developments is long and significant: (1) invasive forms of neurostimulation involving neurosurgical procedures such as deep brain stimulation (Bell et al. 2009); (2) noninvasive forms of neurostimulation that can, in some cases, be developed at home (e.g., do it yourself transcranial direct current stimulation) (Cohen Kadosh et al. 2012; Wexler 2016); (3) imaging technologies for understanding brain function (e.g., functional magnetic resonance imaging (fMRI)) (Poldrack and Farah 2015); (4) various diagnostic procedures for ailments such as Alzheimer's disease or Huntington's disease for which no cures are in sight (Gauthier et al. 2013); (5) new drugs to be assessed and older drugs which can be repurposed for nontherapeutic uses such as cognitive enhancement (Farah 2015); and a great deal more.

There are also lingering healthcare-related issues, such as the profound effects of poverty on mental health and the impact of stigma on people with neurodisabilities. These more clinical concerns represent, for some, perhaps the most fundamental issues to be reckoned with (Racine 2010). In sum, that new neuroscience insights, neurotechnologies, and profound healthcare issues call for ethical attention in the context of basic and clinical neuroscience is uncontroversial. In other words, whether these contributions represent a genuine "neuroscience revolution" (Wolpe 2002) or a mere "evolution" of science and technology does not remove the need to consider what such insights and their implications could mean for society. However, it is much less certain that there is a consensus surrounding one of the primary responses to these issues: the development of a "neuroethics."

Neuroethics, now a vibrant area of academic inquiry, has evolved tremendously since key North American discussions at the turn of the millennium (Marcus 2002; Canadian Institutes of Health Research 2002), expanding upon prophetic, yet often overlooked, prior international discussions (Pontius 1973, 1993; Engelhardt 1977; Cranford 1989; Bernard 1994; Vincent 1995). Since its inception, contemporary neuroethics has raised debates and controversies over its goals, methodologies, and status as a field or discipline. Some celebrate its existence, claiming it promises to reinvigorate a dialogue between disciplines of the humanities and social sciences (e.g., philosophy, anthropology) and the neurosciences (Roskies 2002; Gazzaniga 2005); others criticize this view for endorsing an ethics they feel is rooted too firmly, and wrongly, in biology (Stent 1990; Farah and Wolpe 2004). Still others fault neuroethics for having no clear foundations, other than the development of a rather servile and uncritical ethics that promotes neurotechnology (Parens and Johnston 2006, 2007; De Vries 2005, 2007). Another controversy is whether neuroethics can be defined and should be developed as an interdisciplinary field (Racine 2010), as a full-fledged discipline (Illes and Raffin 2002), or as a subcategory of bioethics (Parens and Johnston 2006, 2007; De Vries 2005, 2007). Ultimately, the question of what neuroethics is, or ought to be, remains heavily debated—perhaps a sign of a healthy and thriving academic endeavor, where different views raise constructive dialogue and reveal underlying tensions.

Debates About Neuroethics: Perspectives on Its Development, Focus, and Future is the first volume entirely dedicated to the exploration of issues associated with the nature of neuroethics itself. It is an effort to reflect on some of the underlying assumptions in neuroethics and the implications of those assumptions with respect to training and education programs, research activities, policy engagement, public discourse, teaching, ethics consultation, and mentoring, to name but a few areas of interest. Indeed, in this edited volume, established and emerging neuroethics leaders take up the pen to express and debate their views about the development, focus, and future of neuroethics. We hope that these exchanges will, without complacency, help to both identify strengths and address weaknesses of the goals and activities embraced by neuroethics thus far. As neuroethics experiences a new level of growth and development, partly due to the creation of large-scale international (e.g., Human Brain Project) and national (e.g., BRAIN Initiative, Brain Canada, Japan Brain/Minds, China Brain Science Project) neuroscience projects, the need to carefully consider interpretive, theoretical, methodological, and programmatic aspects of neuroethics is essential.

Readers will discover three parts in this book, each individually prefaced by a section introduction which gives an overview of the topic and of the contributions found in that section. Part I, *Development and History of Neuroethics*, presents different interpretations of how neuroethics has evolved, as well as different views about its nature as a practical and academic endeavor. Readers will become acquainted with some key historical facts while they reflect on whether or not there has been a need to develop a distinct neuroethics and what that distinctness might imply. Part II, *Focus, Theories, and Methodologies in Neuroethics*, features contributions that offer different views on the theories that have shaped neuroethics, their value, and the guidance they offer. This section stages discussions about relevant theories and methodologies for neuroethics as well as their limitations. Part III, *The Future of Research Programs, Training, and International Neuroethics*, offers a rich set of contributions examining questions associated with the emergence of training programs, trends in neuroethics research, and the topic of the international development of neuroethics. Many examples are discussed, providing the reader with information about these developments, while engaging in self-reflection about the path(s) to take in the future.

In many respects, one of the hallmarks of the contributions in this volume is their diversity: diversity of academic backgrounds (e.g., clinical, philosophical, sociological, historical), diversity of styles (e.g., argumentative or narrative, analytic or synthetic), and diversity of locations (e.g., United States, Canada, Germany, Sweden, Spain, Switzerland, Australia)—although we do acknowledge that even more diversity than offered in this volume is a worthwhile aim to continuously and vigorously pursue. This diversity will be appreciated by readers seeking both to learn about topical neuroethics problems and who wish to reflect on its nature and characteristics—with an invitation to contribute to its development as scholars or as curious readers in the future. As editors, we have encouraged this diversity and solicited experts with different backgrounds to provide their own accounts in their own words, thereby hoping to stimulate reflection and encourage constructive discussion about neuroethics itself. In our opinion, reflecting on what we do in neuroethics and why we do it is fundamental to our work, especially if neuroethics claims to be able help science, healthcare, and policy engage in similar and necessary self-reflection.

References

Bargh JA, Chartrand TL (1999) On the unbearable automaticity of being. Sci Am 462–479

Bell E, Mathieu G, Racine E (2009) Preparing the ethical future of deep brain stimulation. Surg Neurol 72:577–586

Bernard J (1994) La bioéthique. Flammarion, Paris

Canadian Institutes of Health Research (2002) INMHA-CIHR workshop on neuroethics. Institute of Neurosciences, Mental Health and Addiction, Toronto

Cohen Kadosh R, Levy N, O'Shea J, Shea N, Savulescu J (2012) The neuroethics of non-invasive brain stimulation. Curr Biol 22:R108–R111

Cranford RE (1989) The neurologist as ethics consultant and as a member of the institutional ethics committee. Neurol Clin 7:697–713

De Vries R (2005) Framing neuroethics: a sociological assessment of the neuroethical imagination. Am J Bioeth 5:25–27

De Vries R (2007) Who will guard the guardians of neuroscience? Firing the neuroethical imagination. EMBO Rep 8 Spec No:S65–S69
Decety JW, Wheatley T (eds) (2015) The moral brain: a multidisciplinary perspective. MIT Press, Cambridge
Engelhardt TJ (1977) Splitting the brain, dividing the soul, being of two minds: an editorial concerning mind-body quandaries in medicine. J Med Philos 2:89–100
Farah MJ (2015) The unknown of cognitive enhancement. Science 350:379–380
Farah MJ, Wolpe PR (2004) Monitoring and manipulating brain function: new neuroscience technologies and their ethical implications. Hastings Center Rep 34:35–45
Farah MJ, Smith ME, Gawuga C, Lindsell D, Foster D (2009) Brain imaging and brain privacy: a realistic concern? J Cognit Neurosci 21:119–127
Gauthier S, Leuzy A, Racine E, Rosa-Neto P (2013) Diagnosis and management of Alzheimer's disease: past, present and future ethical issues. Prog Neurobiol 110:102–113
Gazzaniga MS (2005) The ethical brain. Dana Press, New York
Greene J, Cohen J (2004) For the law, neuroscience changes nothing and everything. Philos Trans R Soc Lond B 359:1775–1785
Haggard P (2011) Decision tim e for free will. N euron 69:404–406
Harris S (2 012) Free will. Free Press
Illes J, Raffin TA (2002) Neuroethics: an emerging new discipline in the study of brain and cognition. Brain Cognit 50:341–344
Levy N (2007) Neuroethics: challenges for the 21st century. Cambridge University Press, Cambridge
Libet B, Wright EW Jr, Gleason CA (1982) Readiness-potentials preceding unrestricted 'spontaneous' vs. pre-planned voluntary acts. Electroencephalogr Clin Neurophysiol 54:322–335
Libet B, Gleason CA, Wright EW, Pearl DK (1983) Time of conscious intention to act in relation to onset of cerebral activity (readiness-potential). The unconscious initiation of a freely voluntary act. Brain 106(Pt 3):623–642
Marcus SJ (ed) (2002) Neuroethics: mapping the field. Conference proceedings. The Dana Foundation
Mele AR (2014) Free: why science hasn't disproved free will. Oxford University Press, Oxford
Parens E, Johnston J (2006) Against hyphenated ethics. Bioeth Forum [Online]. http://www.bioethicsforum.org/genethics-neuroethics-nanoethics.asp
Parens E, Johnston J (2007) Does it make sense to speak of neuroethics? Three problems with keying ethics to hot new science and technology. EMBO Rep 8 Spec No:S61–S64
Poldrack RA, Farah MJ (2015) Progress and challenges in probing the human brain. Nature 526:371–379
Pontius AA (1973) Neuro-ethics of "walking" in the newborn. Perceptual Motor Skills 37:235–245

Pontius AA (1993) Neuroethics vs neurophysiologically and neuropsychologically uninformed influences in child-rearing, education, emerging hunter-gatherers, and artificial intelligence models of the brain. Psychol Rep 72:451–458

Racine E (2010) Pragmatic neuroethics: improving treatment and understanding of the mind-brain. MIT Press, Cambridge

Roskies A (2002) Neuroethics for the new millenium. Neuron 35:21–23

Stent GS (1990) The poverty of neurophilosophy. J Med Philos 15:539–557

Vincent J-D (1995) Ethics and neuroscience. In: Proceedings. Third session. International Bioethics Committee of UNESCO, Paris. http://unesdoc.unesco.org/images/0013/001323/132339e.pdf

Wegner D (2002) The illusion of conscious will. MIT Press, Cambridge

Wexler A (2016) The practices of do-it-yourself brain stimulation: implications for ethical considerations and regulatory proposals. J Med Ethics 42:211–215

Wolpe PR (2002) The neuroscience revolution. Hastings Center Rep 32:8

Part I

Development and History of Neuroethics

Section Introduction: Development and History of Neuroethics

1

Eric Racine and John Aspler

Abstract

In this introduction to the section on the development and history of neuroethics, we describe several distinct visions of this area of inquiry and its evolution. Accordingly, the following four chapters provide contrasting accounts of neuroethics and its foundations, including in-depth discussions about the merits of its existence, and detailed accounts of the key issues that have driven its development.

There exist several somewhat competing narratives about the history and development of neuroethics (Racine 2010), which makes any attempt to provide a definitive or comprehensive account a daunting task; however, neuroethics should not avoid such debates about its foundations, even if they might never be settled in their entirety. Indeed, perhaps no single account of neuroethics can—or even should—be sought or enforced. However, a common thread shared by most accounts is that neuroethics explores concerns tied to significant developments in neuroscience, neurology, and neurotechnology.

E. Racine (✉)
IRCM, Montréal, QC, Canada

Université de Montréal, Montréal, QC, Canada

McGill University, Montréal, QC, Canada
e-mail: eric.racine@ircm.qc.ca

J. Aspler
IRCM, Montréal, QC, Canada

McGill University, Montréal, QC, Canada
e-mail: John.Aspler@ircm.qc.ca

E. Racine, J. Aspler (eds.), *Debates About Neuroethics*, Advances in Neuroethics, DOI 10.1007/978-3-319-54651-3_1

The diversity of views about neuroethics is thrown into sharp relief in debates about the novelty and identity of neuroethics. Claims that it is an innovative or unique area of inquiry are plentiful, but closer scrutiny of events of fundamental importance to neuroethics suggests that it shares a history with the field of bioethics, which has also been heavily shaped by neuroscience. For example, some of the Third Reich's "medical science" concerned neuroscience research (e.g., EEG experiments) (Shevell 1999). It is no accident that Leo Alexander, Medical Consultant to the Secretary of War and the US Chief Counsel for War Crimes, was a neurologist and psychiatrist (Shevell 1996, 1998). Furthermore, Alexander was a main contributor to the Nuremberg Code, a key document in the development of modern principles of research ethics. As another example, a central event that led to the creation of the National Commission for the Protection of Human Subjects of Biomedical and Behavioral Research in 1974, which produced the influential Belmont report in 1978, was the unhappy legacy of thousands of psychosurgeries conducted in the USA (Gostin 1980). In fact, among the Commission's many reports (e.g., on research with fetuses, with prisoners), one is entirely dedicated to psychosurgery, including recommendations for more restrictive use of the procedure (The National Commission for the Protection of Human Subjects of Biomedical and Behavioral Research 1977). Therefore, there is a strong case to be made for the interwoven relationship between bioethics and neuroethics; neuroscience—and its historical precursors—has been a key part of the history that sparked the development of both bodies of literature.

At the same time, and in spite of these historical links, different scholars have put forward arguments suggesting that neuroethics is a distinct field of scholarship or practice. For example, Adina Roskies historically captured how neuroethics involves an ethics of neuroscience, a neuroscience of ethics and, most importantly, promising synergy at their intersection (Roskies 2002). Others have suggested that this comprehensive view conflates two distinct areas of scholarship (Farah and Wolpe 2004) and that neuroethics should be viewed as a content field of bioethics, i.e., one defined by its focus on specific neurotechnologies and areas of neuroscience (Wolpe 2004). Others still have called upon neuroethics to pay more attention to clinical issues (Fins 2008) in the spirit of the work of esteemed ethicist-neurologists like Cranford (Bernat and Anderson 2006) and Bernat (Bernat 2002). Influenced by pragmatism, one of the coeditors of this volume has put forth that neuroethics has, at its core, the goals of building a community of interest, as well as of increasing knowledge and the capacity to respond to the ethical challenges of neurological and mental health patients, research subjects, and the societal implications of neuroscience more broadly (Racine 2010). Obviously, defining the history, the boundaries, and the foundations of neuroethics is not a straightforward task. It may not be a task that can ever be accomplished by a single scholar, although individual perspectives will offer greater knowledge and reflection about such foundational matters.

This first part of this book revisits key arguments in favor of establishing neuroethics as a distinct field or discipline, while critical contributions scrutinize some of these arguments. In Chap. 2, *Nudging Toward Neuroethics: Prehistory and Foundations*, the famous bioethicist and historian Albert Jonsen, who has been a participant in and a witness of major developments in bioethics, uses his many decades of professional experience as a lens through which to explore how the ethical challenges of twentieth century neuroscience and neurology anticipated the arrival of a "neuroethics."[1] We remain ever grateful for Dr. Jonsen's collaboration and have been profoundly moved by his decision to contribute what he foresees to be one of his last scholarly contributions to the field of bioethics to this volume. In Chap. 3, *The Evolution of Neuroethics,* Walter Glannon, a philosopher and prolific contributor to neuroethics inquiry, provides an overview of key issues that have driven its development throughout the twenty-first century—such as neuroimaging, functional neurosurgery, neuroenhancement, and disorders of consciousness. In Chap. 4, *Toward a Pragmatic Neuroethics in Theory and Practice,* the engaged physician and pragmatist scholar Joseph Fins emphasizes the importance of continuing a shift toward a more clinically grounded neuroethics. He unpacks his argument alongside a moving testimonial in which he reflects on the struggles he and his family faced when a loved one became gravely ill and needed urgent neurological care. Finally, in Chap. 5, the celebrated historian of the human sciences Fernando Vidal is joined by an emerging bioethics scholar, Michelle Piperberg, for *Born Free: The Theory and Practice of Neuroethical Exceptionalism.* They critically examine the arguments and assumptions upon which neuroethics has been built, notably the notion of "neuro-exceptionalism." Reviewing an extensive literature, they find that a great deal of neuroethics scholarship hinges on the reductionist and essentialist belief that we are our brains and that this belief is problematic for several theoretical and practical reasons. Their analysis is likely to provoke further thought, discussion, and debate. Together, these chapters present four distinct perspectives on the history and development of neuroethics; they will engage readers in a first set of important discussions and debates about neuroethics and its identity.

[1] Albert Jonsen once told a group of scholars at Stanford University that neuroethics should be renamed "encephaloethics" (Jonsen 2008). The scholars had been assembled by our friend and colleague Judy Illes to discuss the ethics of fMRI-based detection of consciousness, a topic which Jonsen traced back to early debates about brain death in the 1950s. He suggested that encephalon (from the Ancient Greek *ἐνκέφαλος*, meaning literally what is in one's head) was more precise than the term "neuro" (from the Ancient Greek *νεῦρον,* designating nerves). In his opinion, these debates about brain death marked the beginning of modern discussions about the philosophical and ethical significance of the brain with respect to issues such as death, organ donation, consciousness, and the meaning of life.

References

Bernat JL (2002) Ethical issues in neurology. Butterworth-Heinemann, Boston

Bernat JL, Anderson DC (2006) Ronald E. Cranford, MD (1940-2006). Neurology 67:1338–1339

Farah MJ, Wolpe PR (2004) Monitoring and manipulating brain function: new neuroscience technologies and their ethical implications. Hastings Cent Rep 34:35–45

Fins JJ (2008) A leg to stand on: Sir William Osler and Wilder Penfield's "neuroethics". Am J Bioeth 8:37–46

Gostin LO (1980) Ethical considerations of psychosurgery: the unhappy legacy of the prefrontal lobotomy. J Med Ethics 6:149–156

Jonsen AR (2008) Encephaloethics: a history of the ethics of the brain. Am J Bioeth 8:37–42

Racine E (2010) Pragmatic neuroethics: improving treatment and understanding of the mind-brain. MIT Press, Cambridge

Roskies A (2002) Neuroethics for the new Millenium. Neuron 35:21–23

Shevell MI (1996) Neurology's witness to history: the combined intelligence operative subcommittee and reports of Leo Alexander. Neurology 47:1096–1103

Shevell MI (1998) Neurology's witness to history: part II. Leo Alexander's contributions to the Nuremberg code (1946 to 1947). Neurology 50:274–278

Shevell MI (1999) Neurosciences in the Third Reich: from ivory tower to death camps. Can J Neurol Sci 26:132–138

The National Commission for the Protection of Human Subjects of Biomedical and Behavioral Research (1977) Psychosurgery: report and recommendations. Washington

Wolpe PR (2004) Neuroethics. In: Post SG (ed) The encyclopedia of bioethics, 3rd edn. MacMillan Reference, New York, pp 1894–1898

Nudging Toward Neuroethics: Prehistory and Foundations

2

Albert R. Jonsen

Abstract
As neurological science and its medical application in neurology and neurosurgery evolved during the twentieth century, ethical questions began to emerge about the relationship between mind and body. This chapter outlines how some of the earliest questions, although only slightly noted at the time, anticipated the appearance of a new discipline called neuroethics.

2.1 Introduction

The problem posed to any writer of history is where to begin the narrative. An additional problem posed by this assignment is how to write about "neuroethics," a term and a topic of scholarly interest that appeared only very rarely before the 1990s. I have chosen a starting point and a series of events between the early 1930s and about 2002, the year in which a conference was held at which the word "neuroethics" was coined (Marcus 2002). This choice is somewhat arbitrary but not, I hope, unreasonable. My strategy to reduce arbitrariness is an immodest one: many of these events were ones in which I played a part, sometimes tangential and sometimes substantial. My tactic for writing about "neuroethics" before neuroethics existed accounts for my enigmatic, somewhat silly title, "nudging toward neuroethics."

All the events I describe concern the field of human neurology as it unrolled during the twentieth century; the events I have chosen pose issues that, had there been a "neuroethics," would likely have been topics for that field. Since the field did not

A.R. Jonsen
California Pacific Medical Center, San Francisco, CA, USA
e-mail: arjonsen@aol.com

E. Racine, J. Aspler (eds.), *Debates About Neuroethics*, Advances in Neuroethics, DOI 10.1007/978-3-319-54651-3_2

exist formally, these issues, if they were discussed, were given passing notice or were noted as problems without solutions or as problems to be tolerated as the price of medical progress. One by one, these events "nudged" scholars toward formulating a field of academic discourse that merits the name "neuroethics." This chapter is a "prehistory," reporting how certain events and ideas converge toward a recognizable field of study.

I distinguish and comment on three such events: first, Mapping the Brain, that is, discerning segments of the cerebral cortex that correspond to mental and physical activities; second, the Silent Brain, as in the clinical implications of when no such activities can be discerned; and third, the Brain in Living Motion, i.e., scanning brain activity using advanced neuroimaging. A final short section, Neuroethics, steps up to the edge of the topics treated in the remaining chapters of this book.

2.2 Mapping the Brain

I met Wilder Penfield at the home of a friend in Montreal in 1965. I did not know that evening that I was meeting a world famous pioneer of neurosurgery. He and our host were both board members of the Vanier Institute, founded by Sir George Vanier, Governor General of Canada, to foster family values. Soon after, my host, a professor of medicine at McGill University, informed me of Dr. Penfield's preeminence. My brief meeting with him provides my starting point for this chapter.

Penfield (1891–1976) was born an American and later became a Canadian citizen. After graduating from Princeton (1913), he won a Rhodes scholarship and then completed a medical degree at Johns Hopkins University (1918). During his studies, he was mentored by the leading neuroscientists of the era, Sir Charles Sherrington and Dr. Harvey Cushing. Penfield was invited to join the faculty of McGill University and Royal Victoria Hospital in Montreal, where he created the Montreal Neurological Institute in 1934 (Penfield 1977).

Penfield's primary interest was the treatment of epilepsy: his work centered on finding those segments of the brain that seemed responsible for seizures and surgically excising them. He developed the unique procedure of removing sections of the skull under local anesthesia and probing the cerebral cortex with electrodes to find the offending tissue. Remarkably, he began to discern a wide range of sensorimotor responses to the stimulation. Since the patients were awake, they could report psychological phenomena such as recalled memories, hallucinations, and déjà vu. Based on these experimental findings, Penfield created extensive maps of the brain, designating the loci related to psychological and bodily function. Versions of his maps remain in use today (Penfield 1977).

Penfield deserves a place in the prehistory of neuroethics. His pioneering work in associating brain and behavior anticipates the properly ethical questions that will be raised decades later, namely, questions about whether the functions of the brain involved in motivation and intention, and in making prudential and moral decisions, can be directly discerned and controlled. Several authors have recently noted Dr. Penfield's relevance for neuroethics (Fins 2008; Sedney and Bernstein 2014).

He had, says Joseph Fins, "a rich ethical sensitivity" (2008). That sensitivity was manifested in his consistent concern for the welfare of his patients. He was an active participant in the first of a series of academic conferences during the 1960s that explored the "great issues of conscience in modern medicine" (Fins 2008; Jonsen 1998). During what Penfield called his "second career," he reflected on a most profound ethical problem—one that underlies every neuroethical problem today—the relationship between brain and mind. He realized that this was a problem that called for explicit engagement between philosophy and science (Penfield 1963, 1975).

2.2.1 Electroencephalography

Penfield was certainly aware of the technique called electroencephalography. It had been known since the mid-nineteenth century that the brain emitted electrical currents. Physical methods to trace these currents by placing electrodes on the skull were in use; finding the loci of epileptic seizures was one of the earliest applications. Penfield's clinical explorations went beyond these by directly stimulating the tissue of the open brain. As I shall show in the section entitled The Silent Brain, electroencephalography would lead to major ethical problems, namely, the issues associated with therapeutic efforts to eliminate brain waves thought to be noxious and the implications of the finding that these brain waves are sometimes absent.

2.2.2 Electroconvulsive Therapy

Neurologists in the late nineteenth and early twentieth centuries believed that the electrical properties of the brain could be manipulated to treat psychiatric illness. Although application of external electrical current was occasionally attempted, drugs were more commonly used to induce seizures for depression, schizophrenia, mania, and other conditions. In the 1930s, technology made possible the application of strong electrical currents, and "electroconvulsive therapy" (ECT) became a widely used tool to treat intractable psychiatric disorders. It appeared to have considerable benefit and few side effects (one of these, amnesia, was considered a benefit, since patients could not remember the violence of the treatment). However, as ECT became more widely used, criticism arose.

The debate anticipated a broad ethical problem about informed consent to treatment. ECT was used without consent (indeed, many of the patients were incapable of consent). Also, its violent physical effects frightened psychiatric patients and could be used as "instruments of terror" to control difficult patients. This aspect of ECT was dramatized in Ken Kesey's novel *One Flew Over the Cuckoo's Nest (1962)* and the subsequent film (1975). (The menacing "Big Nurse Ratchet" was played by the charming actress Louise Fletcher, whom I met at the home of her brother, my bioethics colleague John Fletcher.) Many reports were produced by health organizations (e.g., American Psychiatric Association 1990), and many states passed legislation limiting its use. Today, ECT remains in use in restricted situations; in those situations, its therapeutic benefits are acknowledged. (My close friend, Yale surgeon Sherwin Nuland, was cured of paralyzing depression by ECT as he reports in his TED talk (2001).)

2.3 The Silent Brain

The electrical activities of the brain "speak" to those who can read the fluctuations of voltage; what does the silence of those signals mean medically and philosophically? Two historical episodes respond to this question: the first involving a technique to "therapeutically" silence a part of the brain and the second, the implications of a "dead brain," whose signals have "turned off" temporarily or permanently.

2.3.1 Frontal Lobotomy

In the mid-1900s, neurophysiology located major brain functions in distinct segments of the organ. It was hypothesized that treatment for certain dysfunctions might be provided by locating their source in the brain and, in some manner, surgically modifying them. This had, as Penfield demonstrated, been true of epilepsy. Occasional experiments in surgical modulation of brain tissue to treat psychiatric illness had been attempted with little success. Then, in the mid-1930s, a Portuguese neurologist, Egas Moniz (1874–1955), proposed that depression, anxiety, and other debilitating psychological conditions could be addressed by severing the frontal lobes of sufferers from the rest of the brain. This could be done by severing the fibers that connected the two which would leave the patient unharmed except in affect.

Moniz's work (he directed but did not perform the surgery, since he was not a neurosurgeon) quickly caught on; it was replicated and modified around the world; it became known as "ice pick surgery," since fibers were severed by inserting an instrument like an ice pick through the nose into the brain. However, scientific studies to demonstrate efficacy were sketchy and inadequate. The degree of impairment to affect was much more severe than anticipated; many patients (like Rosemary Kennedy, sister of President Kennedy (Larson 2015)) who were "cured" remained for life in institutions.

Frontal lobotomy had fallen into disrepute by the 1960s; there was an unsuccessful movement to recall the Nobel Prize which had been awarded to Moniz in 1949. However, many other forms of brain surgery continued to attempt to treat psychiatric disease. In 1974, when the US Congress established the National Commission for the Protection of Human Subjects of Biomedical and Behavioral Research, one of its mandates was to examine the conditions, if any, under which those procedures could be supported with federal funds. This particular mandate arose from the concerns of African-American congressman Louis Stokes that persons of color were improperly subjected to such procedures.

The Commission did not find this to be the case, but did find that almost all psychosurgery lacked scientific merit; only one procedure, amygdalotomy, seemed useful for intractable obsessive-compulsive disorders. The Commission's report, *Psychosurgery* (1977), was informed by the work of a leading neurosurgeon, Elliot Valenstein, whose book *Brain Control: A Critical Examination of Brain Stimulation and Psychosurgery* (1974) had concluded that contemporary psychosurgery was nothing more than the discredited lobotomy of the previous decades.

I was a member of this National Commission. We commissioners considered our *Psychosurgery* report an almost total discrediting of the practice. However, for some, it was not total enough. The Commission presented its report at a meeting held in San Francisco. A local group of psychiatric patients entitled Network Against Psychiatric Assault (NAPA, an acronym based on the location of a large mental hospital in Napa, California) mounted a demonstration during the meeting and forced its adjournment. As a resident of San Francisco, I was advised to have a police watch at my home for the duration of the Commission meeting, which was moved to the safer confines of the federal building. The episode nudges toward neuroethics because it revealed how many of the issues in neuroscience involve psychiatric disease and that psychiatric disease raises profound questions about human freedom and constraint (National Commission for the Protection of Human Subjects of Biomedical and Behavioral Research 1977).

2.3.2 Persistent Coma: Henry Beecher and Pope Pius XII

I met Dr. Henry Beecher at a meeting of the Institute for Society, Ethics and the Life Sciences in, I believe, 1975. Unlike my meeting with Dr. Penfield several years before, I certainly knew who Dr. Beecher was; everyone working in bioethics knew who he was. In title, he was the Dorr Professor of Anesthesia at Harvard University School of Medicine. He was also a leading research scientist in his field, and when in 1966 he published a paper entitled "Ethics and clinical research" in the *New England Journal of Medicine* (Beecher 1966), he sent shockwaves through the world of medical science. He excoriated a score of prominent researchers for unethical research, primarily for conducting experimental procedures without the consent of the subjects, or for subjecting them to unjustifiable risks. Beecher became infamous among his colleagues and famous in the burgeoning field of bioethics. Dan Callahan, founder of the Institute for Society, Ethics and the Life Sciences, invited him to be a board member. Several years later, I was elected a fellow of the Institute; it was at my first meeting that I encountered Dr. Beecher.

The Institute for Society, Ethics and the Life Sciences (later renamed The Hastings Center) is a major mark in the history of bioethics. In "nudging toward neuroethics," it is helpful to explain what "bioethics" means. Dan Callahan, a newly minted Harvard Ph.D. in philosophy, hoped to stimulate critical examination of developments in biological and medical science. In 1969, he collected a small group of eager young scholars in a small office in Hastings on Hudson, New York, and convinced a larger group of famous scientists and philosophers to serve on an advisory board. Henry Beecher was one of these. This nascent organization began to select issues such as genetics, death and dying, behavior control, and population control. Scholars from around the country were invited to participate in systematic analyses of the ethical problems entailed by scientific progress in these areas. The Institute's journal, *The Hastings Center Report*, began to appear in 1971.

Callahan explained what he meant by the neologism "bioethics" (a word coined but only vaguely defined by the biologist Van Rensselaer Potter (1970)). Callahan noted the widespread interest in moral issues raised by the biomedical sciences; these concerns had been voiced in general, diffuse terms at many conferences (Jonsen 1998). He then proposed that to become a discipline, bioethics must use the traditional modes of philosophical analysis—logic, consistency, careful use of terms, rational justification of argument—supplemented by sensitivity to feelings and emotions, as well as to the political and social influences on behavior (Callahan 1973). Bioethics should move from a random expression of concerns over advances in the life sciences to a structured analysis of its issues. In other words, bioethics should become a discipline. Neuroethics similarly will move in this direction, as other essays in this volume show.

We return to Henry Beecher's place in the prehistory of neuroethics; it has two acts. In the first, Dr. Beecher is listening to Pope Pius XII addressing a World Conference of Anesthesiologists in Rome in 1957 (Kopp 1997). A photo shows him standing next to the Pope who had been invited to cast light on a significant moral issue facing those specialists who directed care in the newly created intensive care units (Pope Pius XII 1959). Advances in respiratory support technology had not only brought great benefits, it had also produced a moral paradox: what was the moral course of action when a patient was highly unlikely to recover from deep coma, yet whose breathing was supported by a respirator? The Pope drew on traditional moral theology to answer this modern question: there was no moral obligation to sustain life by "extraordinary means," that is, means that impose great burdens on self or others without compensating benefit (Pope Pius XII 1958).

The Pope was asked a second question: has death already occurred after deep trauma of the brain, or did death occur only after complete arrest of circulation (the traditional criterion of death)? The Pope declined to answer, saying that this was not a question for moral theology but for medical science. To the dismay—or delight—of his audience, he invited them to invent a clear and precise definition of death. Apparently, Dr. Beecher was delighted, because several years later, he gathered an ad hoc committee of Harvard medical school faculty to do just what the Pope requested. This was the second act in his role as a precursor of neuroethics.

The product of the Harvard ad hoc committee, commonly called The Harvard Report, appeared in the *Journal of the American Medical Association* in 1968 (Harvard Ad Hoc Committee 1968). Considerable debate over the issue had marked the years since organ transplant—particularly heart transplant—had begun. The Harvard Report proposed to settle the debates with a papistic "clear and precise definition." However, it complicated rather than clarified. This was suggested in the very title "A definition of irreversible coma: report of the ad hoc committee of Harvard medical school to examine the definition of brain death." Was irreversible coma equivalent to brain death? Was "brain death" equivalent to the death of the person? In fact, the Harvard Report did not offer any definition of death; it provided only operational criteria for recognizing a presumably irreversible coma.

The phrase in the title, "Brain death," had come into use to designate a neurological fact, namely, the silence of brain waves detected by electroencephalography. The Harvard Report lists "a flat electroencephalogram … as of great confirmatory value" in diagnosing irreversible coma and devotes a long paragraph to its clinical features. This technique, while certainly of "confirmatory value," was of little clinical relevance, since electroencephalography detects a silent cortex but does not reach the lower brain stem, which is crucial in determination of death, as we note in the following section.

The Harvard Report quickly achieved standing as the accepted definition of death in law and in medical practice. But its failure to distinguish firmly between "brain death" and "irreversible coma" caused confusion in the widely publicized case of Karen Ann Quinlan (1975). Karen, a 21-year-old woman, was found in a coma due, apparently, to alcohol and drug ingestion; she was put on a respirator but failed to recover consciousness. After some months, her father requested that the ventilator be turned off and that Karen be allowed to die. The hospital and doctors refused; Mr. Quinlan sought relief from the New Jersey courts. This was granted by the Supreme Court of New Jersey. The judicial ruling reviewed in some detail the contemporary neurological science but in the end relied on precedent: a person has the right to reject medical care and Karen, herself incompetent to do so, could exercise refusal through her father/guardian. Karen's respirator was turned off; however, remarkably, she began to breathe on her own, living another 10 years without any sign of recovery from coma (New Jersey Supreme Court 1975).

2.3.3 The President's Commission: Defining Death

In 1980, the US Congress renewed the National Commission under a new title, The President's Commission for the Study of Ethical Problems in Medicine. The first of its mandates was to "study the ethical and legal implications of the matter of defining death, including the advisability of developing a uniform definition of death" (1981). At this point, the Harvard Report, and a variety of conflicting state laws, had sown confusion. I was one of two bridge commissioners between the National and the President's Commission. After extensive review and consultation, the Commission framed a draft statute called the Uniform Declaration of Death Act in its report, *Defining Death*: "an individual who has sustained either (1) irreversible cessation of circulatory and respiratory functions, or (2) irreversible cessation of all functions of the entire brain, including the brain stem, is dead" (1981). The Commission recommended that this proposed statute be legally enacted in all states; for all practical purposes, it was.

The words "including the brain stem" are an unusual bit of anatomy in a legal statute, yet they are vitally important. They represent the Commission's election of a "whole brain" view of death. If the brain stem no longer functions, the controlling centers for respiration are lost and the integration between all other bodily functions deteriorates. Death occurs then when either the signs of circulation/respiration are gone or when the signs of cerebral and brain stem function disappear: these clinical signs are two sides of the same reality, loss of physical integrity. An appendix, prepared by a group of leading neurologists, stated the clinical tests for determining whether the uniform definition was met: either way, the individual is dead (A Collaborative Study 1977).

This formulation resolves any standing questions about the eventual applicability of the Harvard formulation to Quinlan's case. Quinlan was in a form of "prolonged" or "persistent" coma," which can erase consciousness but leave vegetative function and should not be called death or irreversible coma (Jennett 2002). As the Quinlan court rightly concluded, Karen was not dead: she was a living person with a "silent brain"; the ethical question was whether to allow her to die. This question ignited the burning ethical debate of the next decade and the topic of another Commission report, *Deciding to Forego Life-Sustaining Treatment* (1983).

Behind the definition of death, then, lay the deep ethical question—perhaps the deepest of all ethical questions—what is a person? Should a being who can no longer process the information that allows for communication and collaboration between humans be considered a person, with dignity and rights? A flurry of philosophical opinion attended the Commission's report, but left this deep question, expressed in neurological terms as the "higher brain" formulation of death, open.

2.4 The Brain in Living Motion

Neuroscience is a rapidly evolving field of science in which anatomy, physiology, chemistry, genetics, neurology, psychology, and psychiatry unite to explore the structure and function of nerve cells, singly and woven into the cables and sheets that carry internal and external information throughout the organism. This field of science, in which physical structures and processes are so intertwined with psychological states, touches human experience at almost every point, from emotion to intellection, from sexual attraction to religious meditation, from personality to culture.

2.4.1 Brain Imaging

The rather silly nineteenth-century pseudoscience called phrenology "mapped" the parts of the brain where certain human behaviors were supposed to sit. Bumps on the skull over those brain areas supposedly revealed the skull owner's personality. Antique shops still display plaster busts with sections marked on the skull: combativeness, amiability, amorousness, mirthfulness, etc. The idea that a single spot in the brain performed a single function persisted until it was possible to visualize dynamic activity. Then it became obvious that, although certain large parts of the brain are dominantly associated with certain sorts of activities, every sensation and action and thought sets off intricate connections across the organ.

These advances have been propelled by the techniques of brain imaging. The functioning human brain, encased in a bony citadel, has resisted invasion until recently. Although the occasional bold explorer, such as Wilder Penfield, probed the living brain during surgery, only since the development of imaging technology such as computerized axial tomography (CAT), positron emission tomography (PET), magnetic resonance imaging (MRI), and the more advanced functional MRI (fMRI) have investigators been able to see in intricate detail and vivid color the brain at work. This latter technique, introduced in 1990, is particularly useful; it captures changes in blood oxygen levels, a measure correlated with increased neuronal activity. Neuroscience can chart the intricate cellular changes that affect and reflect behavior. They can go further to pinpoint where and how thought, affection, and action arise and respond to environment and external stimuli.

As research subjects (some healthy, some with neurological conditions) recline in PET and MRI scanners, they perform various simple tasks. For example, subjects may be directed to move a particular finger. When they do, those parts of the brain that recognize the words of the command and those that govern physical movement are activated. When subjects are invited to move fingers at will, quite different parts of the brain, those presumably having to do with choice, jump into sight. These technologies have made possible extensive mapping of the enormously complex geography of the human brain, revealing brain activity associated with sensation, movement, memory, emotion, choice, and thought. "Mapping" may not be the right metaphor, since these studies reveal a constantly moving flow of energy that represents information. They show not a static geography of the brain but a field of vital tissue that resculpts itself as it responds to experience. It is more like a movie than a map.

2.4.2 Phineas Gage and Brain Damage

A story frequently cited in neuroscience literature provides a striking example of the way in which physical brain and psychic life are associated. In 1861, a derelict man, Phineas Gage, died in San Francisco. Thirteen years before, Gage had been the victim of a freak accident. While supervising railroad construction, an accidental explosion blasted an iron bar into his left cheek, through the front of his brain and out the top of his skull. Gage lived, his physical capacities intact and his cognitive facilities unimpaired with one gaping exception: he became incapable of making moral and prudential decisions. Neuroscientist Antonio Damasio and his wife Hanna, a neuroanatomist, obtained Gage's skull and precisely mapped the path of the bar; it ripped through the ventromedial region of his frontal lobe.

Damasio opens his book, *Descartes' Error: Emotion, Reason and the Human Brain*, with Gage's history. He writes,

> Gage had once known all he needed to know about making choices conducive to his betterment. He had a sense of personal and social responsibility … He was well adapted in terms of social convention and appears to have been ethical in his dealings. After the accident, he no longer showed respect for social convention; ethics … were violated, the decisions he made did not take into account his best interest … there was no evidence of concern about his future, no sign of forethought … Intriguing as these questions are, they may not be as important as those which surround Gage's status as a human being. May he be described as having free will? Did he have a sense of right and wrong or was he the victim of his new brain design, such that his decisions were imposed upon him and inevitable. Was he responsible for his acts? … Gage had lost something uniquely human: the ability to plan his future as a social being (Damasio 1994).

After the accident, Gage became irresponsible, disorganized, reckless, although still manifesting intelligence and comprehension. He could not generate the power to move from understanding a moral situation to making a moral choice. It was clear that by ripping through the ventromedial region of his frontal lobes, the bar had not only destroyed tissue but an essential feature of his personality, his moral consciousness, the ability to guide himself through the complex world of right and wrong.

After describing Gage's lesion and subsequent behavior (as well as several of his own patients with similar conditions), Damasio makes a remark that can serve as a theme for neuroethics: "the fact that acting according to an ethical principle requires the participation of simple circuitry in the brain core does not cheapen the ethical principle. The edifice of ethics does not collapse, morality is not threatened and in a normal individual the will remains the will" (1994). This is a bold affirmation. What are we to say about that elemental concept of ethics, free will, in the light of the neurosciences? What do the profound claims, made by philosophers and theologians and humanists about human dignity, and the assumptions of lawyers and judges about culpability and responsibility, mean in the light of comparative and evolutionary neuroscience? How should the ethicist look at daily problems of affirmation or dereliction of moral duty in the light of neuroscience?

2.5 Neuroethics

With the sciences of brain scanning opening up the moral questions of brain behavior, the leaders of the Dana Foundation of New York decided to host a conference exploring these issues. The conference was held in San Francisco in May 2002. The University of California San Francisco was a cosponsor; I was invited to serve as cochairman. Prominent neuroscientists and philosophers (including Antonio and Hanna Damasio) were invited to present papers that would "nudge" the question from somewhat ill-defined moral issues into more precisely formulated ethical arguments that might deserve the name "neuroethics."

The other co-chair was the chairman of the Dana Foundation's Board, journalist William Safire. Known for his popular newspaper column about words, Safire opened the San Francisco conference with the comment, "We need a new word to identify this field. I suggest 'neuroethics'." He ventured a definition: "neuroethics is an explanation of what is right or wrong, good or bad about the treatment of, perfection of, or unwelcome invasion of and worrisome manipulation of the human brain" (Safire 2002). Safire was not an ethicist, and his definition calls for ethicists to provide cogent arguments around his broad invocation of "right and wrong," "good or bad," "unwelcome," and "worrisome."

Several years before the Dana Conference, another journalist produced an excellent layperson's guide to this emerging field. Rita Carter wrote in the introduction of her book, *Mapping the Mind* "The personal, social and political implications of (brain mapping) are awesome, and one of the most serious ethical questions we will face in the new century is deciding how this powerful new tool should be deployed" (Carter and Frith 1998).

2.5.1 Whither Now?

Carter asks a speculative question about the awesome future. But the awesome turns "worrisome" or "unwelcome" in Safire's terms, when someone actually does something that changes the world in which we live (e.g., Einstein's equations informing the atomic bomb). On the very day that I finished writing this chapter, *The New York Times* (Taylow 2016) reported that a Chinese surgeon was planning to perform a "full-body transplant." A paraplegic patient is waiting; he would be decapitated and have his head attached to a different healthy decapitated body. The article was filled with quotations from ethicists in the United States and in China. If the stupendous technical obstacles can be overcome, should it be done? One Chinese commentator said, "ethically, it's impossible." Rita Carter's "new century" is here and so is neuroethics, as the subsequent chapters of the volume will show.

References

A Collaborative Study (1977) An appraisal of the criteria of cerebral death. A summary statement. JAMA 237(10):982–986

American Psychiatric Association (1990) The practice of ECT: recommendations for treatment, training and privileging. Convuls Ther 6(2):85–120

Beecher H (1966) Ethics and clinical research. N Engl J Med 274:1354–1360

Callahan D (1973) Bioethics as a discipline. Hastings Cent Stud 1(1):66–73

Carter R, Frith CD (1998) Mapping the mind. University of California Press, Berkeley

Damasio AR (1994) Descartes' error: emotion, reason, and the human brain. Putnam Publishing, New York

Fins JJ (2008) A leg to stand on: Sir William Osler and Wilder Penfield's "neuroethics". Am J Bioeth 8(1):37–46.

Harvard Ad Hoc Committee (1968) A definition of irreversible coma. Report of the Ad Hoc Committee of the Harvard Medical School to examine the definition of brain death. JAMA 205(6):337–340

Jennett B (2002) The vegetative state: medical facts, ethical and legal dilemmas. Cambridge University Press, Cambridge

Jonsen AR (1998) The birth of bioethics. Oxford University Press, New York
Kesey K (1962) One flew over the Cuckoo's nest. Penguin, New York
Kopp V (1997) Henry Knowles Beecher and the redefinition of death. Bull Anesth Hist 15:6–8
Larson KC (2015) Rosemary: the hidden Kennedy daughter. Houghton Mifflin Harcourt, Boston
Marcus SJ (ed) (2002) Neuroethics: mapping the field. Dana Foundation Press, New York
National Commission for the Protection of Human Subjects of Biomedical and Behavioral Research (1977) Psychosurgery. US Government Printing Office, Washington
New Jersey Supreme Court (1975) In the matter of Karen Quinlan: the complete legal briefs, courts proceedings, and decision in the superior court of New Jersey. University Publications of America, Arlington
Nuland S (2001) How electroshock therapy changed me. TED. http://www.ted.com/talks/sherwin_nuland_on_electroshock_therapy?language=en. Accessed 28 Jul 2016
Penfield W (1963) The second career: with other essays and addresses. Little, Brown and Company, New York
Penfield W (1975) The mystery of the mind. Princeton University Press, Princeton
Penfield W (1977) No man alone: a neurosurgeon's life. Little, Brown and Company, New York
Pope Pius XII (1958) The prolongation of life: an address of Pope Pius XII to an international congress of anesthesiologists. The Pope Speaks 4:393–398
Pope Pius XII (1959) In: Angelini F (ed) Discoursi ai Medici. Orizzante, Rome
Potter VR (1970) Bioethics, the science of survival. Perspect Biol Med 14(1):127–153
President's Commission for the Study of Ethical Problems in Medicine and Biomedical and Behavioral Research (1981) Defining death: a report on the medical, legal and ethical issues in the determination of death. US Government Printing Office, Washington
President's Commission for the Study of Ethical Problems in Medicine and Biomedical and Behavioral Research (1983) Deciding to forego life-sustaining treatment: a report on the ethical, medical and legal issues in treatment decisions. US Government Printing Office, Washington
Safire W (2002) Visions for a new field of neuroethics. In: Marcus SJ (ed) Neuroethics: mapping the field. Dana Foundation Press, New York
Sedney CL, Bernstein M (2014) Wilder Penfield—bioethicist. Can J Neurol Sci 41(2):177–181
Taylow D (2016) Doctor's plans for full-body transplant raises anxiety even in daring China. New York Times
Valenstein ES (1974) Brain control: a critical examination of brain stimulation and psychosurgery. Wiley, New York

The Evolution of Neuroethics

3

Walter Glannon

Abstract

Neuroethics has rapidly developed as a theoretically rich and practically significant field at the intersection of brain science, social science, philosophy, and law. This chapter is an analysis and discussion of the most debated issues in neuroethics as it has evolved over the last 15 years. I take the opportunity to examine whether or not a foundational claim about the duality of neuroethics (i.e., the ethics of neuroscience and the neuroscience of ethics, as well as their interaction) has been a feature of these debates. After noting the most important events in the development of this field, I examine ethical issues in four key areas that form the core of neuroethics research and discourse: (1) neuroimaging, with a focus on incidental findings, brain privacy, and the impact of imaging on normative judgments of moral and criminal responsibility; (2) functional neurosurgery for psychiatric disorders, focused on capacity to consent and patient autonomy; (3) cognitive and moral enhancement; and (4) chronic disorders of consciousness, with particular attention paid to how prognostic uncertainty can impact life-sustaining care. In the case of point four, these issues should be framed by the question of what is in the best interests of patients with these disorders, which is difficult to know given their neurologically compromised condition. In the concluding section, I speculate on some of the new ethical questions that may arise from advances in neuroscience in the future and how these advances may shape the continued evolution of debate in neuroethics.

W. Glannon
Department of Philosophy, University of Calgary, Calgary, AB, Canada
e-mail: wglannon@ucalgary.ca

E. Racine, J. Aspler (eds.), *Debates About Neuroethics*, Advances in Neuroethics,
DOI 10.1007/978-3-319-54651-3_3

3.1 Introduction

Neuroethics is an interdisciplinary field at the intersection of the clinical brain sciences of neurology, psychiatry, and neurosurgery, as well as other areas of medicine and research such as radiology, cognitive psychology, moral philosophy, philosophy of mind, and law. Consideration of the ethical aspects of neuroscience is not just a recent phenomenon. One can find references to "neuroethics" at least as far back as the 1970s (Pontius 1973), but 2002 was an especially significant year in the evolution of this field. Conferences held in San Francisco and London, Ontario addressed a range of ethical and social issues in neuroscience research and practice (Marcus 2002; http://publications.gc.ca/site/eng/329536/publication.html). At the San Francisco conference, *New York Times* columnist William Safire broadly defined "neuroethics" as "the examination of what is right and wrong, good and bad, about the treatment of, perfection of, or unwelcome invasion of and worrisome manipulation of the human brain" (Safire, in Marcus 2002: 3).

Perhaps the most influential event in this same year was the publication of a landmark paper by neuroscientist and philosopher Adina Roskies (2002). She argued that neuroethics should not be described as a subcategory of bioethics because the connections between the brain and behavior, as well as the ability to measure and manipulate the brain, raise a novel set of ethical questions. The distinctive nature of neuroethics is also a reflection of the exponential advances in neuroscience over the last 30 years. For example, functional neuroimaging can display brain activity in real time, which may have implications for control of thought and behavior.

Roskies drew a general distinction between "the ethics of neuroscience" and "the neuroscience of ethics" (2002: 21). The first division of neuroethics includes "the ethics of practice," which takes into consideration the risks and benefits faced by patients and research participants whose brains are imaged, or else altered by surgery, drugs, or electrical or magnetic stimulation. It also includes consideration of whether these individuals have the mental capacity to give informed consent to receive or undergo these interventions. Roskies further divided the ethics of neuroscience into the "ethical implications of neuroscience," which aims to "investigate the implications of our mechanistic understanding of brain function for society" (21). The second division of neuroethics, the neuroscience of ethics, pertains mainly to examining the neurobiological bases of moral reasoning and decision-making in theoretical and applied normative ethics. While acknowledging that "each of these can be pursued independently to a large extent," Roskies noted that "perhaps most intriguing is to contemplate how progress in each will affect the other" (2002: 21; See also Farah 2002; Illes 2006; Glannon 2007; Levy 2007; Racine 2010).

One example of overlap between the ethics of neuroscience and the neuroscience of ethics is when a patient with a psychiatric disorder, such as schizophrenia or major depression, decides whether to accept or refuse an intervention or a treatment. The capacity to weigh potential benefits against the risks of the intervention is necessary for the patient to be able to give informed consent. This can be problematic if the disease being treated involves impairment in this capacity that raises questions about the ability to consent to the intervention. Paying particular attention to the neuroscience of ethics, Roskies asked whether increased knowledge of the brain and how it regulates thought and behavior would change our views about moral and criminal responsibility. In addition, she raised questions about how information from brain scans might be used for lie detection in legal settings and the more general implications of this information for privacy and confidentiality. At a deeper level, Roskies asked whether future developments in neuroscience would cause us to revise our definition of "normal" behavior. She concluded by arguing that neuroethics should not be confined to specialists in neuroscience, philosophy, and law, but should include public debate with broad social participation from all members of our society in discussing the implications of reading and changing people's brains.

In this chapter, I analyze and discuss the four most debated issues in neuroethics as it has evolved over the last 15 years. First, I consider incidental findings and the privacy of brain information as they relate to neuroimaging. Most of this section focuses on the limits of neuroimaging and its implications for behavior control and judgments of moral and criminal responsibility. In section two, I examine functional neurosurgery for psychiatric disorders. I discuss whether patients and research participants with these disorders can give informed consent to undergo this type of neuromodulation and, if undergone, whether they retain their autonomy. Third, I examine cognitive enhancement used to improve the execution of mental tasks outside of therapy, as well as moral enhancement to improve a person's responsiveness to the rights, needs, and interests of others. Fourth, I consider ethical issues in chronic disorders of consciousness, paying particular attention to prognostic uncertainty and what this might suggest about the care patients with these disorders should receive. These last issues should be shaped by the best interests of patients with chronic disorders of consciousness, which can be difficult (and in some cases impossible) to know given their neurologically compromised state. In a concluding section, I speculate on some new ethical questions which advances in neuroscience may generate in the future, and how they may shape the continued evolution of debate in neuroethics

3.2 Neuroimaging

Neuroimaging has certainly been one of the most discussed technologies in neuroethics, and it may be useful to review the features of different structural and functional neuroimaging modalities. The main types of structural brain imaging are computed tomography (CT) and magnetic resonance imaging (MRI). CT scans provide three-dimensional x-rays of anatomical brain features, while MRI measures the alignment of magnetic fields in the brain to generate images of neural structures. In addition, MRI provides higher spatial resolution than CT and can discriminate between gray- and white-matter brain regions.

The main types of functional brain imaging are positron emission tomography (PET) and functional magnetic resonance imaging (fMRI). PET measures brain function in terms of different levels of glucose metabolism. With an intravenously infused radioactive tracer to detect gamma rays, PET generates images by displaying metabolic changes in different brain regions. In contrast, fMRI measures brain function by showing differences in the magnetic properties of oxygenated and deoxygenated blood (Farah and Wolpe 2004; Roskies 2013a, b).

3.2.1 Incidental Findings and Brain Privacy

Beyond safety concerns about exposing patients and research participants to potentially harmful levels of ionizing radiation in PET and especially in CT, one question about imaging in the ethics of neuroscience has been what to do with incidental findings (Illes et al. 2006). These are typically unexpected, asymptomatic brain abnormalities identified in the context of unrelated clinical care or research, such as cysts, aneurysms, or vascular changes. These findings may or may not be indicative of actual or probable disease, or a life-threatening condition, and they often warrant referral to a neurologist or neurosurgeon for follow-up, although in the research context, researchers may not even be trained to identify them at all. Whether people with the training to recognize these findings, such as radiologists and clinical neuroscientists, should disclose information about abnormalities deemed ambiguous, or not clinically significant, may depend on the patient's or subject's medical history, professional guidelines about disclosure, and the unique features of each doctor-patient relationship.

A potentially more serious ethical problem is unauthorized third-party access to neuroimaging data (e.g., insurers, neuromarketers) (Illes and Racine 2005). This would violate the confidentiality shared between a patient and their health-care provider. The information becomes more difficult to control in a research setting. Here, the relationship is between an investigator, or group of investigators, and what may be a large number of research participants generating more data. More fundamentally, it would violate privacy of information about one's brain, which is grounded in the principle that what lies within the skull is inviolable and off-limits to anyone but the person whose brain it is.

Third-party access should be prohibited as a matter of ethical principle based on autonomy and the negative right to noninterference in one's body and brain. As a matter of ethical practice, such access should also be prohibited to prevent third parties from drawing invalid inferences from imaging data to a person's current and future behavior. For example, prospective employers could prejudicially use evidence of clinically ambiguous or benign brain abnormalities to make hiring decisions based on questionable predictions about the risk of the person developing a neurological or psychiatric disorder. Prospective insurers could also draw invalid or questionable inferences from the data in order to deny medical coverage. As one illustration of this problem, in a letter to *Nature*, an individual who volunteered for an MRI brain study reported having to deal with future insurance eligibility after a tumor was discovered in his carotid artery to the left of his brainstem. He emphasized the need to include this and other possible consequences of brain imaging in the process of obtaining informed consent from those who volunteer for this research (Anonymous 2005).

Other third-party users, like neuromarketers, might also gain access to this information and use it to construct erroneous user profiles based on a person's history as a consumer (Ariely and Berns 2010). Among other concerns, some of these profiles could result in an unfavorable credit rating that could have a negative impact on a person's purchasing ability and quality of life. Like the case of the MRI study volunteer just cited, this raises another ethical issue regarding the use of neuroimaging data.

To prevent these different forms of discrimination and harm, privacy laws must be extended to neuroimaging and be duly enforced.

3.2.2 Neuroimaging Limitations and Culpability

Functional imaging has the power to reveal the neural correlates of the cognitive and affective processes that enable moral reasoning and decision-making—an important development in the neuroscience of ethics. However, there are limitations to what neuroimaging can actually reveal about how brain function mediates mental capacities. These limitations warrant greater caution in interpreting imaging data as evidence of moral and criminal culpability. More than neuroscience itself, neuroethics discourse has constructively engaged with the normative challenges surrounding the relationship between the brain and behavior.

The technological aspects and limitations of neuroimaging, and the complex relation between the brain and behavior, should make us circumspect in making these empirical and normative judgments. For example, the blood oxygenation level dependent (BOLD) signal in fMRI measures hemodynamic changes, not changes in neural activity (Roskies 2013a). Blood flow usually lags several seconds behind the neuronal activity that produces it and thus is not a direct measure of this activity. There are also problems with the signal-to-noise ratio in fMRI. This is a measure of how much relevant information (signal) is corrupted by junk information (noise). The ratio is too low in a single scan for it to have neurophysiological value. For this reason, images from fMRI have to be averaged over many brains from many studies to have any statistical significance. In addition, the fMRI signal cannot distinguish between excitatory and inhibitory neural activity. This makes it difficult to know whether brain activity indirectly displayed by scans has an enabling or disabling effect on cognitive, affective, and motivational states at the mental level.

Images from functional neuroimaging scans are not "snapshots" of events or processes occurring in the brain. They are more accurately described as scientific constructs than real-time indices of neural activity (Illes and Racine 2005). There is thus an inferential distance between images of metabolism and blood flow in the brain and what actually occurs at the neural level (Roskies 2008, 2013a). There is also the issue of task dependency, focusing on local neural activity associated with the performance of a specific cognitive task. This is often overlooked and yet can limit the validity of generalized claims about global neural activity (Bell and Racine 2009). Even if imaging could provide an accurate account of brain activity, there would be an additional inferential distance between this activity and our thoughts and actions. Although structural and functional brain scans may show correlations between neural activity and behavior, correlation is not causation. The absence of a causal connection between the brain and behavior indicates that the former cannot completely account for the latter. A recent meta-analysis of fMRI studies raises additional questions about the accuracy of fMRI and further widens the inferential gap between data from brain imaging and brain activity. The analysis showed a false positive rate of up to 70% among researchers interpreting data from fMRI results (Eklund et al. 2016). The authors pointed out that spatial autocorrelation in statistical analyses of imaging data can lead investigators to "find" brain activity where it does not exist. The meta-analysis also reinforces skepticism about the claim that brain function or dysfunction alone can offer a satisfactory explanation of whether or to what extent we can control our behavior.

Many claims about the brain mediating or determining choice and action are based on localized anatomical and functional properties of particular brain regions. Yet the reasoning and decision-making associated with behavior control consists of sensorimotor, cognitive, affective, and volitional capacities mediated by a distributed network of interacting brain regions and pathways. These include (but are not limited to) the ventromedial prefrontal and orbitofrontal cortex (vmPFC and OFC), the anterior cingulate cortex (ACC), limbic structures (e.g., the amygdala), and subcortical structures (e.g., the striatum, the cerebellum).

The neural circuits mediating prudential and moral reasoning and decision-making are domain-general rather than domain-specific (Decety et al. 2012; Pessoa 2013; Decety and Cowell 2015). This complicates any effort to associate a structural feature or pattern of activity in a particular brain region with an action or pattern of behavior. Even if there were a strong correlation between activity in a particular brain region and a person's decision to act, this would not imply that the activity caused the decision. It is possible that the conscious decision causes the change in unconscious brain activity rather than the other way around. It cannot be known whether changes in blood flow and glucose metabolism as indications of brain activity cause or are caused by the mental process of forming and executing an intention in action.

By this same line of reasoning, imaging showing brain abnormalities in regions associated with reasoning and decision-making cannot prove that a criminal offender could not control his actions and be responsible for them. The problem of inferential distance between brain-imaging data and behavior partly explains why the evolution of neuroscience has not resulted in a significant change in the assessment of normative judgments of moral and criminal responsibility. In cases where imaging can inform these judgments, it is more often less advanced structural imaging than more advanced functional imaging that provides the relevant information.

Neuroimaging displaying anatomical abnormalities affecting regions mediating rationality and inhibitory mechanisms may clarify questions about impulsivity, for example, when behavioral criteria alone are ambiguous and inconclusive.

Ultimately, neuroimaging will, at most, supplement and not supplant behavioral evidence of the mental capacities necessary for responsibility (Kamm 2009; Schauer 2009; Vincent 2011; Jones et al. 2013; Glannon 2014). Normative questions about how a person should or can be expected to act are conceptually different from empirical questions about how the brain functions. What *ought* to be the case cannot be reduced to, explained away by, or directly inferred from what *is* the case. Normative questions about whether a person can control her actions have been and will likely continue to be based on assessments of mental capacities grounded primarily in observable behavior rather than structural and functional features of the brain (Morse 2006, 2011).

More sophisticated imaging techniques may show stronger correlations between neural circuits and the mental capacities necessary for prudential and moral reasoning and decision-making. They may have better spatial and temporal resolution and be able to reveal subtle brain abnormalities at cellular and subcellular levels (Roskies 2013b). These techniques may include advanced diffusion tensor imaging (DTI), which can detect changes in white-matter volume and connectivity; voxel-based morphometry (VBM), which can detect gray-matter changes; diffusion functional MRI (dfMRI); and diffusion spectrum imaging (DSI). Still, these advances may not provide a conclusive answer to the question of how discrete anatomical and functional abnormalities in local neural circuits might affect the distributed neural networks that enable rational and moral agency. What matters for judgments of both behavior control and moral and criminal responsibility is not brain processes as such. Rather, what matters is whether or to what extent these processes enable or disable the mental capacities necessary for control and responsibility. While neuroimaging can be one component of an assessment of control, in most cases one's behavior will provide the most reliable evidence of whether one has or is impaired in these capacities.

These considerations underscore the limitations of functional and structural neuroimaging. Brain scans can elucidate the neurobiological basis of reasoning and decision-making, but cannot determine whether or to what extent the brain determines thought and behavior.

3.3 The New Psychosurgery

The first era of psychosurgery began in Switzerland in the 1890s, with the work of Gottlieb Burckhardt (Joanette et al. 1993). But perhaps the most significant portion of this era spanned roughly 20 years in the first half of the twentieth century. Portuguese neurologist Egas Moniz and neurosurgeon Almeida Lima hypothesized that psychiatric disorders such as schizophrenia were caused by abnormalities in white-matter tracts in the frontal cortex and their projections to limbic structures

(Pressman 1998). Creating lesions in these tracts to relieve psychiatric symptoms was the ill-conceived rationale for the prefrontal leucotomy they practiced in the 1930s and 1940s. American neurologist Walter Freeman and neurosurgeon James Watts performed approximately 3500 lobotomies in the United States between the 1930s and 1960s. These procedures failed to achieve their therapeutic goals and often resulted in significant cognitive, affective, and motor impairment in many patients. This resulted in widespread public criticism and a backlash leading to some ethical reflection. The advent of antidepressant and antipsychotic drugs in the 1950s, together with the growing recognition of patients' rights and investigators' duty to protect them, coincided with the demise of these crude procedures.

Functional neurosurgery can be described as the second era of psychosurgery. This technique consists in electrical or magnetic stimulation or ablation of neural tissue to modulate function in brain regions implicated in neurological and psychiatric disorders. Although it began in the late 1980s, functional neurosurgery did not fully develop until the first decade of the twenty-first century. In 1987, Alim-Louis Benabid and co-investigators used high-frequency electrical deep brain stimulation (DBS) of the central thalamus without ablating any neuronal tissue to control motor symptoms in a patient with Parkinson's disease (Benabid et al. 1987; Benabid 2007). It was also discovered that the technique could have modulating effects on mood and motivation. This led some psychiatrists to use it as an experimental treatment for major depressive disorder (Mayberg et al. 2005; Lozano et al. 2008), obsessive-compulsive disorder (Mallet et al. 2008), and other treatment-refractory psychiatric conditions. MRI-guided stereotactic neurosurgical techniques and more precise targeting of dysfunctional neural circuits have improved symptoms for many patients and reduced the incidence of neurological and psychiatric sequelae.

Functional neurosurgery is now an established treatment for movement disorders such as Parkinson's disease, dystonia, and essential tremor; however, it remains experimental and investigational for psychiatric disorders. In some respects, "psychosurgery" may be an appropriate term, not only for disorders of mood and motivation but also for movement disorders such as Parkinson's, which often involve cognitive and affective symptoms in addition to motor symptoms.

Functional neurosurgery is different from the more traditional structural neurosurgery consisting of resecting tumors, clipping intracranial aneurysms, and other procedures. The most frequently used functional neurosurgical technique is DBS. In addition to its neuromodulating effects, DBS can be used as a research tool to probe brain regions mediating motor, affective, cognitive, and volitional capacities associated with reasoning and decision-making (Lozano and Lipsman 2013). Still, most of the ethical issues surrounding DBS pertain to its neuromodulating effects. Improvements in the safety of functional neurosurgery in general have enabled investigators to meet their ethical obligation of nonmaleficence in protecting patients and research participants from harm (Beauchamp and Childress 2012: ch. 5). Nevertheless, the technique has resulted in adverse effects—including hypomania, mania, and compulsive behavior in some cases (Muller and Christen 2011; Christen et al. 2012). These effects may result from not stimulating targeted circuits with the requisite precision, overstimulating them, or from expanding effects on other neural circuits. More precise stimulation of circuits at the right frequency has reduced the incidence of neurological and psychological sequelae.

Even with a better safety profile and reduction of risk, there remain ethical questions about whether the psychiatric conditions for which DBS is used compromise a patient or participant's capacity to give informed consent (Rabins et al. 2009; Lipsman et al. 2012).These specific questions are an extension of the more general question of capacity assessment in medicine (Appelbaum and Grisso 1988; Appelbaum 2007). There are also questions about whether a stimulating device implanted in one's brain undermines autonomy, which I will elaborate on below. These two ethical questions blur the distinction between the ethics of neuroscience and neuroscience of ethics because the neural circuit targeted by the intervention is in many cases the source of the capacity to freely decide to undergo it. And yet, it is the dysfunctional nature of this source that justifies using DBS to modulate and restore it to normal function.

3.3.1 Informed Consent

The cortical, limbic, or subcortical circuits targeted by DBS for psychiatric disorders are associated with the cognitive and affective capacities necessary for consent (Glannon 2010; Skuban et al. 2011). These circuits are dysfunctional in these disorders, and the mental capacities they mediate may be impaired by them. They may interfere with the patient's or research subject's ability to process information about the potential benefits and risks of this intervention in the brain.

Mental impairment caused by these disorders comes in degrees. Being affectively and cognitively impaired does not necessarily undermine a person's capacity to rationally process information about DBS, or to voluntarily choose or reject its implantation and activation. In conditions that are refractory to other interventions, this capacity would not only include consideration of the potential benefits and risks of the procedure; it would also include a comparative assessment of the probability of adverse neurological and psychiatric events *with* DBS against the continuation of mental suffering and possible suicide *without* DBS. The risk of suicide and the unremitting harm from a treatment-refractory condition can justify participation in first-in-human and proof-of-principle trials, as well as more advanced DBS trials, even when they entail some risk of harm to the patient or research subject (Lipsman et al. 2012).

Still, the severity, chronicity, and treatment-resistant nature of some psychiatric and neurological disorders may cause desperation for symptom relief—interfering with a rational assessment of risk in DBS (Ford 2009). Some patients with severe depression may also have a disregard for their well-being and may fail to adequately appreciate or even ignore the risks. They may perceive it as a treatment of last resort and have unreasonable or unrealistic expectations about its effectiveness. These expectations may be more likely in DBS than in pharmacological of psychological therapies because patients may believe that stimulating neuronal circuits will get at the "root" of the problem (Dunn et al. 2011; Lipsman et al. 2012). This is an example of the potential power of brain-based explanations of thought and behavior, which can be misleadingly reductive in suggesting that psychiatric disorders can be explained entirely in terms of localized dysfunction in neural circuits or particular nodes of these circuits. In addition, if the outcome fails to meet their expectations, these beliefs could set patients up for psychological harm (over and above what they have already experienced from the disorder).

In these respects, the nature of psychiatric disorders could make patients and research participants vulnerable in clinical trials testing the safety and efficacy of DBS, raising concerns about their capacity to provide informed consent (Bell et al. 2014). When the patient's decisional capacity comes into question, a family member acting in the patient's best interests could give proxy consent. This may occur in psychiatric disorders such as anorexia nervosa (Lipsman and Lozano 2014; Maslen et al. 2015). Otherwise, some form of third-party involvement in the consent process may be required when a patient's decisional capacity and psychological preparedness for the procedure are in question (Laxton et al. 2010) A clinical psychologist or psychiatrist not directly involved in the patient's care or the research protocol could confirm that the patient had sufficient mental capacity to make a rational and voluntary decision to agree or decline to participate in DBS research or treatment.

3.3.2 Autonomy

The idea that a device can control one's behavior outside of one's conscious awareness might suggest that they are not the author or source of their mental states and actions. It seems that the actions one performs while undergoing DBS are not autonomous but generated and sustained by something alien to them. Autonomy (*autos* = self + *nomos* = rule or law) consists of two general mental capacities: competency and authenticity (Dworkin 1988; Taylor 1991; Mele 1995). Competency involves the cognitive and affective ability to critically reflect on the mental states that issue in one's actions. Authenticity involves the ability to identify with or endorse these mental states following a period of critical reflection. The process of critically reflecting on and identifying with one's mental states and actions is what makes them one's *own*.

DBS does not undermine behavior control and autonomy—it supports them by modulating dysfunctional neuronal circuits that ordinarily generate and sustain thought and action. The device does not interfere with but enables critical reflection on and identification with one's mental states. By modulating dysfunctional neural circuits, DBS allows the formation and execution of conscious intentions in promoting and facilitating effective agency. Insofar as the stimulating device enables the subject to have the mental states she wants to have and perform the actions she wants to perform, she can identify it as her own, as an integrated feature of her brain-mind. It is the brain disease, not the brain implant that impairs or undermines autonomy.

All behavior is regulated by a balance of interacting conscious and unconscious mental and neural processes. When it operates safely and effectively, DBS restores this balance through its modulating effects on the brain-mind. Being a "passive recipient" of the effects of DBS does not imply that the subject has no control of her behavior. Although it operates outside of conscious awareness, the device does not replace but restores the patient as the source or author of her actions (Lipsman and Glannon 2013). This highlights Roskies' definition of the "neuroscience of ethics," where the ethical issues pertain to the neurobiological basis of reasoning and decision-making necessary for autonomy and informed consent.

3.4 Enhancement: Cognitive and Moral

3.4.1 Cognitive Enhancement

Cognitive enhancement refers to interventions in the brain that improve information processing, attention, and concentration for more effective execution of cognitive tasks. Its aim is to raise certain cognitive functions above levels considered normal for humans (Juengst 1998). While transcranial magnetic stimulation (TMS) and transcranial direct current stimulation (tDCS) have been used to enhance some cognitive functions, most cognitive enhancement has been in the form of psychotropic drugs (Young et al. 2010; Iuculano and Cohen Kadosh 2013). For example, methylphenidate and dextroamphetamine can increase circulating levels of dopamine in the brain and enable people to narrow their focus, sustain attention, and execute cognitive tasks more effectively (De Jong et al. 2008). Modafinil can alter sleep-wake cycles and enable people to remain focused and continue performing tasks despite sleep deprivation (Volkow et al. 2009). Propranolol can have an indirect cognition-enhancing effect by reducing the body and brain's sympathetic nervous system response to perceived threatening stimuli. This last drug enables those who are anxious about musical performance or public speaking to avoid being distracted by a rapid heartbeat or other physical symptoms and remain attentive to the desired action (Elliott 2008; Glannon 2011: 129–133). In this section, I will provide specific examples of the use of cognition-enhancing drugs and explain their effects in more detail.

The ability to alter the brain to improve normal cognitive functions initiated a debate on the ethical reasons for and against neuroenhancement roughly 12 years ago (Farah et al. 2004). This debate is as lively today as it was then (Jotterand and Dubljevic 2016). Studies indicate that a significant number of students and professionals take drugs to work more efficiently and productively for longer periods (Wilens et al. 2008; Heinz et al. 2012). In 2008, the journal *Nature* published the results of an informal survey of its readers (Maher 2008). One-third of the 1400 respondents said that they had used psychotropic drugs for the off-label non-therapeutic purpose of enhancing their cognitive functions. This has been a significant development in the ethics of neuroscience because it involves taking psychotropic drugs, not to treat neuropsychiatric diseases, but to improve normal mental capacities (Tome 2005). Some of the main ethical issues involve weighing the benefits against the risks of enhancement and whether it should be regulated (Forlini et al. 2013).

Some studies suggest that methylphenidate enhances executive functions on novel tasks, but impairs these functions on tasks that have already been learned (Farah et al. 2004). In a recent experiment using transcranial electrical stimulation to test the learning and application of mathematic information, researchers found that stimulating an area of the subjects' prefrontal cortex impaired learning new information, but enhanced the application of what was learned. Stimulating an area of the parietal cortex had the opposite effect, impairing the ability to apply the new information (Iuculano and Cohen Kadosh 2013). More troubling is the finding that chronic use of methylphenidate to raise levels of dopamine in the brain could result in addiction in some susceptible individuals (Heinz et al. 2012).

Taking modafinil once to remain awake for many hours in order to perform a cognitively demanding task may not have any adverse physiological effects; however, repeated use of the drug could cause hyper-activation of the stress response and disturb the feedback loop that regulates stress and relaxation responses in the sympathetic and parasympathetic nervous systems (Volkow et al. 2009). Moving between sleep and attention states is an adaptation to the environment. If the brain is in a constant state of attention, then some regions mediating cognition may degenerate from being overloaded with information (Zhang et al. 2014). In this regard, unnecessary wakefulness can be more problematic than unnecessary sleep. Constant manipulation of circadian rhythms and sleep-wake cycles with modafinil could cause these cycles to become dysfunctional and interfere with one's ability to adapt to one's natural and social environment (Zhang et al. 2014). Like methylphenidate, some studies have shown that modafinil can increase circulating levels of dopamine in the brain and can lead to addiction in susceptible individuals (Volkow et al. 2009).

In addition, most studies indicate that those with greater cognitive capacity tend to benefit less from cognition-enhancing drugs, while those who are worse off on this scale tend to benefit more. All of the studies also suggest that enhancement may involve trade-offs: improving some cognitive functions through drugs or stimulation may impair others (Farah et al. 2004; De Jong et al. 2008). Moreover, chronic use of these drugs may have deleterious effects on normal brain functions. Any use of "cognitive enhancement" thus warrants qualification, both in terms of net improvement in overall cognitive function and the risk of adverse physiological effects.

Assuming that cognition-enhancing drugs were safe and effective, some would object to enhancement on the grounds that it would exacerbate social and economic inequality. For example, enhancement could work against a system of justice based on fair equality of opportunity to obtain social goods (Rawls 1971; Daniels 1985).

It may seem unfair that those who are naturally cognitively better off could use drugs to enhance their capacities to gain an even greater competitive advantage in securing positional goods, such as an elite university education or lucrative jobs. However, these capacities are already enhanced when parents arrange for private tutors for their children and send them to private schools, or when athletes employ private trainers. Drug-induced cognitive enhancement is only one example of a broader spectrum of enhancement.

The ability to improve one's mental capacities pharmacologically may also reflect income inequality. Those who lack the financial means to purchase these drugs do not have the same opportunity as others who can afford them. The cognitively worse off may also lack the sophistication to persuade physicians to prescribe enhancers for them, quite apart from whether this practice would be consistent with medical professional codes of ethics (Forlini et al. 2013; Synofzik 2009). This may generate an unjust situation in which not everyone who could benefit from the drugs would have access to it. But if cognition-enhancing drugs were available to all, and if the positive effects of the drugs were greater among the cognitively worse off than among the cognitively better off, then enhancement practiced on a broad scale would not likely result in greater social inequality. The issue is not whether some people are better or worse off than others as such, but where their cognitive functions lie with respect to an absolute baseline of these functions. There would probably be no levelling down of the better off either if the drugs resulted in modest improvement in their mental capacities without significant adverse effects. Yet because differences in cognitive ability are only one aspect of opportunity and well-being, pharmacological enhancement of this ability would not support the stronger claim that it would reduce social inequality. The point here is not that enhancement should not be permitted, but that its overall social effect would likely be limited.

If one decides to take a drug to enhance one or more of one's cognitive capacities, social and individual factors may influence whether or to what extent one's goal is achieved. These may involve the circumstances in which one takes the drug and whether one makes good use of the increased capacity. Psychopharmacological enhancement is just one form of enhancement designed to increase individual and collective levels of well-being. Whether it has these effects will depend on a complex set of biological and nonbiological factors, such as drug metabolism and the social circumstances in which one lives.

3.4.2 Moral Enhancement

Moral enhancement aims to improve people's capacity to recognize reasons for respecting the rights, needs, and interests of others and to react to these reasons by performing appropriate actions (Fischer and Ravizza 1998). The moral enhancement debate extends and further develops the cognitive enhancement debate; however, it also raises a distinct and more controversial set of ethical questions. Citing threats to human survival via the potential use of weapons of mass destruction, the effects of climate change, and the failure of education and political institutions to defuse these threats, Ingmar Persson and Julian Savulescu have defended what they call “moral bioenhancement” (2008, 2012; Douglas 2008). They claim that this could be a complement to traditional means of motivating people to act morally in addressing and ideally resolving the “predicament of humankind” (2012: 2). Morally enhanced actions would balance deontological considerations of treating individuals as ends in themselves and not merely as means (Kant 1785/1998) with consequentialist considerations of acting to bring about outcomes that decrease harm and increase human welfare (Mill 1863/1987). In addition to the threats mentioned by Persson and Savulescu, moral enhancement would aim at preventing or reducing the incidence of harm from genocide, ethnic conflict, civil war, and economic inequality. Questions such as whether psychotropic drugs could enhance our moral disposition and whether the voluntary or compulsory use of these drugs could be justified are further examples of overlap between the ethics of neuroscience and the neuroscience of ethics. An answer to the first question would depend on an answer to the second, insofar as moral enhancement would be achieved through pharmacological means.

Neuroscientific studies have shown that rational and moral decision-making are enabled by both cognition and emotion—as integrated and interacting systems rather than separate processes mediated by distributed networks in the brain (Decety et al. 2012; Decety and Cowell 2015; Liao 2016). To be effective then, any form of moral enhancement would have to target networks mediating cognitive-emotional processes. The use of certain psychotropic drugs has been proposed as a possible way of modulating these networks and improving moral behavior. But there are reasons to be skeptical of the idea that administering or taking a psychotropic drug targeting neural circuits could have this effect.

One study has shown that the selective serotonin reuptake inhibitor (SSRI) citalopram increases harm aversion in healthy subjects. The main author of this study, neuroscientist Molly Crockett, claimed that “these findings have implications for the use of serotonergic agents in the treatment of antisocial and aggressive behavior” (Crockett et al. 2010: 17437). She is more circumspect about the behavior-modifying potential of these drugs in a more recent paper: “Most neurotransmitters serve multiple functions and are found in many different brain regions … serotonin plays a role in a variety of other processes … and there are at least 17 different types of serotonin receptors that produce distinct effects on neurotransmission. Thus, interventions … may have undesirable side effects, and these should be considered when weighing the costs and benefits of the intervention” (Crockett 2014: 370). Differences among people in how neurotransmitters influence the activity of brain circuits and how altering neurotransmitter levels might influence this activity suggest that there would be different effects on the behavior of healthy people taking SSRIs, including no effects at all. How increasing levels of these substances would influence the cognitive-emotional processing in moral reasoning is unclear (Crockett 2016; Oliveira-Souza et al. 2016). While the activity of neurotransmitters plays a critical role in regulating activity in these circuits and the mental processes they sustain, neural and mental functions are influenced not just by neurotransmitters, but by the effects of genetic, endocrine, immune, and environmental factors on these substances as well. The capacity for moral sensitivity depends on more than the function of a particular neurotransmitter. Increasing harm aversion would not imply a corresponding increase in the capacity to be more attentive to the needs of others. For example, it would not necessarily enhance the motivation to act in response to these needs along with awareness and may incline one to refrain from acting in this way by reducing the motivation to perform beneficial actions. Specifically, stimuli perceived as aversive would more likely inhibit one from performing these actions than incline one to perform them.

Other researchers have claimed that the neuromodulating effects of the neuropeptide oxytocin may have the greatest potential as a morality-enhancing agent. Oxytocin plays a critical role in social cognition and is associated with inhibition of the fear response to social stimuli. Increasing levels of oxytocin in the brain through intranasal administration could reduce fear and increase trust and social cooperation. Yet any positive social effects of oxytocin may be more local than global and limited to particular groups, with negative effects more likely to occur outside the scope of these groups (Bartels 2012). Studies show that this neuropeptide facilitates social bonding, but also produces non-prosocial effects that may have evolved to promote offspring survival (Hurlemann and Scheele 2016). Specifically, oxytocin may promote antisocial rather than prosocial behavior on a broad scale by strengthening a person's bonding and identification with an in-group, as well as the perception of those in out-groups as competitors or threats. It could promote, rather than prevent or reduce, aggression between individuals and groups.

It would be an oversimplification to claim that increasing the level of a psychoactive substance alone would make one less self-regarding and more other-regarding. Moral behavior is a function of multiple biological, psychological, and environmental factors including brain function, incentives, and social cues, not just neural chemicals and circuits. It is a disposition that depends on factors both inside and outside of the brain. The idea that pharmacological neuromodulation alone could enhance moral behavior fails to appreciate the complexity of human moral psychology (Agar 2014; Wiseman 2016).

Another challenge with moral enhancement would be reaching public consensus on which moral theory should drive it. Such a theory would have to accurately reflect the psychology behind our actions, as well as the reality of how we act. But we do not have a very good understanding of what motivates people's behavior. Most people are not altruistic and do not typically act from other-regarding interests. The most plausible theory might be one based on a lower common behavioral denominator—rational self-interest—and thus be some version of social contract theory (Hobbes 1651/1987; Gauthier 1986; Harris 2016: ch. 11). It might be based on rational choice and promote social cooperation for mutual benefit. Each person could give up some of their self-interest by cooperating, but all would be better off by doing this. This would be consistent with deontological considerations of individual rights and consequentialist considerations of acting to bring about the best outcomes.

Assuming that there could be public consensus on adopting a social contract theory as a foundation for a moral bioenhancement program, though, there would be problems with implementing and sustaining cooperation. If less than a critical mass of people decided to cooperate by enhancing their moral sensitivity, then the goals of reducing collective harm and increasing well-being would not be achieved. Many would refuse to enhance for a variety of reasons. In a scenario where a majority enhanced and a threshold of cooperation had been reached, some would calculate that they could refuse to enhance and free ride off the cooperation of others without sacrificing any self-interest. Those who engaged in the most harmful behavior and had the greatest need for moral enhancement would take advantage of this situation for their own maleficent ends. In light of the likelihood of voluntary enhancement failing to achieve its goals, Persson and Savulescu state that moral bioenhancement should be obligatory: "safe, effective moral enhancement would be compulsory" (2008: 173, 2012a, b).

Even if one believed that safe and effective moral bioenhancement should be compulsory for all citizens in a liberal democratic society, there would remain the daunting task of enforcing compliance. This would require different levels of coordinated social and political action. It would also assume the moral integrity and public acceptance of those empowered with overseeing this task. This may assume too much. As John Harris asks, "who guards the guardians?" (2016: 105). More fundamentally, assuming that compulsory moral bioenhancement significantly reduced harm, it would come at what many would consider an unacceptable cost: it would leave no space for freedom. For some, the magnitude of actual and potential harm resulting from voluntary action might be significant enough to justify imposing limits on our actions. For others, no amount of harm from these actions could justify such limits. Compulsory moral bioenhancement would undermine the freedom of choice necessary for moral responsibility, praise, and blame. Without the ability to choose among different courses of action—good and bad, right and wrong—we would no longer be moral agents. Some might equate moral bioenhancement with moral progress insofar as it reduced harm among humankind. Yet harm reduction alone would not be sufficient for a robust conception of moral progress. Indeed, depriving people of their agency through compulsory moral bioenhancement would eliminate the "moral" in moral progress. Our actions would not be our own, but the products of brain-altering interventions forced on rather than chosen by us. The main ethical question is whether individual freedom to perform harmful, even evil, actions could be overridden by collective interest in reducing and preventing harm. There may be an intractable conflict between individual and collective values on this question. This turns on a larger question at the intersection of the ethics of neuroscience and the neuroscience of ethics: whether or how we should intervene in the brain to modify human behavior.

3.5 Chronic Disorders of Consciousness

3.5.1 Diagnostic Criteria

The most broadly discussed area in the ethics of neuroscience in the last 16 years has been chronic disorders of consciousness. These disorders result from different types of brain injury and involve ethical questions related to diagnosis, prognosis, and which forms of care would be in patients' best interests (Fins 2005, 2015). There are three chronic disorders of consciousness: coma, the vegetative state (VS), and the minimally conscious state (MCS). These are distinct from shorter-duration disorders of consciousness such as epilepsy or longer-duration disorders of consciousness such as hemispatial neglect—conditions which neuroethics has long neglected given its perhaps narrow conception of these kinds of disorders.

Following severe brain injury from trauma, anoxia due to cardiac arrest, or infection, some people lose consciousness and fall into a coma. This state is characterized by a lack of arousal and awareness in the form of complete unresponsiveness (Plum and Posner 1966). Brainstem reflexes and autonomic functions are intact, but integrated cortical function is absent. Some comatose patients regain full arousal and awareness, usually within 2–4 weeks of brain injury. Others eventually lose all brain functions and die. Still others progress from coma to a VS in which they show arousal in sleep-wake cycles, but are not aware of their self or their surroundings (Jennett and Plum 1972; Multi-Society Task Force on PVS 1994). The VS has also been described as unresponsive wakefulness syndrome (UWS) (Laureys and Boly 2012). Many neurologists consider a persistent VS to become a permanent VS 3 months after an anoxic brain injury or 12 months after a traumatic brain injury. Permanent VS patients have no possibility of recovering any degree of awareness. Some persistently vegetative patients progress to the MCS, in which they have varying degrees of awareness. This depends on the extent of intact thalamic-cortical connectivity and axonal connectivity in white-matter tracts of the brain (Fins 2015: ch. 8). The MCS is characterized by overt unresponsiveness despite intermittent evidence of awareness (Giacino et al. 2002; Bernat 2009). Awareness may be confirmed by inconsistent but reproducible neural responses to stimuli such as verbal commands through EEG or fMRI. Emergence from the MCS is defined as the recovery of reliable and consistent responses displaying the ability to communicate with others.

The MCS became a diagnostic category in 2002 (Giacino et al. 2002). Distinguishing the MCS from the VS was significant because the greater degree of integrated brain function in minimally conscious patients suggests a more favorable prognosis and possible recovery of some cognitive and motor functions (Fins 2015: ch. 8). The incidence of misdiagnosing minimally conscious patients as vegetative is alarmingly high—around 40% (Bernat 2009). Adrian Owen and co-investigators drew attention to this problem in a study using fMRI to show that a woman with a traumatic brain injury presumed to be vegetative was covertly aware (Owen et al. 2006). Her minimally conscious state was confirmed by activation in cortical regions when she was asked to imagine performing certain activities around her home. The diagnostic difference between the VS and the MCS has prognostic implications that bear on decisions about care for patients in the second group and the need for research that might lead to therapeutic interventions.

3.5.2 Prognostic Uncertainty

A mistaken diagnosis of a patient as permanently vegetative rather than minimally conscious could lead to withdrawal of life-sustaining artificial nutrition and hydration. By causing the patient's death, this action could preclude interventions that might restore a greater degree of cognitive and motor functions. These interventions could include psychotropic drugs or deep brain stimulation to unlock thalamic-cortical connectivity disrupted by brain injury.

Nevertheless, the general prognosis of these patients is fraught with uncertainty (Bernat 2016). The period over which a patient may recover cognitive and motor functions can vary from weeks to months to years. There is considerable variability among these patients in the extent of recovery (Fins 2015: ch. 8). The degree of intact neuronal connectivity alone may not be predictive of long-term outcomes. In fact, while there have been a few cases of patients receiving the drug zolpidem or DBS who have recovered a significant degree of cognitive and motor functions, there have been no cases of patients who have recovered these functions to pre-injury levels (Luaute et al. 2010; Sen et al. 2010; Magrassi et al. 2016). More specifically, some studies have shown that thalamic stimulation has resulted in some improvement in arousal, according to the Coma Recovery Scale, and reduced limb spasticity in some patients in the VS and MCS. In a recent study, however, "none of these patients returned to a fully conscious state" (Magrassi et al. 2016: e1). In 2010, the authors of one review of DBS for disorders of consciousness concluded: "Current research is extremely limited both in the number of patients studied, as well as the clinical situations in which to best consider the application of DBS therapy. Hence, no definitive conclusions can be made based on current data regarding whether to attempt DBS therapy, or even in which patients it would be most beneficial" (Sen et al. 2010: 5). It cannot be predicted whether different forms or targets of neuromodulation might restore a significant degree of cognitive and motor control for these patients in the near future.

One question that has been raised in the neuroethics literature is whether it is better to be minimally conscious than vegetative (Kahane and Savulescu 2009; Wilkinson and Savulescu 2013). Some studies have indicated that MCS patients can experience pain, which is less likely among VS patients because they have less neural integration in regions constituting the "pain matrix" (Boly et al. 2008). Even if analgesics can control pain, MCS patients may have a level of awareness high enough to be conscious of their mental and physical limitations and functional dependence on others, which could cause them to suffer. Whether they are better off may depend on whether interventions could increase axonal and thalamic-cortical connectivity and result in meaningful recovery of cognitive and motor functions and emergence from the MCS. This could offset any pain or suffering they might experience while minimally conscious.

Because of their inability to communicate verbally or gesturally, it can be difficult if not impossible to know what the experience of being covertly aware is like for these patients or their quality of life. When they are in this state, it cannot be known what their best interests are or which treatments would be consistent with these interests. EEG and fMRI may reveal that they are aware, but not what it is like for them to be in this state. Brain-computer interfaces (BCIs) have enabled some patients with locked-in syndrome to express their attitudes about quality of life (Birbaumer et al. 2008, 2014). It may also be possible for them to express wishes about treatment, including life-sustaining treatment. But these patients are fully conscious, with most if not all of their cognitive capacities intact. This may enable them to process the semantic information necessary to communicate through a device when they are unable to do this by ordinary means. In contrast, the cognitive impairment in most MCS patients may preclude them from communicating through a BCI. They might not be able to express their wishes about life-sustaining treatment or give informed consent to continue or discontinue it. Theoretically, a surrogate could give proxy consent on the patient's behalf, though the high stakes of withdrawing treatment may rule out anything other than a clear expression of an informed and considered judgment from the patient. Still, assuming that they are acting in their best interests, surrogates could give proxy consent to allow these patients to participate in research aimed at developing therapies for them (Farisco et al. 2014).

Given the severity of their injury and the judgment that they are worse off than other patients with other conditions, some would claim that priority in medical research should be given to conducting a sufficient number of clinical trials to test the safety and efficacy of potential therapeutic interventions for MCS patients. The claim would be driven by a principle of justice whereby priority should be given to the worse off (Rawls 1971: 40–45). It would be supported by the severely debilitating consequences of anoxic and traumatic brain injury and the large number of patients with this disorder of consciousness.

It has been estimated that there are between 250,000 and 300,000 patients with disorders of consciousness lingering in long-term care facilities in the United States alone (Fins 2015). Those with more integrated brain activity and the potential for some degree of functional recovery may be appropriate candidates for clinical trials. These might involve attempts to replicate a technique used in an earlier case, where stimulation of the central thalamus to "unlock" the mesolimbic circuit and thalamic-cortical connections in an MCS patient restored some degree of cognitive and motor function (Schiff et al. 2007; Fins 2015: ch. 15). This potential therapy could justify providing more funding for the research. Priority to the medically worse off is not absolute, however. It is conditional on outcomes of interventions aimed at meeting the greater needs that justify priority. Whether more resources should go into research testing potential therapies for patients with chronic disorders of consciousness than research for other conditions would depend on the probability of the research leading to significant functional recovery in these patients. A claim of priority for this group could be weaker if the research did not yield promising results. The strength or weakness of the priority claim would also depend on how outcomes of research into potential therapies for MCS compared with outcomes of research into therapies for other medical conditions. These are open questions that need to be addressed at both medical professional and health policy levels.

3.6 Conclusion: Future Directions

Ethical questions generated by the ability to measure and intervene in the brain will continue to be debated by neuroscientists, legal theorists, social scientists, philosophers, and the general public. New technologies may generate new questions beyond those I have discussed in the four previous sections about neuroimaging, functional neurosurgery for psychiatric disorders, cognitive and moral enhancement, and chronic disorders of consciousness. The ethical questions raised by these technologies can be framed in terms of the duality between the ethics of neuroscience and the neuroscience of ethics.

Arguably, the most rapidly advancing and ethically challenging area of neuroscience is neuromodulation in general and neural prosthetics in particular. Unlike invasive forms of neuromodulation, such as DBS, noninvasive neuromodulating techniques can change neural tissue and circuits while avoiding the risk of intracranial surgery. These techniques may lead to more favorable risk-benefit ratios for patients and research participants. They include optogenetics, which uses light to activate or inhibit targeted cells in the brain (Deisseroth 2015), and focused ultrasound as a "cleaner" form of lesioning brain tissue (Lipsman et al. 2013). Yet because these are novel interventions, it will be some time before the actual benefit and long-term risk can be known.

Existing techniques such as transcranial magnetic stimulation (TMS), repeated transcranial magnetic stimulation (rTMS), and transcranial direct current stimulation (tDCS) involve less risk. But the current from these techniques may not penetrate deeply enough into all subcortical regions implicated in neurological and psychiatric disorders (Rossi et al., 2009). Microelectrode chips implanted in motor and parietal cortices may enable people with extensive paralysis to translate neural signals into moving a computer cursor or robotic limb (Hochberg et al. 2006, 2012). As mentioned in the preceding section, BCIs may enable individuals who are verbally or behaviorally nonresponsive to communicate their wishes about care. These techniques may enable some patients to recover some degree of agency lost through brain injury. Like DBS, though, they also raise questions about autonomy and behavior control. The ethical concern is not limited to the question of whether the person or a device controls behavior. It is possible that the information on a microelectrode array in the motor cortex could be hacked by an external source. This could disrupt the intended function of the device and undermine the shared behavior control between the person and the device. Consistent with the approach to existing experimental and investigational forms of neuromodulation, policies have to be implemented and enforced to protect patients and research participants from potential and actual harm in novel brain interventions.

A more general philosophical concern is that development of more sophisticated neural prosthetics could gradually replace natural circuits in the brain and transform it into a completely artificial organ. This could turn us into complete machines or cyborgs in a transhuman world. But the expectation that future devices may seamlessly integrate with the brain and body does not imply that they will replace them. Neural prosthetics will likely continue to compensate for neural dysfunction while supplementing rather than supplanting normal functioning neural circuits. Any fear that these devices might cause us to lose our personhood or humanity would be unfounded.

Nevertheless, complete replacement of a natural brain with an artificially constructed one is a theoretical possibility. The Human Brain Project is a large collaborative endeavor whose initial aim was to achieve a multi-level integrated understanding of neural structure and function through simulation of the entire brain (Markram et al. 2011), though it has had a number of organizational problems (Fregnac and Laurent 2014). For this and other reasons, the Project now has the more modest goal of developing platforms for neurocomputing and neurorobotics research and development. In addition, the Human Connectome Project aims to create a comprehensive network map of brain circuitry and connectivity through different imaging modalities (Fornito et al. 2015). There will likely be other initiatives attempting to reconstruct the brain. But it is questionable whether an artificially constructed brain could replicate the complexity of the central nervous system and how its natural structure and function are shaped by dynamic interaction between and among genetics, epigenetics, bodily systems, and the environment. It is also questionable whether an artificial brain could replicate the mental capacities that drive the main questions in the ethics of neuroscience and neuroscience of ethics.

Persons are constituted by their brains, but are not identical to them. The contents of our mental states are shaped not only by our brains, but also by the natural and social contexts in which we are embedded. Ethical questions about how neural prosthetics will affect persons need to be informed by factors both inside and outside of the brain. These questions in turn should inform the development of technologies designed to gain a better understanding of the differences between normal and diseased brains, reduce harm, and improve the quality of our lives.[1]

[1] I am grateful to Eric Racine and John Aspler for many helpful comments and suggestions on an earlier draft of this chapter.

References

Agar N (2014) A question about defining moral bioenhancement. J Med Ethics 40:369–370
Anonymous (2005) How volunteering for an MRI scan changed my life. Nature 434:17
Appelbaum P (2007) Assessment of patients' competence to consent to treatment. N Engl J Med 357:1834–1840
Appelbaum P, Grisso T (1988) Assessing patients' capacities to consent to treatment. N Engl J Med 319:1635–1638
Ariely D, Berns G (2010) Neuromarketing: the hope and hype of neuroimaging in business. Nat Rev Neurosci 11:284–292
Bartels A (2012) Oxytocin and the social brain: beware the complexity. Neuropsychopharmacology 37:1795–1796
Beauchamp T, Childress J (2012) Principles of biomedical ethics, 7th edn. Oxford University Press, New York
Bell E, Racine E (2009) Enthusiasm for fMRI often overlooks its dependence on task selection and performance. Am J Bioeth 9(1):3–13
Bell E, Racine E, Chiasson P, Dufourcq-Brana DL, Dunn LB, Fins J et al (2014) Beyond consent in research: revisiting vulnerability in deep brain stimulation for psychiatric disorders. Camb Q Healthc Ethics 23:361–368
Benabid A-L (2007) What the future holds for deep brain stimulation. Expert Rev Med Devices 4:895–903
Benabid A-L, Pollak P, Louveau A, Henry S, de Rougemont J (1987) Combined (thalamotomy and stimulation) stereotactic surgery of the VIM thalamic nucleus for bilateral Parkinson disease. Appl Neurophysiol 50:344–346
Bernat J (2009) Chronic consciousness disorders. Annu Rev Med 60:381–392
Bernat J (2016) Prognostic limitations of syndromic diagnosis in disorders of consciousness. AJOB Neurosci 7(1):46–48
Birbaumer N, Murguialday A, Cohen L (2008) Brain-computer interfaces in paralysis. Curr Opin Neurol 21:634–638
Birbaumer N, Gallegos-Ayala G, Wildgruber M, Silvoni S, Soekadar S (2014) Direct brain control and communication in paralysis. Brain Topogr 27:4–11
Boly M, Faymonville M-E, Schnakers C, Peigneux P, Lambermont B, Phillips C et al (2008) Perception of pain in the minimally conscious state with PET activation: an observational study. Lancet Neurol 7:1013–1020
Christen M, Bittlinger M, Walter H, Muller S (2012) Dealing with side effects of deep brain stimulation: lessons learned from stimulating the STN. AJOB Neurosci 3(1):37–48
Crockett M (2014) Moral bioenhancement: a neuroscientific perspective. J Med Ethics 40:370–371
Crockett M (2016) Morphing morals: neurochemical modulations of moral judgment and behavior. In: Liao, pp 237–245
Crockett M, Clark L, Hauser M, Robbins T (2010) Serotonin selectively influences moral judgment and behavior through effects on harm aversion. Proc Natl Acad Sci U S A 107:17433–17438
Daniels N (1985) Just health care. Cambridge University Press, New York
De Jong R, Bolt I, Schermer M, Olivier B (2008) Botox for the brain: enhancement of cognitive, emotional and pro-social behavior and blunting of unwanted memories. Neurosci Biobehav Rev 32:760–776
Decety J, Cowell J (2015) The equivocal relationship between morality and empathy. In: Decety J, Wheatley T (eds) The moral brain: a multidisciplinary perspective. MIT Press, Cambridge
Decety J, Michalska K, Kinzler K (2012) The contribution of emotion and cognition to moral sensitivity: a neurodevelopmental study. Cereb Cortex 22:209–220
Deisseroth K (2015) Optogenetics and the circuit dynamics of psychiatric disease. J Am Med Assoc 313:2019–2020
Douglas T (2008) Moral enhancement. J Appl Philos 25:228–245

Dunn L, Holtzheimer P, Hoop J, Mayberg H, Roberts L, Appelbaum P (2011) Ethical issues in deep brain stimulation research for treatment-resistant depression: focus on risk and consent. AJOB Neurosci 2(1):29–36

Dworkin G (1988) The theory and practice of autonomy. Cambridge University Press, New York

Eklund A, Nichols T, Knutsson H (2016) Cluster failure: why fMRI inferences for spatial extent have inflated false-positive rates. Proc Natl Acad Sci U S A 113:7900–7905

Elliott C (2008) In defense of the beta-blocker. The Atlantic, Aug 20. http://www.theatlantic.com/magazine/archive/2008/08/in-defense-of-the-beta-blocker/606/

Farah M (2002) Emerging ethical issues in neuroscience. Nat Neurosci 5:1123–1129

Farah M, Wolpe PR (2004) Monitoring and manipulating brain function: new neuroscience technologies and their ethical implications. Hastings Cent Rep 34(3):35–45

Farah M, Illes J, Cook-Deegan R, Gardner H, Kandel E, King P et al (2004) Neurocognitive enhancement: what can we do and what should we do? Nat Rev Neurosci 5:421–425

Farisco M, Evers K, Petrini C (2014) Biomedical research involving patients with disorders of consciousness: ethical and legal dimensions. Annali Dell Istituto Superiori Di Sanita 50:221–228

Fins JJ (2005) Rethinking disorders of consciousness: new research and its implications. Hastings Cent Rep 35(2):22–24

Fins JJ (2015) Rights come to mind: brain injury, ethics, and the struggle for consciousness. Cambridge University Press, New York

Fischer JM, Ravizza M (1998) Responsibility and control: a theory of moral responsibility. Cambridge University Press, New York

Ford P (2009) Vulnerable brains: research ethics and neurosurgical patients. J Law Med Ethics 37(1):73–82

Forlini C, Gauthier S, Racine E (2013) Should physicians prescribe cognitive enhancers to healthy individuals? Can Med Assoc J 185:1047–1050

Fornito A, Zalesky A, Breakspear M (2015) The connectomics of brain disorders. Nat Rev Neurosci 16:159–172

Fregnac Y, Laurent G (2014) Neuroscience: where is the brain in the human brain project? Nature 513:27–29

Gauthier D (1986) Morals by agreement. Oxford University Press, New York

Giacino J, Ashwal S, Childs N, Cranford R, Jennett B, Katz D et al (2002) The minimally conscious state: definition and diagnostic criteria. Neurology 58:349–353

Glannon W (2007) Bioethics and the brain. Oxford University Press, New York

Glannon W (2010) Consent to deep brain stimulation for neurological and psychiatric disorders. J Clin Ethics 21:104–111

Glannon W (2011) Brain, body and mind: neuroethics with a human face. Oxford University Press, New York

Glannon W (2014) The limitations and potential of neuroimaging in the criminal law. J Ethics 18:153–170

Harris J (2016) How to be good: the possibility of moral enhancement. Oxford University Press, Oxford

Heinz A, Kipke R, Heimann H, Wiesing U (2012) Cognitive neuroenhancement: false assumptions in the ethical debate. J Med Ethics 38:372–375

Hobbes T (1651/1987) In: Oakeshott M (ed) Leviathan. Basil Blackwell, Oxford

Hochberg H, Serruya M, Friehs G, Mukand J, Saleh M, Caplan A et al (2006) Neuronal ensemble control of prosthetic devices by a human with tetraplegia. Nature 442:164–171

Hochberg H, Bacher D, Jarosiewicz B, Masse N, Simerai J, Vogel J et al (2012) Reach and grasp by people with tetraplegia using a neurally controlled robotic arm. Nature 485:372–375

Hurlemann R, Scheele D (2016) Dissecting the role of oxytocin in the formation and loss of social relationships. Biol Psychiatry 79:185–193

Illes J (ed) (2006) Neuroethics: defining the issues in theory, practice and policy. Oxford University Press, New York

Illes J, Racine E (2005) Imaging or imagining? A neuroethics challenge informed by genetics. Am J Bioeth 5(2):5–18

Illes J, Kirschen M, Edwards E, Stanford L, Bandettini P, Cho M et al (2006) Incidental findings in brain imaging research. Science 311:783–784
Iuculano T, Cohen Kadosh R (2013) The mental cost of cognitive enhancement. J Neurosci 33:4482–4486
Jennett B, Plum F (1972) Persistent vegetative state after brain damage: a syndrome in search of a name. Lancet 177:734–737
Jones O, Wagner A, Faigman D, Raichle M (2013) Neuroscientists in court. Nat Rev Neurosci 14:730–736
Joanette Y, Stemmer B, Assal G, Whitaker H (1993) From theory to practice: the unconventional contribution of Gottlieb Burckhardt to psychosurgery. Brain Lang 45:572–586
Jotterand F, Dubljevic V (eds) (2016) Cognitive enhancement: ethical and policy implications in international perspectives. Oxford University Press, Oxford
Juengst E (1998) What does enhancement mean? In: Parens E (ed) Enhancing human traits: ethical and social implications. Georgetown University Press, Washington, pp 29–47
Kahane G, Savulescu J (2009) Brain damage and the moral significance of consciousness. J Med Philos 34:6–26
Kamm FM (2009) Neuroscience and moral reasoning: a note on recent research. Philos Public Aff 37:330–345
Kant I (1785/1998) Groundwork of the metaphysics of morals. Gregor M (trans and ed). Cambridge University Press, New York
Laureys S, Boly M (2012) Unresponsive wakefulness syndrome. Arch Ital Biol 150:31–35
Laxton A, Tang-Wai D, McAndrews M, Zumsteg D, Wennberg R, Keren R et al (2010) A phase 1 trial of deep brain stimulation of memory circuits in Alzheimer's disease. Ann Neurol 68:521–534
Levy N (2007) Neuroethics: challenges for the 21st century. Cambridge University Press, Cambridge
Liao SM (ed) (2016) Moral brains: the neuroscience of morality. Oxford University Press, Oxford
Lipsman N, Glannon W (2013) Brain, mind and machine: what are the implications of deep brain stimulation for perceptions of personal identity, agency and free will? Bioethics 27:465–470
Lipsman N, Lozano A (2014) Targeting emotional circuits with deep brain stimulation in refractory anorexia nervosa. Neuropsychopharmacology 39:250–251
Lipsman N, Giacobbe P, Bernstein M, Lozano A (2012) Informed consent for clinical trials of deep brain stimulation in psychiatric disease: challenges and implications for trial design. J Med Ethics 38:107–111
Lipsman N, Schwartz M, Huang Y, Lee L, Sankar T, Chapman M et al (2013) MR-guided focused ultrasound thalamotomy for essential tremor: a proof-of-concept study. Lancet Neurol 12:462–468
Lozano A, Lipsman N (2013) Probing and regulating dysfunctional circuits using deep brain stimulation. Neuron 77:406–424
Lozano A, Mayberg H, Giacobbe P, Hamani C, Craddock R, Kennedy S (2008) Subcallosal cingulate gyrus deep brain stimulation for treatment-resistant depression. Biol Psychiatry 64:451–461
Luaute J, Maucourt-Boulch L, Tell F, Quelard T, Sarraf J, Iwaz D et al (2010) Long-term outcomes of chronic minimally conscious and vegetative states. Neurology 75:246–252
Magrassi L, Maggioni G, Pistarini C, Perri C, Bastianello S, Zippo A et al (2016) Results of a prospective study (CATS) on the effects of thalamic stimulation in minimally conscious and vegetative state patients. J Neurosurg 125(4):972–981
Maher B (2008) Poll results: look who's doping. Nature 452:674–675
Mallet L, Polosan M, Nematollah J, Baup N, Welter M-L, Fontaine D et al (2008) Subthalamic nucleus stimulation in severe obsessive-compulsive disorder. N Engl J Med 359:2121–2134
Markram H, Meier K, Lippert T, Grillner S, Frackowak R, Dehaene S et al (2011) Introducing the human brain project. Proc Comput Sci 7:39–42
Maslen H, Pugh J, Savulescu J (2015) The ethics of deep brain stimulation for the treatment of anorexia nervosa. Neuroethics 8:215–230

Mayberg H, Lozano A, Voon V, McNeely H, Seminowicz D, Hamani C et al (2005) Deep brain stimulation for treatment-resistant depression. Neuron 45:651–660
Mele A (1995) Autonomous agents: from self-control to autonomy. Oxford University Press, New York
Mill JS (1863/1987) In: Ryan A (ed) Utilitarianism and other essays. Penguin, London
Morse S (2006) Brain overclaim syndrome and criminal responsibility: a diagnostic note. Ohio State J Crim Law 3:397–412
Morse S (2011) Lost in translation? An essay on law and neuroscience. In: Freeman M (ed) Law and neuroscience: current legal issues, vol 13. Oxford University Press, Oxford, pp 529–562
Muller S, Christen M (2011) Deep brain stimulation in Parkinsonian patients—ethical evaluation of cognitive, affective and behavioral sequelae. AJOB Neurosci 2(1):3–13
Multi-Society Task Force on PVS (1994) Medical aspects of the persistent vegetative state. N Engl J Med 330:1572–1579
Oliveira-Souza R, Zahn R, Moll J (2016) The neuropsychiatry of moral cognition and social conduct. In: Liao, pp 203–236
Owen A, Coleman M, Boly M, Davis M, Laureys S, Pickard J (2006) Detecting awareness in the vegetative state. Science 313:1402
Persson I, Savulescu J (2008) The perils of cognitive enhancement and the urgent imperative to enhance the moral character of humanity. J Appl Philos 25:162–177
Persson I, Savulescu J (2012) Unfit for the future: the need for moral enhancement. Oxford University Press, Oxford
Pessoa L (2013) The cognitive-emotional brain: from interactions to integration. MIT Press, Cambridge
Plum F, Posner J (1966) The diagnosis of stupor and coma. F.A. Davis Co., Philadelphia
Pontius A (1973) Neuroethics of walking in the newborn. Percept Mot Skills 37:235–245
Pressman J (1998) Last resort: psychosurgery and the limits of medicine. Cambridge University Press, New York
Rabins P, Appleby B, Brandt J, DeLong M, Dunn L, Gabriels L et al (2009) Scientific and ethical issues related to deep brain stimulation for disorders of mood, behavior and thought. Arch Gen Psychiatry 66:931–937
Racine E (2010) Pragmatic neuroethics: improving treatment and understanding of the mind-brain. MIT Press, Cambridge
Rawls J (1971) A theory of justice. Belknap Harvard Press, Cambridge
Roskies A (2002) Neuroethics for the new millennium. Neuron 35:21–23
Roskies A (2008) Neuroimaging and inferential distance. Neuroethics 1(1):19–30
Roskies A (2013a) Brain imaging techniques. In: Morse S, Roskies A (eds) A primer on criminal law and neuroscience. Oxford University Press, New York, pp 37–74
Roskies A (2013b) Other neuroscientific techniques. In: Morse and Roskies, pp 75–88
Rossi S, Hallett M, Rossini P, Pascual-Leone A (2009) Safety, ethical considerations and application guidelines for the use of transcranial magnetic stimulation in clinical practice and research. Clin Neuropsy 120:2008–2039
Safire W (2002) Visions of a new field of 'neuroethics'. In: Marcus S (ed) Neuroethics: mapping the field. Dana Press, New York, pp 3–9
Schauer F (2009) Neuroscience, lie-detection and the law. Trends Cogn Sci 14:101–103
Schiff N, Giacino J, Kalmar K, Victor J, Baker K, Gerber M et al (2007) Behavioral improvements with thalamic stimulation after severe traumatic brain injury. Nature 448:600–603
Sen A, Peter B, Campbell P, Yadla S, Jallo J, Sharan A (2010) Deep brain stimulation in the management of disorders of consciousness: a review of physiology, previous reports and ethical considerations. J Neurosurg 29:4–5
Skuban T, Hardenacke K, Woopen C, Kuhn J (2011) Informed consent in deep brain stimulation—ethical considerations in a stress field of pride and prejudice. Front Integr Neurosci 5:7
Synofzik M (2009) Ethically justified, clinically applicable criteria for physician decision-making in psychopharmacological enhancement. Neuroethics 2(2):89–102
Taylor C (1991) The ethics of authenticity. Harvard University Press, Cambridge

Tome A (2005) Listening to the past: history, psychiatry and anxiety. Can J Psychiatr 50:373–380
Vincent N (2011) Neuroimaging and responsibility assessments. Neuroethics 4(1):35–49
Volkow N, Fowler J, Logan J, Alexoff D, Zhu W, Telang F et al (2009) Effects of modafinil on dopamine and dopamine transporters in the male human brain; clinical implications. J Am Med Assoc 301:1148–1154
Wilens T, Adler L, Adams J, Sgambati S, Rostroson J, Sawtelle R et al (2008) Misuse and diversion of stimulants prescribed for ADHD: a systemic review of the literature. J Am Acad Child Adolesc Psychiatry 47:21–31
Wilkinson D, Savulescu J (2013) Is it better to be minimally conscious than vegetative? J Med Ethics 39:557–558
Wiseman H (2016) The myth of the moral brain: the limits of moral enhancement. MIT Press, Cambridge
Young L, Camprodon J, Hauser M, Pascual-Leone A, Saxe R (2010) Disruption of the right temporoparietal junction with transcranial magnetic stimulation reduces the role of beliefs in moral judgments. Proc Natl Acad Sci U S A 107:6753–6758
Zhang J, Zhu Y, Zhan G, Fenik P, Panossian L, Wang M et al (2014) Extended wakefulness: compromised metabolics in and degeneration of locus ceruleus neurons. J Neurosci 34:4418–4431

Toward a Pragmatic Neuroethics in Theory and Practice

4

Joseph J. Fins

Abstract

The evolution of neuroethics over the past decades has been marked by a penchant for the speculative and distance from the lived reality of patients, families and clinicians confronting severe neuropsychiatric illness and disability. In this chapter, I seek to ground neuroethics more closely to clinical reality by discussing the normative challenges posed by stroke, a common malady which has been mostly overlooked in the annals of neuroethics. Through a thick description of a clinical narrative, a pragmatic neuroethics is articulated.

4.1 Introduction

Recently my mother-in-law had a stroke, a malady that obviously affects the brain, but is not a topic typically the purview of neuroethics. Although about 17 million people have a stroke each year, with 6 million deaths, and 33 million survivors worldwide (Feigin et al. 2014), neuroethics has placed its focus elsewhere, considering the more esoteric topics of neuroimaging, neuroprosthetics, and speculative issues like enhancement and the national security applications/implications of advanced technologies. I, myself, in writing about disorders of consciousness, and

J.J. Fins, M.D., M.A.C.P.
Division of Medical Ethics, Weill Medical College of Cornell University, New York Presbyterian Weill Cornell Medicine, 435 East 70th Street, Suite 4-J, New York, NY 10021, USA

The Consortium for the Advanced Study of Brain Injury, Weill Cornell Medical College and The Rockefeller University, New York, NY, USA

Solomon Center Distinguished Scholar in Medicine, Bioethics and the Law, Yale Law School, New Haven, CT, USA
e-mail: jjfins@med.cornell.edu

E. Racine, J. Aspler (eds.), *Debates About Neuroethics*, Advances in Neuroethics,
DOI 10.1007/978-3-319-54651-3_4

the minimally conscious state (MCS) in particular, could be accused of writing about a condition on the demographic margins (Giacino et al. 2014), though, truth be told, no one has cared enough to make a truly systematic count in the United States, although smaller studies have been performed elsewhere with suggestive data on related conditions (Fins et al. 2007a, b; Stepan et al. 2004; Løvstad et al. 2014). But none of these are definitive or the basis for health policy efforts to meet the needs of this population (Fins et al. 2007b).

But in my work on MCS, I have always tried to be true to its clinical reality (Schiff and Fins 2007; Fins and Schiff 2002; Fins 2005b, 2015b), distinguishing the real from the imagined (Fins 2007b), and the known from the imagined, even as I looked toward future innovation (Schiff et al. 2007, 2009). I have looked for historical antecedents to neuroethics in the works of pioneering figures like William Osler and Wilder Penfield (Fins 2008a, b). I have maintained that there is an older clinical neuroethics that has been forgotten, and which should be restored to prominence, notwithstanding the important contributions of pioneering scholars like Cranford (1998) and Bernat (2008) and more recent work by Ford (Boissy et al. 2008), Racine (2013), and Gillett, the latter of whom has written prudently about the ethics of decompressive hemicraniectomy following malignant stroke (Honeybul et al. 2015). Like these colleagues, I have sought to locate ethical reflection both within clinical reality and in this dedication to a grounded and pragmatic clinical reality (Fins 2005a, 2008c, 2015a). I have focused on the lived experience of patients and their families, rather than on more speculative concerns (Fins and Schiff 2002; Fins and Hersh 2011), which brings me back to my mother-in-law's stroke.

Her case, which I share with the permission of her daughter, was chock full of neuroethics questions, both basic and speculative, and yet our field has not delved into the profound ethical questions that arise in the case of what might appear a rather pedestrian clinical case: a conventional stroke. So, in homage to my mother-in-law, and all the many millions who need our care and attention, I will share her story.

Using narrative methods in medical ethics (Brody 1988; Charon 2008; Fins 2013), I will discuss why this case is important and illustrative of a clinical and grounded neuroethics that has escaped us. I hope to use it to prompt the further development of a pragmatic neuroethics (Fins et al. 1997, 2012; Fins 2005a, 2008c, 2015a; Racine 2008; Brendel 2004), which could go deeper than the traditional formulations of clinical ethics, consumed as it often is with formulaic questions of the withholding and withdrawing of life-sustaining therapy, futility, and the like (Racine 2007). We need to broaden our focus to more fully capture the needs of the clinic. We need to expand the scope beyond the usual confines of clinical ethics (Moeller et al. 2012; Nilson et al. 2008) and broaden it to deeper questions of personal identity and relationality (Noddings 2013).

As we shall see through this case vignette, these are the key questions that arise when a mind is transformed by ischemia and hemorrhage, and language and self are lost. Could anything be more profound, more devastating, and a more appropriate topic for neuroethics? And given the demographics of cerebral vascular accidents, is there much that is more important to neuroethics or to patients and families touched by neuropsychiatric disease? To help better integrate practical realities of care and ethical theory, let me share this narrative. The goal is to further the conversation at the interstices where engaged clinical care requires both immersion in the real and speculation about meaning. This integrated, pragmatic approach is that to which we should aspire.

4.2 Preludes and Prevention

The morning started off like any other. My wife, Amy, was visiting her mother, a robust 88. She was preparing to move to another apartment that she was to purchase. To make a purchase like that so late in one's life was a bold move, but that was Helen. She was determined and forward-looking and always planning for the future. Amy had brought muffins and Helen was hungry. They ate together in the breakfast nook, and then Helen began to develop aphasia, an inability to express her thoughts in fluid and cogent language.

It was not the first time. Over the past several years, she had had several transient ischemic attacks—or mini-strokes—that presented with blood pressure spikes and language difficulties. When her blood pressure was controlled, her speech would return to normal without incident. Head scans, echocardiograms, Holter monitoring, and electroencephalogram were all unrevealing, and the etiology of her events became a diagnosis of exclusion. We were greeted with clinical "happy talk" that she had a "good-looking brain" for someone her age. I interpreted that as good gray matter and not the deep sulci one sees in patients with Alzheimer's disease. Great, I thought. But that isn't the question. No one suspected cognitive impairment, but her doctors told us what the tests could explain, not that which they couldn't.

My family members—and Helen in particular—were reassured, I suspect by the negative test results, but I was worried about a Sword of Damocles which always hovered over us because we remained ignorant of the causes of Helen's mishaps. Statistically, my mother-in-law was still at risk. If we didn't understand what had happened, we could not truly prevent a recurrence. I worried that it was only a matter of time before a more life-altering, or life-threatening, event would occur.

And here is our first question for a clinical neuroethics. How do we deal with the uncertainty of Bayesian theory and population-based recommendations about preventive treatment with anticoagulation, statins, and the like (Goodman 2002)? That is, how do we deal with probabilities of risk in clinical practice, and how do we translate data from large clinical trials to the needs of unique individual patients? When we evaluate patients who are symptomatic and need evaluation, what is the sensitivity and specificity of our testing? What did a negative test really mean? That nothing had overtly changed, but the underlying pathophysiologic predicate for a recurrence remained? These are questions of translational medicine that require a plurality of approaches and fonts of medical knowledge (Solomon 2014; Fins 2014) and which appeal to anthropology and what has been described as ordinary ethics (Lambek 2010; Zizzo et al. 2016).

Most critically, they require careful explication. When clinicians speak of a negative test, especially one that deals with the complexity of the human brain and its centrality to who we are and will be, their words need to be laced with humility and compassion. Humility, because no test can really provide either the reassurance that all will remain well or compassion, since that news can be terribly disquieting. Few doctors can get this balance right and that is also a question for neuroethics. How do we educate and inform patients about the uncertainty attendant to stroke and other complex illnesses, or conditions, and do so in a humane and constructive manner that minimizes crises and disputes (Loupis and Faux 2013)?

This is an important responsibility for our field because nothing really quelled the anxiety that lurked in the background for Helen and our family. Every time my mother-in-law searched for a word, something we all do, she was rushed to the hospital. But the accumulated episodes dulled the sensibilities of her doctors, most of whom were generalists and not neurologists, who perhaps became sanguine about the lurking risks of a cerebral vascular accident. They were providing all the known preventive measures, but failed to fully respond to the emotional stress engendered by each event. This too is another opportunity for neuroethics: the enduring responsibility not to turf or abandon patients whose trajectories are uneven or unpredictable, whether it be in the outpatient clinic or the transitions that follow during hospitalization (Back et al. 2009).

And then there is the clinical reality of primary care providers who develop a negative countertransference (i.e., unreflected upon and unconscious negative emotions) toward patients and/or families (Goldberg 2000), as recurrent events test their patience and the limits of science and prevention. Although all knew *in theory* why her probability was heightened every time she had an event, her doctors remained without any clear-cut or meaningful actionable intelligence with which to respond. They became frustrated and distant.

As a member of a cohort, she was receiving all that modern, but still primitive, medicine could do to prevent thromboembolic stroke. She was placed on clopidogrel, a blood thinner which makes platelets less sticky, but could not tolerate antihypertensive medication because her baseline blood pressures were routinely in the low to normal range. So she was protected from clot, but still vulnerable to elevations in blood pressure.

It struck me as highly ironic and indeed distressing that with all the recent media hype about precision medicine (Joyner 2016), we were still waging battles over the effects of aspirin, warfarin, or one of the newly marketed antiplatelet agents. While there is no shame in academic debate, the reality presented to the public implies a degree of confidence and indeed hubris about the state of our emerging knowledge. What was being celebrated on TV was not so precise or tailored to Helen's specific needs or the biology of her vasculature. Stroke prevention remained rather crude and, in principle, was not much more evolved than what I had learned in medical school three decades earlier. Yet, one would not know this if you watched the evening news and direct-to-consumer advertisements of anticoagulants which give viewers a false sense of security.

The marketing of these drugs is another real-world ethics issue, not only because of the influence of pharma on practice, but on perhaps the more nefarious consequence of the messaging and the power of the crafted imagery that go into these direct-to-consumer ads (Biegler and Vargas 2016): if you take these drugs, you won't have a stroke, you'll be fine, and you can go golfing with a pro who also suffers from atrial fibrillation, an abnormal heart rhythm that affects the upper chambers of the heart. The message is that modern medicine has reached its apogee and all will be well. That is, until you or your loved one is not. And when that happens, will it be harder to accept the frailty of old age or infirmity? After all, the incessant ads have painted a different picture of an active old age without infirmity, with treatments that are increasingly less burdensome.

Despite what I saw each night advertised on the news, I was still worried—even if marketers sought to reassure the laity, and sell the public their products. But I worried for all those with a history of mini-strokes, who remained at increased risk because of their underlying pathophysiology. At an individual level, if we didn't understand what had happened, could we truly prevent a recurrence?

So yet another question for clinical neuroethics: how do we deal with this uncertainty at the intersection of the doctor-patient relationship and what is now euphemistically called population medicine? As a member of a cohort, Helen was getting what she was supposed to receive, all those population-based recommendations about treating with anticoagulation, statins, and the like. But I was skeptical about its effect on preventing what I thought would eventually take place.

4.3 Watershed Decisions

So I wasn't entirely surprised to receive a call from Amy that her mother was having word-finding problems. We had been there before, but this time the symptoms seemed to linger and were more intense. Although it was a Sunday morning, I was in the office when I received the call about 20 minutes into the onset of symptoms. Amy put Helen on the phone and she was clearly aphasic. In response to questions like how she felt, she replied, "I need directions," a non sequitur and many more like that. I told Amy she needed to take her mom to the emergency room at the nearby hospital. Unlike past episodes, we couldn't wait this one out. It seemed, at least on the face of it, more serious.

Helen had been under a lot of stress the past weeks. I had seen her the week before, and she was worried about her finances, her ability to make the move, and all the logistics that it entailed. Although we don't know precisely how stress or inflammation, its stigmata at the more molecular level, play into stroke, the relationship of mind to brain and how one's psychological state affects underlying neural mechanisms is another rich area of neuroethics, Cartesian dualism notwithstanding, if we ground one's state of mind in forces beyond biological reductionism.

Certainly, the stress Helen was experiencing, whether it was more psychological or physiological, increased the probability of an event that morning. Like the run up to the move and the closing on the apartment, her event had been brewing, how precisely we just don't know. But this does raise questions for neuroethics too in the realm of determinism, that is, was this event destined, even fated. Who would have ever thought that something as quotidian as a stroke could raise all these foundational issues for philosophy?

When Helen arrived at the emergency room, she was expeditiously triaged because her blood pressure was hovering at 200 mm Hg systolic and she remained aphasic. The doctor in the emergency room was very responsive and very concerned. He immediately ordered a CT scan, which revealed that the stroke was not hemorrhagic and that the ischemia that was causing the aphasia was likely due to thrombosis, or a clot.

The treatment for thrombotic stroke, especially when it is available early in the course of the event, is to treat with a clot-buster so as to restore flow to the area(s) in the brain deprived of blood. So the emergency room doctor contacted an academic neurologist who was involved with clinical trials for tissue plasminogen activator (TPA), an agent which helps to degrade thrombin, an essential ingredient of clot formation (The National Institute 1995; del Zoppo 2004).

Using telemedicine, the neurologist, at a sister hospital 5 miles away, assessed the situation and made a recommendation for TPA. He dutifully explained the basis for the therapy as proportionate given the clear progression of the stroke against the likelihood of bleeding which he characterized as a 6% risk. In general, it was all proper and fulfilled the broad mandate of informed consent about the usual criteria explicating the rationale, risks, benefits, and alternatives. Nonetheless, we were not told that the best centers in the country only achieve a success rate of just 35% in ischemic stroke when the territory involved is supplied by the internal carotid or middle cerebral artery—and when care is provided at leading centers (McKhann 2016; González et al. 2013). But something else was still missing.

In the world I grew up in, doctors did not practice from a distance, and neurologists saw patients in person, actually took a history, and examined them *in person* before venturing to make a treatment decision. Perhaps doctors examining patients is an archaic notion; perhaps I am entrenched in an old reality and a philistine. Who could be against telemedicine as a vehicle to generalize expertise, especially when a recent analysis of outcomes suggests that outcomes in telemedicine-guided IV thrombolysis had similar hemorrhagic complication and mortality rates and no difference in functional communication (Kepplinger et al. 2016)? But I was skeptical about its use at a major academic medical center with two campuses with large hospitals and a newly formed medical school in a densely populated suburb. I wondered how economics has altered clinical practice and how a venture like telemedicine, perhaps well suited to rural and underserved regions, has been co-opted as a cost-saving device for large medical centers that should have the wherewithal to have a neurologist on staff and present to oversee care.

So here is another theme for a clinically grounded neuroethics: how healthcare economics balances access to specialized neurological care against quality of care. Society may not be able to have it both ways, but when it comes to a specialized practice like neurology, there is something to be said about keeping the purview of specialized medicine *specialized.* For example, the quality of brain death assessment differs when done by neurologists versus nonspecialists. Variability in institutional policies about who can make these determinations has been associated with differential error rates in brain death assessment in a national survey. The rigor of the clinical evaluation of brain death is greater when it is done by a specialist than by a generalist, with the latter having higher error rates with their evaluations (Greer et al. 2016).

Beyond the clinical contributions of a locally present practitioner, there is the human dimension of having that skilled interlocutor present when families are asked to make a momentous decision about care in an expeditious manner. All the more so when treatment decisions involved a drug like TPA, which, in breaking up a clot, can also unleash bleeding. Having this conversation by Skype was made harder because of time pressures. The need for a decision was rather urgent: the longer one waits, the firmer the architecture of the clot matrix and the less likely it is that TPA will help unweave the thrombosis.

So the most momentous decision a daughter could make for her mother was made during a quick Skype encounter. It seemed unfair and rather callous. At the very least, you would like to be able to take the measure of the man (or woman) who is telling you to take a risk in the service of health, to entrust your mom to his judgment and the proportionate recommendation to avoid one peril (progression of the thrombotic stoke) versus the possibility of iatrogenic bleeding, bleeding which was a consequence of the treatment.

Given this, neuroethics could also explore the role of the doctor-surrogate relationship in times of pivotal decision-making. This was a topic that I addressed in my book on disorders of consciousness, *Rights Come to Mind: Brain Injury, Ethics and the Struggle for Consciousness* (Fins 2015b). I interviewed over 50 family members of patients who had severe brain injury, and there is no substitute for clinicians with skilled and empathic communication skills, in the sense of being present and face to face with patients and families. When a serious brain insult occurs, surrogates are suddenly thrust into a decision-making role for which most are unprepared. The shock is sudden and wholly unexpected. Lee Woodruff, the wife of the ABC News anchor Bob Woodruff who sustained a serious brain injury during his coverage of the War in Iraq, titled their memoir *In an Instant* to capture how suddenly and irrevocably their lives had been changed (Woodruff and Woodruff 2006).

Whether from a traumatic brain injury or an encroaching ischemic stroke, in a moment one's security and place in the world as a wife, or a daughter, is displaced. The burdens which were shared are now assumed without the counsel of the one person in the world who had always provided advice and love. So instead of making decisions together, surrogates are left to contend with decisions that they themselves need to make in an instant. These choices can be highly technical and hard to understand and process under normal circumstances, much less moments of intense stress when family members are worried about survival and a bereft future. This is hardly the time for a phone discussion, if it can be avoided. Personal contact and solidarity, which may have its basis in neuro- or sociobiology (Ferrari 2014), is necessary because it is more humane and compassionate.

With the decision to proceed made, the ER physician administered the TPA, and Amy was gratified that her mother's speech improved after about a half hour. Her word finding was better, as was her prosody. It looked like the TPA had gotten the better of the nefarious clot lodged somewhere upstream of Broca's area. We were all delighted. Helen was once again showing the resilience that had marked her life and career and becoming herself again.

But soon after those few precious moments of celebration, Helen started to become agitated and uncomfortable. Amy told the ER doctor who quickly ordered another scan to confirm what was already feared: whatever good that had occurred in the speech areas was now overtaken by new areas of bleeding elsewhere in the brain in her frontal cortex, bilateral temporal lobes, and her occipital lobe. She was transferred to the ICU to better manage her blood pressure and to receive clotting factors to reverse the bleeding across her brain.

Clinically, she became stuporous, barely arousable, and nonverbal. She lay in her bed, fidgeting and reaching for what was not there, turning from side to side unable to find comfort. It was distressing to watch, and progressively so, as we learned that subsequent CT scans showed that the bleeding was continuing and was now accompanied by edema—or swelling—as well.

We began to wonder if this was all for naught. Where was this heading? And in the immediate aftermath of her move to the ICU, there was a lot of second guessing. Suppose we had done nothing, if Amy and Helen had waited it out at her apartment. Would the aphasia have passed like it always had in the past? Suppose we had decided against the TPA? What then? It was a dialogue that still swirls in our heads as consequential and all too sad. In speaking with Amy and her brothers, I tried to piece together our logic so as to assuage the guilt we all felt for acquiescing to the recommendation. It was essentially a question of causality and responsibility. Whatever the decision and our felt responsibility, the outcome would be dictated by the biology of stroke and the pharmacology of the TPA. That was the logic, but decisions to proceed with treatment always seem to sting more than ones of omission when complications occur, notwithstanding these analytics.

I tried to console my family members by recapitulating the process of our decision. The stroke was progressing and Helen was getting worse, and it seemed that the ischemic stroke constituted an enduring threat if we had done nothing. And we were right in giving the TPA, I thought, because Helen's speech *did* improve after treatment. That suggested that there was a diagnosis (a clot) amenable to the treatment (the TPA) that was indicated. Because the progression of the ischemic stroke was pretty clear and the risk of bleeding potentially serious but less probable, the right action was to administer the TPA because it was the only course that could prevent the morbidity that Helen had always feared and that we hoped to prevent. It was our only hope. Sadly, any success was quickly overtaken by hemorrhagic complications.

As a physician son-in-law, I entered into these discussions because no one else did. I thought it necessary, notwithstanding the conflicted role I had as a nonobjective family member. Although I have had countless conversations like this with other families as a physician and clinical ethicist, I would have preferred if someone else had engaged in this retrospective analysis with us. I had hoped that the consulting neurologist who recommended the TPA might have made the brief journey to visit us and reassure our family as I was compelled to do. TPA has evolved as the standard of care, although the literature suggests it is frequently, and inappropriately, underutilized (Adeoye et al. 2011). It is hard to confront failure, even when a virtuous effort was made. Maybe that is why we never saw him after the complications set in. It was disheartening and made all that followed all the more difficult.

I wonder if he felt any physicianly obligation to us given his rather virtual contact with Helen and our family. In other contexts, a lack of follow-up might be construed as a violation of the ethical principle of non-abandonment, but that sense of obligation may only be perceived as operative if a doctor appreciates that a doctor-patient relationship has been established, and that the person who received care was more than an image on a screen: another theme for neuroethics in this age of telemedicine.

4.4 Apprehending Aphasia

Our time in ICU was one of uncertainty about how to interpret Helen's condition and gauge what she was experiencing. Initially, she was in a deep stupor and barely responsive. All we had to assess her condition were a series of scans which showed progression of the bleeding and a penumbra of edema—or shadows of swelling—around the areas of hemorrhage. We were adjusting to this new reality, with her surrogates thinking about how to fulfill Helen's prior wishes.

But then, after about 72 h, she seemed to wake up and rally, even speaking a bit, although her aphasia remained and her memory and recognition of family members faltered. She was there but not herself. Still, her speaking at all was quite unexpected, even to me, although I should have known better. I had seen patients improve after the edema recedes and intracranial pressure is reduced on key structures, but this was no ordinary patient. It was Amy's mom. And I of course was not her doctor.

As a family member, I forgot what I should have known. Imagine the surprise of other families, laying crepe and preparing for the worst (Siegler 1975), who return on the third-day post-event to find their loved one sitting up and trying to speak. It was predictable, if not knowable, but no one had provided a roadmap of what might happen and that this could be expected. And it was unfortunate, because after resigning ourselves to an inevitable decline, we were both happy to see some recovery and terrified about the degree of morbidity it might portend.

Her new found speech prompted all kinds of questions. What does an aphasic patient know and experience? She seemed to understand more than she could express, as would be expected based on her initial presentation. But she had additional pathology in her temporal and frontal lobes. She asked Amy, for example, "how is your man?" Amy responded, "You mean, Joe?" Helen replied, "Yes. What does he do?" Amy told her I was a doctor and Helen replied, "That's nice, dear."

It is a simple exchange but prompts us to ask: did she forget me or my name? Did she not recall what I did? How much of herself was gone? Since we can never know the consciousness of the other and only infer it through language, aphasia is a special challenge. Aphasia could be primary and stem from the actual seat of consciousness, the loss or cognition or memory, or it could be that the thinking is clear but that the expression of the thought is not. It is a question that gets to the heart, or maybe the essence, of personal identity. The later explication, where it is an issue of transmission and not the generation of the thought, is less an assault on personhood, suggesting that the self has been preserved.

I recall talking to my father who had transient aphasia a decade ago and recovered to tell the tale, as it were. He vividly recalled being tested for object naming. He was asked by a neurologist to name a "pen." He later told me that he knew it was a pen but could not find the word. He even knew it was a Mont Blanc pen but could not find that proper name, so he said "Germany" in response to the neurologist's query, knowing the pen's origins. The neurologist might have thought the patient confused, but his inner monologue—how he spoke to himself—was intact. He knew what he wanted to say and was struggling with a way to express what was known, understood but frustrated by an inability to engage in output.

There was discordance between what was happening on the inside and what was expressed, much as has been demonstrated with neuroimaging in patients who are in a state of non-behavioral MCS. In such cases, the patient appears vegetative (i.e., in a state of wakeful unresponsiveness) (Jennett and Plum 1972), but actually responds volitionally on neuroimaging when asked to imagine doing a task (Owen et al. 2006). Such cognitive-motor dissociation is a pressing issue in the realm of disorders of consciousness (Schiff 2015), and aphasia could be seen as a parallel scenario where fractured speech may not wholly represent what is going on inside a patient's head. The parallels are striking and suggest a broadening of a categorization in neuroethics that distinguishes phenotypic behaviors and what might be thought of as the underlying analogue to a neurological genotype grounded in intact neural circuitry that remains totally or partially entrapped within the confines of the brain (Fins 2007a). Later into recovery, this distinction can have a bearing on the legal standing and rights of people with aphasia (Morris et al. 2014). This bears on the rights of the disabled and the need to recognize the saliency of nonverbal communication, as noted in the UN Convention on the Rights of the Disabled, which asserts that there are many forms of communication beyond spoken language. It can also and must include: "display of text, Braille, tactile communication, large print, accessible multimedia as well as written, audio, plain-language, human-reader and augmentative and alternative modes, means and formats of communication, including accessible information and communication technology." To disregard these means of communication is to discount the other and to devalue personhood (UN Convention on the Rights of the Disabled 2016).

4.5 Prognostication and the Uncertain Self

Covert cognitive abilities, hidden under a cloak of aphasia, also have a latent potentiality, which is governed by a temporal coefficient. We had already seen how the nonverbal state of deep stupor had been eclipsed by an increased level of arousal with productive speech following the retreat of the edema. The state change imposed by the edema was temporary. How much more improvement could or might occur as blood was resorbed and essential structures and spindly circuits were decompressed and again able to function? Although much of this would be biologically determined, much of what happens next will be determined by what we do, the brain's resilience, and the uncertain prospects of rehabilitation.

Our challenge was that predicting this future was as multivariate as the initial challenges posed by the ischemic stroke. Our challenges were now compounded by perhaps more consequential insults to the frontal cortex and occipital and temporal lobes from the iatrogenic bleeding. And paradoxically, given the early improvement in the aphasia following the initial delivery of the TPA, it might be that the burden of the aphasia might have been ameliorated, so that the burden of hemorrhage might in the end be greater than that of ischemia.

The question was fundamental to one of personal identity. How much of yourself do you have to lose before you cease to be you? Helen, after her rally, had known lesions in areas of her brain associated with language, memory, judgment, and vision. She was able to perceive pain and discomfort, and while she did not always recognize family members who came into her room, she was comforted by their presence nearly until she lost consciousness. So who exactly was the person in the sickbed? Philosophers like Derek Parfit might argue for discontinuity of the self, and there might be a rationale for such an assertion as Helen ceased to be the vibrant (and vivacious) presence we had known and loved, but if she was not there, and not Helen, then why did she brighten when reminded her daughter was in the room?

It is hard to answer this question in the abstract, as most philosophes might, and wrong to be overly dichotomous about the presence or the absence of the self. Like most complex constructs, the self exists on a continuum. In those days after her rally, Helen was weakened and debilitated. Was she her usual self? No, but she was still there as a distinctive human presence who, with assistance, recognized those she loved and who was loved in return by those who loved her. Not perfect continuity, but enough to prompt questions about how to best express our love and maintain fidelity to her prior wishes.

4.6 Preferences, Personal Identity, and Loss

As Helen regained consciousness, we found ourselves in the prototypic Ulysses dilemma, the philosophical trope from the Homeric epic in which the protagonist is heading back home to both his beloved Ithaka and his queen Penelope after the Trojan War. Captaining his vessel, Ulysses instructs his crew to bind him to the masts lest their passage by the seductions of the Sirens cause them to shipwreck before reaching home (Homer 1963). No matter what he says during that passage, he tells his crew to in essence be mutinous and disobey any instructions he might give later in their voyage. A sage leader, he feared that the seductions would alter his state and have him make choices that were inauthentic. The true Ulysses wanted to return home despite any future protestations to the contrary. His instructions, in essence an advance directive, to borrow modern nomenclature, pitted past and considered preferences against current desires.

We too were at sea on our voyage with Helen. Over the course of 18 years, she had completed at least three advance directives stipulating that she would never want artificial nutrition or hydration if she could not return to what she felt was a meaningful recovery, which she deemed as the ability to communicate and live life independently. It was clear enough when she was in a stupor and could not communicate, but less so when she regained speech and indicated that she was hungry, posing the challenge framed by the Ulysses dilemma: past considered stated preferences versus present desires.

Once again, as in the case of the Helen of Troy, another Helen is at the center of an epic question. It forced us to consider questions of personal identity and authenticity. Past and present preferences were seemingly mutually exclusive, so we needed to determine which was more reflective of the "true" Helen. The enduring Helen was the one who consistently opted against being fed under these circumstances pointing to a decision to withhold feedings. Yet it was not that simple, as the current Helen was having the perception of hunger, an uncomfortable experience. Either choice felt like a betrayal. Frankly, we loved *both* Helens.

Moving beyond dichotomous thinking into a more pragmatic space where decisions can be made on a continuum, we decided to respect Helen's prior wishes by not inserting artificial means of nutrition, like a percutaneous gastrostomy tube or the more temporary nasogastric tube. (The latter choice was especially ill advised, as it would not necessarily satiate hunger pangs and brought its own misery and discomfort.) We did decide to offer what are euphemistically referred to as "pleasure feeds," so that patients who are hungry can eat something if they so desire or indicate. This is a palliative (the word derives from "to cloak" or "disguise") (Fins 1992) that allows for the relief of hunger, but does not meet overall caloric needs. The downside is a prolongation of the dying process and the vigil that well-intentioned family members have to endure.

By foregoing artificial and invasive interventions, we were able to honor Helen's long-standing preferences. By providing a modicum of feeding, we sought to palliate the distress of hunger. Like most ethical choices, our approach sought a middle ground and compromise, appreciating that there is more complexity in lived experience and clinical practice than in rarefied ethical theory.

Her children sought to balance the directions she had given prospectively. As surrogates invested with the authority to make judgments when their mother lost decision-making capacity, they were guided by her clear prior wishes. Put more formally, her durable power of attorney sought to be true to substantive wishes, while at the same time not failing to utilize the discretionary moral authority that came with that singular (and therefore empowering) designation.

As I have written, designated surrogates are governed by both contractual and covenantal sources of moral authority based on their knowledge of substantive prior wishes and the fact that they have been chosen (Fins 1999). My empirical research exploring the balance between the contractual and covenantal (Fins et al. 2005;

Hinderer et al. 2015)—or substantive and discretionary—moral authority coalesces into the tripartite invocation of fidelity, wisdom, and love (Fins and Maltby 2003). Fidelity meant keeping true to one's promises, in our case not providing artificial nutrition and hydration, as doing so would violate consistent prior wishes. Wisdom, however, compelled us to not narrowly construe the situation and be aware that the withholding of food had the potential to prompt discomfort, if not suffering. This same pursuit of wisdom prompted a compromise and the heuristic of what are now euphemistically called "pleasure feeds."

In toto, our decisions were motivated by love and the pursuit of beneficence, even altruism, as none of Helen's children wanted to say goodbye. Every moment was precious, even as their mother deteriorated and drifted away. What they did, they did for her, in an act reminiscent of Augustine's definition of true love in *The Confessions*, in which he speaks of how those who abandon themselves, and self-interest, can have true love nurtured in its place (Augustine 1978). To paraphrase a comment that the esteemed theologian Sydney Callahan once made to me, again invoking Augustine: true love is the forgetting of oneself—something echoed by Hannah Arendt in her final work, *Love and Saint Augustine* (Callahan circa 2003; Arendt 1996). And true love was expressed during the hard vigil that followed as all relationality was lost and the end slowly came.

4.7 Why the Disconnect Between Neuroethics and Clinical Ethics?

As difficult and singular an experience as this was for our family, this was not an extraordinary case. The clinical and narrative elements are repeated every time a parent has a stroke, is anticoagulated, and has complications, and children have to make gut-wrenching choices about ongoing care. If prevalence were a marker of utility, cases like this one would dominate the neuroethics landscape. But they don't, despite the philosophical richness that they embody.

And this begs the question, why? Why has neuroethics—and even bioethics—been averse to the commonplace cases, which, as illustrated here, are hardly common or simple in the pragmatic, and even existential, questions that they pose? While leading voices in the palliative care community have usefully sought to articulate core competencies for stroke neurologists and to create protocols for difficult conversations about goals of care and decisions near the end of life (Creutzfeldt et al. 2015), few have paid attention to the deeper ethical questions of personal identity and more practical ethical challenges posed by acute stroke, although it has been addressed in the context of dementia (Dietz et al. 1998; Dresser 2009; Jennings 2003).

One promising exception is the early work by Spokoyny and colleagues who are developing an advance directive designed for those who are at risk of recurrent stroke (Spokoyny et al. 2015). Their COAST (Coordinating Options for Acute Stroke Therapy) instrument seeks to take account of the temporal pressures posed by the need for emergent decision-making about thrombolytics and emerging endovascular therapies by surrogates who may not know the patient's prior wishes. The goal is to ease both surrogate burden and to expedite timely decisions in order to maximize the efficacy of therapy. This tool is currently being piloted.

And this raises the larger question about the relationship between clinical care and neuroethics writ large, not just the question of stroke, but rather the overall clinical utility of neuroethics to the practice of medicine, neurology, neurosurgery, and psychiatry. It seems to me that the field is missing a tremendous opportunity to be meaningfully engaged in the clinical realm and in translational research, an area which the President's Brain Initiative is beginning to recognize, with a recent call for advice on future funding priorities (NIH Brain Initiative's RFI 2016). This is a welcome development, but for it to be sustained against all the other many claims on neuroethics, it is useful to deconstruct why neuroethics and clinical practice seem to have gone their separate ways.

It is all paradoxical and begs for an explanation. Indeed, as one who has sought to straddle both the world of neuroethics and clinical ethics, I have been struck by the lack of overlap of these two communities. So instead of clinical ethicists, or their first cousins palliative care clinicians, attending neuroethics gatherings, I am in the midst of neuroscientists, lawyers, and philosophers.

I am not complaining. I revel in the intelligence and enthusiasm of my friends in neuroethics, but I do miss any sense of the clinical enterprise, the granular reality of caring for patients with neuropsychiatric disorders that afflict millions, but yet occupy a small part of their work space. Having said this, clinical ethics is not without its own blame. As the field has evolved, the clinical practice of ethics consultation has been more focused on the quotidian questions of life and death and ethics at the end of life which can become rather formulaic, informed by governing law on decisions to withhold or withdraw life-sustaining therapy. Deeper questions, of concern to families, and of genuine philosophical import, too often are displaced by the exigencies of clinical need.

Another explication for the disconnect between clinical ethics and neuroethics is one of provenance: the origins of neuroethics did not come out of a clinical space. This demographic is reminiscent of the early days of bioethics itself, when it was dominated by theologians and philosophers, rather than mostly non-physician ethicists, whose interests were less focused on the clinic (Jonsen 1998). The notable exceptions were Jonsen, Siegler, and Winslade who launched clinical ethics as field both independently and through their joint scholarship (Fins and Gracia 2016a, b; Jonsen et al. 1982). Jonsen's fellow casuist and friend, Stephen Toulmin, suggested that this engagement of theory with clinical practice did nothing less than save the life of ethics as an academic discipline (Toulmin 1982). These pioneers laid the groundwork for the following generation of bioethics scholars. And with that transition, deontological reasoning under the general rubric of principlism was supplemented by more pragmatic considerations driven by clinical practice and medical research.

But the pioneers in neuroethics were more like their predecessors in bioethics. Again, most did not have roots in the evolving field of clinical ethics, which involved the rather "pedestrian" questions that come up in routine practice. Most of the early commentators were ethicists, philosophers, and if they had a background in science at all, they were more likely cognitive neuroscientists or psychologists than physicians (Rees and Rose 2004).

And so it comes as no surprise that the early definitions of neuroethics, which have been so formative, did not have a clinical orientation. In fact, the tone was just the opposite, neglecting, even fearing what might be done to or upon the human brain, even if it was a brain afflicted by illness or altered by injury.

So neuroethics evolved as a wary ethic, worried about the advent of new technology and not its therapeutic or explanatory powers to quell human suffering and illness. Just consider the definition of neuroethics offered by the late William Safire, who, as president of the Dana Foundation, had tremendous influence on the early history of the field. In his introduction to Dana's *Neuroethics: Mapping the Field* San Francisco meeting in 2002, Safire notably described neuroethics as the "examination of what is right and wrong, good and bad about the treatment of, perfection of, or unwelcome invasion of and worrisome manipulation of the human brain" (Safire 2002a, b).

It is worth deconstructing the words of Safire—the wordsmith, author of *On Language* (Safire 1980) and long-time columnist for the *New York Times* (Safire 2004)—because they have, I believe, had an outsized influence on the field, at least for readers in the United States. Even though others may have originated the term (Cranford 1989; Pontius 1993) (a point that Safire later conceded in 2005) (Safire 2005b), his definition has influence because of his reach as a popular commentator, who wrote at least four *New York Times* Op-Eds about neuroethics (Safire 2002a, b, 2003, 2005a).

In contrast, one of the earliest articulations of "neuroethics" came from the late neurologist, Ron Cranford in *Neurological Clinics*. He envisioned a singular role for neurologists in discerning ethical dilemmas *in practice* given their vantage point as practicing neurologists. James Bernat, in his magisterial text *Ethical Issues in Neurology*, noted that Cranford envisioned these "neuroethicists" would "...help analyze neuroethical dilemmas from their unique perspective *bridging the gap between clinical neurology and clinical ethics* [emphasis added]" (Bernat 2008).

Sadly, Cranford died too young (Pearce 2006), and his vision for neuroethics neither became the predominant one nor a corrective to unfortunate distortions which I believe redirected the trajectory of neuroethics in a way that has made it more speculative and less relevant to real patients and the needs of the clinic. Instead, for the aforementioned reasons, Safire's definition held early sway on the field when it was nascent. His definition implies that there is a neuroscience of morality. Notwithstanding some musings on the effect of oxytocin on trust, neuroscience hasn't done much to advance a deep understanding of the human condition in the intervening decade and a half since the San Francisco meeting. Shakespeare and the humanities still have a lock on that still rather elusive question (Fins et al. 2013).

Safire's manifesto also helped to seed the perception, at least in the United States, of what has since been called neuroexceptionalism, the notion that the ethics of brain research somehow should be different and singular as an area of inquiry, sequestered from reflections about other areas of the body. Again, in his Dana address, Safire asserts that "… the specifics of brain science hits home as research in no other organ does," asserting that because our brains make us individuals, and our livers do not, that neuroethics deserves a special status.

Although the singularity of our neuro-identities may seem to justify such a demarcation, a closer comparison of our neurological versus our genetic identities indicates that each is unique and that the special status of neuroethics versus genethics versus "quotidian" bioethics is not justified. Despite this flaw in his argument, Safire's call for a special status for neuroethics has persisted and helped to separate it from the broader field of bioethics and most notably from clinical ethics.

This separation from the world of clinical need, and importantly those who work in *clinical* ethics, has distanced neuroethics from the clinic, despite its true origins there as envisioned by Cranford. This sort of argumentation has bred skepticism about clinical care and translational research, even excluding the goal of "medical cure" as a constituent element of neuroethics.

This becomes apparent if we consider Michael Gazzaniga's definition of neuroethics. It at once dismisses "medical cure" and embraces a broader role for the nascent discipline addressing philosophical quandaries. Gazzaniga writes that:

> … I would define neuroethics as the examination of how we want to deal with social issues of disease, normality, morality, lifestyle and the philosophy of living *informed by our understanding of underlying brain mechanisms.* It is not a discipline that seeks resources for medical cure, but one that places personal responsibility in the broadest social and biological context. It is—or should be—an effort to come up with a brain-based philosophy of life (Gazzaniga 2005).

Safire's and Gazzaniga's definitions set the stage for a nervous neuroethics, focused beyond the parochial confines of the clinic and wary—even dismissive—of clinical engagement.

Let us return to the second part of Safire's definition, the "…good and bad about the treatment of, perfection of, or unwelcome invasion of and worrisome manipulation of the human brain." First, when is treatment, which is presumably vetted as opposed to research in equipoise, bad? And why is the therapeutic linked to the pursuit of perfection or presumably to enhancement? The conflation of restorative or regenerative work with enhancement has demonized the former and given scholarly legitimacy to the latter, even though efforts at enhancement remain in the realm of science fiction. Both legitimate therapeutic efforts and enhancement are saddled with the presumption that somehow these "invasions" will be "unwelcome" and "worrisome" and done by force without consent.

The message was clear: notwithstanding pressing clinical need, engaging the needs of patients with disabling neuropsychiatric disorders would be done at one's own peril. Work in this area, which he categorically assumes will constitute an "unwelcome invasion and worrisome manipulation," will only lead to recriminations and controversy (Schiff et al. 2007). So, instead of concern for a practical or clinical neuroethics, which could make a difference in the real world and be prudential, these commentators in neuroethics departed from the Cranford school. These more modern scholars warned of threats to personal identity, neuroprosthetic postmodern cyborgs, threats to neural integrity, and use of these innovations by national security agencies—which could be more proximate and of concern (Farah and Wolpe 2004; Farah et al. 2004; Moreno 2006). They were more concerned about the imagined than the real, and this displaced grounded clinical issues to the periphery of the field.

4.8 What Is to Be Done?

So what can be done to further harmonize clinical ethics and neuroethics in a pragmatic manner which acknowledges the value of both theory and practice? First, we need to validate the richness of clinical narratives, such as the one presented here, which, while presenting a common occurrence, contains questions of philosophical depth worthy of the most analytic voices in neuroethics. Like a day in Joyce's Dublin (Joyce 1986), there is often much latent content beneath a thick, if literary, description of ordinary life (Schwaber 1999).

But to do this in a manner that is both true to reality and which is capable of transcending the general confines of clinical ethics, we need to envision a more integrative approach which represents the convergence of clinical ethics, palliative medicine, and neuroethics. Scholars—and practitioners—from these domains need to collaborate with each other and help to generate a new form of exchange which is truly able to have practice inform theory and have theory, in turn, transform practice. As Racine has observed in his volume, *Pragmatic Neuroethics* (Racine 2010), achieving this confluence would reflect the essence of classical pragmatism and demonstrate its utility.

Acknowledgments I am grateful to Amy B. Ehrlich for editorial comments and permission to write about her mother, Helen Ehrlich Saltzman, to whose memory this piece is dedicated. I am also grateful to Jennifer Hersh for editorial assistance and to the Jerold B. Katz Foundation for its support of my work.

References

Adeoye O, Hornung R, Khatri P, Kleindorfer D (2011) Recombinant tissue type plasminogen activator use for ischemic stroke in the United States: a doubling of treatment rates over the course of 5 years. Stroke 42:1952–1955

Arendt H (1996) In: Vecchiarelli Scott J, Chelius Stark J (eds) Love and Saint Augustine. University of Chicago Press, Chicago

Augustine (1978) Confessions. Pine-Coffin RS (trans) Book VII(18). Penguin Books, New York

Back AL, Young JP, McCown E, Engelberg RA, Vig EK, Reinke LF et al (2009) Abandonment at the end of life from patient, caregiver, nurse, and physician perspectives: loss of continuity and lack of closure. Arch Intern Med 169(5):474–479

Bernat J (2008) Ethical issues in neurology, 3rd edn. Wolters Kluwer Lippincott Williams & Wilkins, Philadelphia

Biegler P, Vargas P (2016) Feeling is believing: evaluative conditioning and the ethics of pharmaceutical advertising. J Bioeth Inq 13(2):271–279

Boissy AR, Ford PJ, Edgell RC, Furlan AJ (2008) Ethics consultations in stroke and neurological disease: a 7-year retrospective review. Neurocrit Care 9(3):394–399
Brendel D (2004) Healing psychiatry: a pragmatic approach to bridging the science/humanism divide. Harv Rev Psychiatry 12(3):150–157
Brody H (1988) Stories of sickness. Oxford University Press, New York
Callahan S (circa 2003) Comment to the author on the philosophy of St. Augustine on true love
Charon R (2008) Narrative medicine: honoring the stories of illness. Oxford University Press, New York
Cranford RE (1989) The neurologist as ethics consultant and as a member of the institutional ethics committee. The neuroethicist. Neurol Clin 7(4):697–713
Cranford RE (1998) The vegetative and minimally conscious states: ethical implications. Geriatrics 53(Suppl 1):S70–S73
Creutzfeldt CJ, Holloway RG, Curtis JR (2015) Palliative care: a core competency for stroke neurologists. Stroke 46:2714–2719
del Zoppo GJ (2004) Thrombolysis: from the experimental findings to the clinical practice. Cerebrovasc Dis 17(Suppl 1):144–152
Dietz SM, Brakman SV, Dresser R, Astrow AB (1998) The incompetent self: metamorphosis of a person? Hastings Cent Rep 28(5):4–5
Dresser R (2009) Substituting authenticity for autonomy. Hastings Cent Rep 39(2):3
Farah MJ, Wolpe PR (2004) Monitoring and manipulating brain function: new neuroscience technologies and their ethical implications. Hastings Cent Rep 34(3):35–45
Farah MJ, Illes J, Cook-Deegan R, Gardner H, Kandel E, King P et al (2004) Neurocognitive enhancement: what can we do and what should we do? Nat Rev Neurosci 5:421–425
Feigin VL, Forouzanfar MH, Krishnamurthi R, Mensah GA, Connor M, Bennett DA, et al; Global Burden of Diseases, Injuries, and Risk Factors Study 2010 (GBD 2010) and the GBD Stroke Experts Group (2014) Global and regional burden of stroke during 1990–2010: findings from the Global Burden of Disease Study 2010. Lancet 383(9913):245–255
Ferrari PF (2014) The neuroscience of social relations. A comparative-based approach to empathy and to the capacity of evaluating others' action value. Behaviour 151(2–3):297–313
Fins JJ (1992) Palliation in the age of chronic disease. Hastings Cent Rep 22(1):41–42
Fins JJ (1999) From contract to covenant in advance care planning. J Law Med Ethics 27:46–51
Fins JJ (2005a) Clinical pragmatism and the care of brain injured patients: towards a palliative neuroethics for disorders of consciousness. Prog Brain Res 150:565–582
Fins JJ (2005b) Rethinking disorders of consciousness: new research and its implications. Hastings Cent Rep 35(2):22–24
Fins JJ (2007a) Border zones of consciousness: another immigration debate? Am J Bioeth 7(1):51–54
Fins JJ (2007b) A review of mind wars: brain research and national defense by Jonathan D. Moreno. New York: Dana Press, 2006. JAMA 297(12):1382–1383
Fins JJ (2008a) A leg to stand on: Sir William Osler and Wilder Penfield's "neuroethics". Am J Bioeth 8(1):37–46
Fins JJ (2008b) "Humanities are the hormones": Osler, Penfield and "neuroethics" revisited. Am J Bioeth 8(1):W5–W8
Fins JJ (2008c) Neuroethics and disorders of consciousness: a pragmatic approach to neuropalliative care. In: Laureys S, Tononi G (eds) The neurology of consciousness, cognitive neuroscience and neuropathology. Academic Press/Elsevier, London, pp 234–244
Fins JJ (2013) Disorders of consciousness and disordered care: families, caregivers and narratives of necessity. Arch Phys Med Rehabil 94(10):1934–1939
Fins JJ (2014) A review of making medical knowledge by Miriam Solomon. Oxford University Press, New York. Notre Dame Philosophocal Reviews. 7 Mar 2016. http://www.ndpr.nd.edu/news/65137-making-medical-knowledge/?utm_source=feedburner&utm_medium=feed&utm_campaign=Feed%3A+PhilosophicalReviews%2FNews+(Notre+Dame+Philosophical+Reviews. Accessed 14 Aug 2016
Fins JJ (2015a) Neuroethics and disorders of consciousness: a pragmatic approach to neuropalliative care. In: Laureys S, Tononi G, Gosseries O (eds) The neurology of consciousness, cognitive neuroscience and neuropathology, 2nd edn. Elsevier, Amsterdam, pp 241–255

Fins JJ (2015b) Rights come to mind: brain injury, ethics and the struggle for consciousness. Cambridge University Press, New York

Fins JJ, Gracia GD (2016a) Entrevista a Albert R. Jonsen. Eidón 44:67–86

Fins JJ, Gracia GD (2016b) Entrevista a Mark Siegler. Eidón 45:64–83

Fins JJ, Hersh J (2011) Solitary advocates: the severely brain injured and their surrogates. In: Hoffman B, Tomes N, Schlessinger M, Grob R (eds) Transforming health care from below: patients as actors in U.S. health policy. Rutgers University Press, New Brunswick, pp 21–42

Fins JJ, Maltby BS (2003) Fidelity, wisdom, and love: patients and proxies in partnership, interactive workbook and educational video. Fan Fox and Leslie R. Samuels Foundation, New York

Fins JJ, Schiff ND (2002) The minimally conscious state: definition and diagnostic criteria. Neurology 59(9):1473

Fins JJ, Bacchetta MD, Miller FG (1997) Clinical pragmatism: a method of moral problem solving. Kennedy Inst Ethics J 7(2):129–145

Fins JJ, Maltby BS, Friedmann E, Green M, Norris K, Adelman R et al (2005) Contracts, covenants and advance care planning: an empirical study of the moral obligations of patient and proxy. J Pain Symptom Manag 29(1):55–68

Fins JJ, Master MG, Gerber LM, Giacino JT (2007a) The minimally conscious state: a diagnosis in search of an epidemiology. Arch Neurol 64(10):1400–1405

Fins JJ, Schiff ND, Foley KM (2007b) Late recovery from the minimally conscious state: ethical and policy implications. Neurology 68:304–307

Fins JJ, Dorfman GS, Pancrazio JJ (2012) Challenges to deep brain stimulation: a pragmatic response to ethical, fiscal and regulatory concerns. Ann N Y Acad Sci 1265:80–90

Fins JJ, Pohl B, Doukas D (2013) In praise of the humanities in academic medicine: values, metrics and ethics in uncertain times. Camb Q Healthc Ethics 22:355–364

Gazzaniga M (2005) The ethical brain. The Dana Press, New York, pp xiv–xxv

Giacino JT, Fins JJ, Laureys S, Schiff ND (2014) Disorders of consciousness after acquired brain injury: the state of the science. Nat Rev Neurol 10:99–114

Goldberg PE (2000) The physician-patient relationship: three psychodynamic concepts that can be applied to primary care. Arch Fam Med 9(10):1164–1168

González RG, Furie KL, Goldmacher GV, Smith WS, Kamalian S, Payabvash S et al (2013) Good outcome rate of 35% in IV-tPA-treated patients with computed tomography angiography confirmed severe anterior circulation occlusive stroke. Stroke 44(11):3109–3113

Goodman KW (2002) Ethics and evidence-based medicine: fallibility and responsibility in clinical science. Cambridge University Press, New York

Greer DM, Wang HH, Robinson JD, Varelas PN, Henderson GV, Wijdicks EF (2016) Variability of brain death policies in the United States. JAMA Neurol 73(2):213–218

Hinderer KA, Friedmann E, Fins JJ (2015) Withdrawal of life sustaining treatment: patient and proxy agreement a secondary analysis of "contracts, covenants, and advance care planning". Dimens Crit Care Nurs 34(2):91–99

Homer (1963) The Odyssey. Fitzgerald R (trans) Book XII, Anchor Books, Garden City

Honeybul S, Ho KM, Gillet G (2015) Outcome following decompressive hemicraniectomy for malignant cerebral infarction: ethical considerations. Stroke 46(9):2695–2698

Jennett B, Plum F (1972) Persistent vegetative state after brain damage. A syndrome in search of a name. Lancet 1(7753):734–737

Jennings B (2003) Hospice and Alzheimer disease: a study in access and simple justice. Hastings Center Rep Suppl:S24–S26

Jonsen A (1998) The birth of bioethics. Oxford University Press, New York

Jonsen AR, Siegler M, Winslade WJ (1982) Clinical ethics. MacMillan, New York

Joyce J (1986) Ulysses. Vintage, New York

Joyner MJ (2016) Precision medicine, cardiovascular disease and hunting elephants. Prog Cardiovasc Dis 58(6):651–660

Kepplinger J, Barlinn K, Deckert S, Scheibe M, Bodechtel U, Schmitt J (2016) Safety and efficacy of thrombolysis in telestroke: a systematic review and meta-analysis. Neurology 87(13): 1344–1351

Lambek M (2010) Ordinary ethics: anthropology, language, and action. Fordham University Press, New York
Loupis YM, Faux SG (2013) Family conferences in stroke rehabilitation: a literature review. J Stroke Cerebrovasc Dis 22(6):883–893
Løvstad M, Andelic N, Knoph R, Jerstad T, Anke A, Skandsen T et al (2014) Rate of disorders of consciousness in a prospective population-based study of adults with traumatic brain injury. J Head Trauma Rehabil 29(5):E31–E43
McKhann G (2016) New stroke therapy. Brain in the news. The Dana Foundation. http://www.dana.org/Brain_in_the_News/New_Stroke_Therapy/. Accessed 28 Aug 2016
Moeller JR, Albanese TH, Garchar K, Aultman JM, Radwany S, Frate D (2012) Functions and outcomes of a clinical medical ethics committee: a review of 100 consults. HEC Forum 24(2):99–114
Moreno JD (2006) Mind wars: brain research and national defense. The Dana Press, New York
Morris K, Ferguson A, Worrall L (2014) A qualitative study of the legal and social justice needs of people with aphasia. Int J Speech Lang Pathol 16(6):541–551
NIH Brain Initiative's RFI (2016) Guidance for opportunities in neuroethics (Notice Number: NOT-MH-16-014)
Nilson EG, Acres CA, Tamerin NG, Fins JJ (2008) Clinical ethics and the quality initiative: a pilot study for the empirical evaluation of ethics case consultation. Am J Med Qual 23(5):356–364
Noddings N (2013) Caring: a relational approach to ethics and moral education, 2nd edn. University California Press, Berkeley
Owen AM, Coleman MR, Boly M, Pickard JD, Tshibanda L, Owen AM et al (2006) Willful modulation of brain activity in disorders of consciousness. Detecting awareness in the vegetative state. Science 313(5792):1402
Pearce J (2006) Ronald E. Cranford, 65, an expert on coma, is dead. The New York Times. http://www.nytimes.com/2006/06/03/science/03cranford.html. Accessed 15 Nov 2016
Pontius AA (1993) Neuroethics vs neurophysiologically and neuropsychologically uninformed influences in child-rearing, education, emerging hunter-gatherers, and artificial intelligence models of the brain. Psychol Rep 72:451–458
Racine E (2007) HEC member perspectives on the case analysis process: a qualitative multi-site study. HEC Forum 19(3):185–206
Racine E (2008) Which naturalism for bioethics? A defense of moderate (pragmatic) naturalism. Bioethics 22(2):92–100
Racine E (2010) Pragmatic neuroethics: improving treatment and understanding of the mind-brain. The MIT Press, Cambridge
Racine E (2013) Pragmatic neuroethics: the social aspects of ethics in disorders of consciousness. Handb Clin Neurol 118:357–372
Rees D, Rose S (2004) The new brain sciences: perils and prospects. Cambridge University Press, New York
Safire W (1980) On language. Times Books, New York
Safire W (2002a) The but-what-if factor. The New York Times. http://www.nytimes.com/2002/05/16/opinion/the-but-what-if-factor.html. Accessed 15 Nov 2016
Safire W (2002b) Introduction: visions for a new field of "neuroethics". In: Marcus SJ (ed) Neuroethics: mapping the field. The Dana Press, New York
Safire W (2003) The risk that failed. The New York Times. http://www.nytimes.com/2003/07/10/opinion/the-risk-that-failed.html. Accessed 15 Nov 2016
Safire W (2004) The right word in the right place at the right time: wit and wisdom from the popular "on language" column in *The New York Times Magazine*. Simon & Schuster, New York
Safire W (2005a) Chimera. N Y Times Mag. http://www.nytimes.com/2005/05/22/magazine/chimera.html. Accessed 15 Nov 2016
Safire W (2005b) Isms and phobias. The New York Times. http://www.nytimes.com/2005/05/15/magazine/isms-and-phobias.html. Accessed 15 Nov 2016
Schiff ND (2015) Cognitive motor dissociation following severe brain injuries. JAMA Neurol 72(12):1413–1415

Schiff ND, Fins JJ (2007) Hope for "comatose" patients. Cerebrum 2003; 5(4): 7–24. Reprinted as: Schiff ND, Fins JJ. Hope for "comatose" patients. "A cerebrum classic". In: Read CA (ed) The Dana Foundation's Cerebrum 2007. Dana Press, New York, pp 185–203
Schiff ND, Giacino JT, Kalmar K, Victor JD, Baker K, Gerber M et al (2007) Behavioral improvements with thalamic stimulation after severe traumatic brain injury. Nature 448(7153): 600–603
Schiff ND, Giacino JT, Fins JJ (2009) Deep brain stimulation, neuroethics and the minimally conscious state: moving beyond proof of principle. Arch Neurol 66(6):697–702
Schwaber P (1999) Cast of characters: a reading of Ulysses. Yale University Press, New Haven
Siegler M (1975) Pascal's wager and the hanging of crepe. N Engl J Med 293(17):853–857
Solomon M (2014) Making medical knowledge. Oxford University Press, New York
Spokoyny I, Cederquist L, Clay B, Meyer BC (2015) Coordinating Options for Acute Stroke Therapy (COAST): an advance directive for stroke. J Clin Ethics 26(3):206–211
Stepan C, Haidinger G, Binder H (2004) Prevalence of persistent vegetative state/apallic syndrome in Vienna. Eur J Neurol 11(7):461–466
The National Institute of Neurological Disorders and Stroke rt-PA Stroke Study Group (1995) Tissue plasminogen activator for acute stroke. N Engl J Med 333(24):1581–1587
Toulmin S (1982) How medicine saved the life of ethics. Perspect Biol Med 25(4):736–750
UN Convention on the Rights of the Disabled (2016) http://www.un.org/disabilities/convention/conventionfull.shtml. Accessed 15 Nov 2016
Woodruff L, Woodruff B (2006) In an instant. Random House, New York
Zizzo N, Bell E, Racine E (2016) What is everyday ethics? A review and a proposal for an integrative concept. J Clin Ethics 27(2):117–128

Born Free: The Theory and Practice of Neuroethical Exceptionalism

5

Fernando Vidal and Michelle Piperberg

Abstract

Since it first appeared in the public eye in the early 2000s, neuroethics has acquired all the sociological features that define a discipline, such as international societies, university chairs, journals, and academic programs. An important element of its rapid development as a discipline was the claim that it should be autonomous from the field that could claim to be its "parent discipline," namely, bioethics. This position gave rise to debate. Theoretical questions of the debate may remain open; on the ground, however, neuroethics won. It was born and has remained free. This chapter examines the ultimate foundation of neuroethics' claim to autonomy, namely (in Adina Roskies' words), the "peculiar relationship between our brains and our selves," and how it functions in the discipline's theory and practice.

5.1 Introduction

The history of neuroethics as a recognized academic endeavor is a success story. When the Decade of the Brain opened in 1990, it did not exist; by the middle of the following decade, it could be considered an established field. There has been

F. Vidal (✉)
Catalan Institution for Research and Advanced Studies (ICREA), Barcelona, Spain

Center for the History of Science, Autonomous University of Barcelona, Bellaterra, Barcelona, Spain
e-mail: fernando.vidal@icrea.cat

M. Piperberg
Independent Scholar, Barcelona, Spain
e-mail: michellepiperberg@gmail.com

E. Racine, J. Aspler (eds.), *Debates About Neuroethics*, Advances in Neuroethics, DOI 10.1007/978-3-319-54651-3_5

oscillation between calling it a "field" and a "discipline," as in the following passage from the introduction to the foundational *Neuroethics: Mapping the Field*:

> … we need a scholarly *discipline* of neuroethics. Thinking in this new *field* must be based on scholarship, with rich connections to philosophy, psychology, law, and other fields. We need people who are *professionally committed* to thinking about these problems … The scholarly *discipline* of neuroethics should become an arena in which the best methods of philosophical analysis—that is, rigorous, disciplined thinking—are brought to bear on these questions (Hall 2002a; different in Hall 2002b: 292; our emphasis).

The same oscillation appears in later literature, it is not problematic, and in any case the distinction is not important for our argument. Neuroethics is a field of research and knowledge. Sociologically, it has the institutionalized and professionalized features that usually characterize disciplines: a recognizable community, journals, institutions, conferences, and training programs. We shall use both terms indistinctly.

More relevant is the fact that, while neuroethics coexists peacefully with bioethics and has been sometimes characterized as a bioethical "subfield" or "subdiscipline," it has constructed its identity largely by defending the claim that it must be autonomous. This claim has been labeled "neuroethics exceptionalism" or "neuroethical exceptionalism" (Wilfond et al. 2005: 20; Stahnisch 2011: 156).

The exceptionalist claim is not merely a tool for discipline building. It expresses the foundations of the neuroethical project via the intermediary affirmation of "neuroexceptionalism" (Schick 2005; Tovino 2007; Wachbroit 2008). Here, "neuro" refers to the neurosciences—specifically, to the exuberantly publicized conviction that they, in particular via brain imaging technologies, pose unparalleled ethical challenges. For example, some would argue that they can yield information typically deemed private and can illuminate the nature of personhood in ways that challenge established beliefs. This conviction is "intermediary" because it is itself based on a further premise, according to which we are essentially our brains. This premise does not mean that we are exclusively that organ or its associated systems, but implies that the fundamental level of analysis of human features and behaviors is the neuroscientific. Of course, the more "basic" genetic level is recognized as crucial. Neuroethics, however, presupposes that the neural aspects of human nature are most directly relevant to many of the questions raised in the Western philosophical and ethical traditions, including issues of personhood and personal identity. Similarly, even though neuroethics forums may include discussions of more contextual understandings of personal identity and their relation to neuropsychiatric illness or brain interventions, such discussions have not in our view affected the discipline's foundations.[1]

Indeed, the basic justification for the neuroethical project has not changed since, in his Introduction to *Neuroethics: Mapping the Field*, the *New York Times*, journalist and Dana Foundation chairman William Safire (2002: 7) explained that whereas

[1] For an example of such discussions, see Baylis (2015), which is the introduction to five chapters on neuroethics and identity in Springer's *Handbook of Neuroethics*.

"one person's liver is pretty much like another's," the brain is "central to our being" and "the organ of individuality." We are not interested here in examining how and in what sense such statements are justified, but in their metamorphosis into norms and expectations.

Philosopher Adina Roskies, in her influential article "Neuroethics for the new millennium" (2002: 21), argued that "the justification for identifying and promoting neuroethics as a new important field" that should "not be merely a subdivision of bioethics" is derived from the "intimate connection" between brain and behavior, from the "peculiar relationship between our brains and our selves" and from "the intuition that our ever increasing understanding of the brain mechanisms underlying diverse behaviors has unique and potentially dramatic implications for our perspective on ethics and social justice." The dozens of identical or equivalent statements that could be quoted convey the same message: "New ethical issues are arising as neuroscience gives us unprecedented ways to understand the human mind and to predict, influence, and even control it" (Farah 2010: 2). In sum, neuroethics' self-appointed role derives from the belief that "[b]y shedding light on the brain, [neuroscience] illuminates our prized rationality, our creativity, our capacity to produce and appreciate art, even our capacity for awe and transcendence" (Clausen and Levy 2015: v).

The authors quoted above leave key terms such as *intimate connection*, *peculiar relationship*, *intuition*, *understanding*, *underlying*, *implications*, *unprecedented*, and *illuminate* vague and unexplained. Such vagueness fulfills an important function. The repetition of their pronouncements turns them into the *performative* basis of the neuroethical enterprise. They are not performative in the same sense as classical performative utterances such as "War is declared," which begins war by declaring it. They are performative because they help bring about and sustain the reality they supposedly describe. Neuroethics depends on considering those utterances as true, and leaving them vague is crucial for conveying the impression that they are indeed true. Reciprocally, its effectively autonomous existence, sustained by equally performative claims such as "Well-prepared neuroethicists are needed more than ever" (Fischbach and Mindes 2011: 370), seems to prove that the quoted statements are empirically substantiated truth claims.

A recent article in the *Stanford Encyclopedia of Philosophy* explains that neuroethics "is concerned with the ethical questions that attend the development and effects of novel neurotechnologies, as well as other ethical and philosophical issues that arise from our growing understanding of how brains give rise to the people that we are and the social structures that we inhabit and create" (Roskies 2016). Nevertheless, as Eric Racine (2010: 75, 76) suggests when he points out that "no one really considers neuroethics to be focusing solely on an organ" and that "few people if anyone really believe that neuroethics is an ethics of the brain per se," discussing the neuroethical project in terms of "reductionism" or "neuroessentialism" is misleading.

Neuroethics indeed provides examples of conspicuously antireductionist positions. Nevertheless, even elaborate arguments, such as those in Walter Glannon's (2011) *Neuroethics With A Human Face* or Neil Levy's (2007) *Neuroethics*, leave intact the conviction that neuroimaging provides insights into the mind (not just the brain) and that, if "the neuro-essentialist notion that the brain defines who we are might seem a bit far-fetched," the brain "is the biological substrate of central human characteristics" and there "is no aspect of our lives that neuroscience cannot, in principle, help to illuminate" (Clausen and Levy 2015: v). But there are many ways of understanding how the brain is *substrate* and how neuroscience *illuminates*.

What, in short, has justified the use of the prefix *neuro* and allowed neuroethics to be born free? One answer could be "exaggerating how much scientific research can tell us about who we are" and "the fantasy of one all-encompassing explanation for complex human traits" (Parens and Johnson 2007: 562). A more specific one would be: the conviction that "brains give rise to the people that we are and the social structures that we inhabit and create" (Roskies 2016: 3) and that "our brains make us the kinds of creatures we are, with our values and goals" (Levy 2007: 8). Such claims, as this article will try to document, embody the ultimate justification for neuroethics, as well as its ideological and anthropological foundations ("ideological" in the general sense of constituting a set of ideas and beliefs and "anthropological" in the sense that they concern the human).

5.2 A Future-Oriented Discipline

From the beginning, neuroethics has been associated with a "new millennium" characterized by the momentous consequences of neuroscientific knowledge for the self and society. The common vision is that, as ideas about the mind and the nervous system "are being revised (if not abandoned) in favor of new notions and novel ways of thinking about the human condition," neuroethics develops "a canon that seeks to (1) identify the humanistic importance of neuroscience…, and (2) provide reflective direction to the pace and trajectory of technological progress" (Giordano 2010: xxviii). The assumption that neuroscience prompts radical reconceptualizations of what it means to be human may express a normative expectation; sociologists, however, have shown that such an expectation has not been realized empirically as far as people's self-perceptions are concerned (O'Connor and Joffe 2013). Yet, even if it does not extend beyond the universe of specialists engaged in various "neuro" disciplines, it has performative efficacy.

Sociologists Ilina Singh and Nikolas Rose noticed that "neuro-ethical anxieties have become part of the very problem they seek to address" and foster "a culture of hype and hope, futurology and fear" (Singh and Rose 2006: 100). Neuroethics indeed advances by announcing tensions that neuroscientific progress ought to generate and offering itself as the best way to manage them: "In the age of neuro-everything, as we enter novel and daring new territories, neuroethicists will be needed to formulate many essential questions, and provide much-needed guidance in addressing difficult and ethically challenging problems" (Fischbach and Mindes 2011: 370). Such claims reflect the alliance between neuroethics and neuroscience, as well as the former's affinity with new disciplines of allegedly humanistic relevance, which (from neuroanthropology to neurotheology) have prospered since the

Decade of the Brain, thanks to the availability of neuroimaging (Littlefield and Johnson 2012; Vidal and Ortega 2017).

Described from the outside with respect to those disciplines, the task of neuroethics is to establish that they are promising endeavors.[2] From the inside, they embody Judy Illes' depiction of neuroethics as a way of "empowering brain science" that proceeds by "moving age-old debates about mind and brain towards modern theoretical discussions about the understanding of human behavior enabled by advances in neurosciences" (Illes 2010: 1294). Both inside and outside the field, neuroethics reproduces the "proleptic structure" that has characterized the brain sciences since phrenology announced new foundations for managing the individual and society (Hagner and Borck 2001: 508–509).[3] "Proleptic" derives from "prolepsis," the representation of future developments as if already existing or accomplished. Such structure defines topics of investigation, shapes public understanding of the brain, sustains future expectations, and identifies the required experts.

By generating both a general vision and concrete research, neuroethics has become central to that structure. For example, the director of the Foresight Lab that runs within the Ethics and Society Subproject of the European *Human Brain Project* (Changeux et al. 2014, 2016) explained "some of the fundamental questions" such an initiative poses:

> What do we know about how the human brain functions? What more do we need to know? What might we do with that knowledge? What would be the social implications of a greater knowledge of the brain and the associated increase in our capacity to intervene in the organ that so many believe to be the seat of consciousness, cognition, intention, and of our private selves? What would be the ethical challenges in gaining that knowledge and in using it? (Rose 2014: 1212).

The modal and the conditional verb tenses would usually suggest mere possibilities. However, when embedded in a major technology-driven, publicly financed multinational enterprise of great cost and high symbolic and political value, they acquire the consistency of the real and prove the need for neuroethics. This process is the heart of neuroethics product positioning. Thus (in one instance among many), the *Oxford Handbook of Neuroethics* devotes its Foreword to arguing in favor of the "proactive approach," underlines the significance of "*emerging* issues" and of "*potential* pressure points for neuroscience-society tension," assumes that some neuroscientific insights "are *likely* to threaten core values and beliefs," and then asks, "How should the scientific community prepare for the *likelihood* that neuroscience advances will result in increasing tension with the rest of society?" (Leshner 2011: v, vii, xi, ix, our emphasis).

Yet the most significant aspect of the proactive outlook is that it translates into research that consolidates its objects. For example, a survey conducted among

---

[2] In contrast, critics (e.g., Fitzpatrick 2012) argue that those "neuro" disciplines are intrinsically inadequate to illuminate the complex human phenomena they claim to study.

[3] The futuristic mode is not exclusive to the neurosciences; see, for example, Mulkay (1993) on the "rhetorics of hope and fear" in the early 1990s British debate on research with human embryos.

health providers and patients diagnosed with major depressive disorder found "high receptivity to brain scan for treatment tailoring and choice, for improving understanding of and coping with disease, and for mitigating the effects of stigma and self-blame" (Illes et al. 2008: 107). Neither at the time of the survey nor as we write these lines has there been neuroimaging for depression treatment "tailoring and choice," and it is far from clear that there ever will be such a possibility (Vidal and Ortega 2012; Ferrari and Villa 2016 for a recent overview).[4] The survey authors acknowledged that, with regard to depression, there is no "translation" of neuroimaging research to the clinic; they nonetheless insisted that "rapid innovation" was taking place along the "trajectory of discovery to implementation," and such an assertion inspired their calls for "the development of responsible social and public policies in response to new diagnostic and prognostic capabilities for the benefit of patients and their families" (Illes et al. 2008: 107)—in short, for giving neuroethicists a crucial social role.

The "capabilities" in question lie entirely in a hypothetical future; the survey authors, however, treated them as fully plausible and claimed that identifying them early on "can maximize benefit and prevent false hope, mitigate hype, and curtail their premature use and even misuse in the private sector" (Illes et al. 2008: 112). The intention was no doubt good. Nonetheless, by adhering to the belief that "fMRI promises significant benefit to the diagnostic process for major depression" and by confidently announcing the realization of that promise, they bolstered confidence in presumptive futures and accomplished the opposite of what they allegedly pursued. By the same token, they reinforced their own legitimacy, and that of the research neuroethics is called to critically examine. Such dynamics illustrate neuroethics' "self-referential methodological structure": its goal is to assess the practices and consequences of scientific activities whose results and assumptions are an essential component of itself (Hoyer 2010: 38, 132). Though seldom as clearly as when Shook and Giordano (2014: 1) write, "Neuroethics applies cognitive neuroscience for prescribing alterations to conceptions of self and society, and for prescriptively judging the ethical applications of neurotechnologies," neuroethicists generally consider the constitutive relationship between neuroscience and their discipline as one of the latter's strengths, rather than as a feature that prevents it from distancing itself from its objects.

While the proactive and self-referential dispositions are integral to neuroethics, the discipline would not prosper if it were a purely "speculative philosophy" (Fins 2008: 38). Neuroethics of course involves conceptual discussion, but the heart of its professional activity is empirical research, as demonstration of its real-world relevance. To our knowledge, the most focused example of an argument for such relevance concerns privacy in connection with neuroimaging research about individual differences in personality and intelligence. On the basis of 16 published studies, it concluded "that the use of imaging to gather information about an individual's psychological traits is already possible, but to an extremely limited extent" (Farah et al.

[4] Lu (2016), a radically optimistic overview of recent research about "tailoring treatment by scanning the brain," is written entirely with modal verbs (*could*, *may*, etc.).

2010: 119). In a broader discussion, the article's first author argued that neuroethics as a whole had moved from being a "predominantly anticipatory field" to a focus on actual developments and explained that transformation by "the rapidly evolving state of neuroscience itself" (Farah 2011: 761). This is not the place to examine the proposed examples—only to indicate that the claim that "neuroscience is calling into question our age-old understanding of the human person" (Farah 2011: 776) is offered as an observation that accounts for the existence of neuroethics, when it actually constitutes the discipline's fundamental postulate.

5.3 Neuroethical Exceptionalism

Although many presentations since "Neuroethics for the new millennium" (Roskies 2002) have referred to two complementary endeavors—the neuroscience of morality and the ethics of neuroscience—*neuroethics* is in practice mostly reserved for the latter category—i.e., for the project of examining the implications of brain science for society or the actual and anticipated ethical, social, political, and legal consequences of neuroscientific knowledge and its uses, including the ethics of research and its clinical biomedical applications. As highlighted in bibliometric analyses (Buniak et al. 2014; Leefmann et al. 2016), and as obvious from the tables of contents of existing handbooks and journals, the field is characterized by a great variety of topics and approaches. Its scope can nonetheless be structured in four basic broad areas: (1) brain science and the self, (2) brain science and social policy, (3) ethics and the practice of brain science, and (4) brain science and public discourse (Lombera and Illes 2009). The question, debated in the early years of neuroethics and of durable theoretical interest, is: why should neuroethics be considered a separate discipline? This question arises especially in light of the existence of bioethics as a professional field with occasionally contested subspecialties, and given that ELSI (Ethical, Legal and Social Implications) programs can be set up in connection with any scientific area. At stake here is not whether neuroethics "has the right" to exist, but rather the reasons put forward for why ELSI in brain science presents challenges so unique that a new profession is required to deal with them.

Of course, like any other process of boundary making and discipline formation, the development of neuroethics involves sociological dimensions (Conrad and De Vries 2011). Situating neuroethics in the contexts of expectations and ideals of progress, of "biocapitalism," and of the global politics of science, Armin Hoyer (2010) argued that its main interest does not reside in its intellectual contents, but in the process by which neuroethicists have become recognized experts for mediating between, on the one hand, the public sphere (from lay audiences to policy-makers) and on the other hand neuroscientific research and diverse "neuro" discourses and practices. Caragh Brosnan (2011: 289) emphasized the extent to which "neuroethicists draw on expectational discourses about the future of neuroscience" and thus "actively align themselves with neuroscience, rather than maintaining the critical distance necessary to act as an ethical watchdog." Raymond De Vries (2007: S68) noted the absence of social scientists and other human science scholars within neuroethics who might contribute critical perspectives, as well as the dearth of discussion on socioeconomic issues, and the structural problems derived from the discipline's institutional organization (e.g., the mission of the Dana Foundation, a major philanthropic advocate of neuroethics, is to support brain research).

Both De Vries and Hoyer also emphasized the connections between the rise of neuroethics and the history and sociology of bioethics. This is particularly significant, since neuroethicists, though sometimes describing their discipline as a subspecialty of bioethics, derive it directly from neuroscientific advances and their transformative potential at the social and individual level.[5] However, the conversation that took place in 2005 around Judy Illes and Eric Racine's *American Journal of Bioethics* target article "Imaging or Imagining? A Neuroethics Challenge Informed by Genetics" shows that neuroscientific progress is not by itself the ultimate reason for advocating neuroethical exceptionalism. Since this is the first time in which such exceptionalism was the focus of attention, that conversation must be seen as a landmark in the history of the field.

The article explains that the specific task of neuroethics is to deal "proactively" and "responsibly" with "the inevitable and omnipresent working hypothesis … that the mind is the brain" and to interpret data so as "to untangle what we image from what we imagine" (Illes and Racine 2005: 10, 12). The data in question are those generated by functional neuroimaging, whose nature and capabilities are depicted as "the model for neuroethical discussions" and as "the principal reason that traditional bioethics analysis… will not suffice as a guide" (7, 6). This is the familiar "technology story" that also characterizes self-presentations of bioethics (De Vries 2007: S66): "advanced capabilities for understanding and monitoring human thought and behavior enabled by modern neurotechnologies have brought new ethical, social and legal issues to the fore front" (Illes and Racine 2005: 5). The article is dominated by the perception of a "new era" defined by "bold new findings and claims" (15) and includes such assertions as: fMRI "poses pivotal challenges to thought privacy" (11), or neurotechnologies "*will* fundamentally alter the dynamic between personal identity, responsibility and free will in ways that genetics never has" and "*are* challenging our sense of personhood and providing new tools to society for judging it" (14, our emphasis). Those assertions suggest that the identification of the mind and the brain does not function as a "working hypothesis" or a theoretical statement, but as an empirical matter of fact.

Several commentaries on *Imaging or Imagining?* discuss said assertions, though only one touches upon their constitutional function for neuroethics. Doucet (2005: 30) points out, "One of the tasks of neuroethics should be to help neuroscientists recognize the limits of their discipline and their natural inclination," and refers to an earlier article where the same authors note that "the brain is used implicitly as a shortcut for more global concepts such as the person, the individual or the self" (Racine, Ofek and Illes. 2005: 160). Questioning the need for bioethics subdisciplines, Wilfond et al. (2005: 21) remark that "treating certain topical areas as new

[5] Bioethics has adopted this narrative; most handbooks published since the mid-2000s make room for neuroethics (though only sometimes in the form of a dedicated chapter).

and exceptional encourages the misguided impression that some information, such as genetic sequences or brain images, is inherently unique because it defines the essence of life and personal identity." Evers (2005: 31) notes that Illes and Racine misleadingly use "brain maps" as an equivalent of "thought-maps." Only Buford (2005: 34) directly questions the assumption "that the mind literally *is* the brain" and argues that "the neurosciences and neurotechnologies are unlikely to have *any* metaphysical implications robust enough" to support the authors' claims about the link between brain and self. He thus opposes not only Illes and Racine but also the editorialist who introduces the debate by claiming that "the simple fact that the brain is the origin of the mind, and therefore in many ways the seat of our humanity, confers very special status on it as an organ" (Leshner 2005: 1). The fact that criticism such as Buford's has not affected the fundamental assumptions of neuroethics, as expressed in the last quotation, illustrates the extent to which the discipline can safely open itself to contrary views.

5.4 "The Idea That a Person Is His or Her Brain"

Given the crucial role played within neuroethics by claims about the impact of neuroscience on notions and practices of the self and personhood, one could imagine that the latter would make up core areas of neuroethical investigation. That, however, is not the case. As of 2009, 60% of publications in the field were devoted to the ethics and practice of brain science; policy issues (21%), self and personhood (16%), and scientists' public discourse (3%) followed at a considerable distance (Lombera and Illes 2009: 61 and 62, Fig. 2). The most recent bibliometric analysis confirms that trend: "the mainstream of neuroethics research is tackling with ethical questions within the established theoretical frameworks of bioethics" (Leefmann et al. 2016: 12). The most frequent topics do not concern the vigorously publicized transformative potential of neuroscientific research (e.g., cognitive enhancement, the modulation, or erasure of memories), but rather medical and healthcare issues related to psychiatric and neurodegenerative diseases (Leefmann et al. 2016: 13; see also Racine and Illes 2008, where all the subjects sketched are "conventionally" bioethical).

The fact that "almost all topics discussed within neuroethics today have been present in debates in medical ethics, pharmacology or philosophy of mind" seems to call into question "the potential of neuroethics to establish itself as a discipline of its own" (Leefmann et al. 2016: 17). Neuroethics, however, has already institutionalized itself. Yet that does not depend on its dealing with new ethical or conceptual problems, but on perpetuating the idea that we are essentially our brains and on acting as an efficient backing for the ever-growing number of projects that depend on it ideologically. Rather than multiplying the examples, let us consider what, to our knowledge, is the most focused defense of neuroethical specificity.

In a chapter of the *Penn Center Guide to Bioethics*, Martha Farah seeks to differentiate the neuroethical issues that are also "familiar bioethical issues, which pertain to the brain as well as other organ systems" from those "that arise in connection with the brain, because it is the organ of the mind." These are "unique to neuroethics" and include "the social and ethical implications of brain imaging, psychopharmacology, brain stimulation, and brain-machine interfaces, as well as the ways in which our advancing understanding of mind-brain relations calls into question basic assumptions about what it means to be a person" (Farah 2009: 72).

The relevant "familiar bioethical issues" in neuroethics are those in which the fact that the brain is centrally involved does not make a fundamental difference. For example, predictive tests for incurable neurodegenerative diseases raise issues related to privacy rights and quality of life akin to those raised by genetic tests. Similarly, fMRI or transcranial magnetic stimulation raises questions of risk, safety, and decision-making that one would ask about other biomedical research methods. The same can be said about incidental findings from research brain scans. Topics less discussed in neuroethics, but which also overlap with bioethics, include the application of stem cell research for neurological disease, future genetic technologies (since those "for selecting or altering the traits of a child are likely to include mental traits such as intelligence and personality" Farah 2009: 74), the use of information about brain development in arguments for and against abortion, or the regulation and safety of neuropsychiatric drugs.

In turn, the "uniquely neuroethical issues," those that "emerge primarily because of the very special status of the brain, compared to other organs, in human life," fall into four main categories: (1) functional neuroimaging, (2) psychopharmacology, (3) brain stimulation and brain prostheses, and (4) brains and persons, i.e., the "challenges to our concepts of personhood posed by our growing understanding of the neural bases of behavior, personality, consciousness, and states of spiritual transcendence" (Farah 2009: 75).

The specifically neuroethical issues related to *psychopharmacology* concern less therapeutic applications than the use of substances "to change or enhance normal people's cognitive abilities and moods" (Farah 2009: 77). The augmentation of attention and memory, the weakening of unwanted memories, and the improvement of mood raise questions about safety, fairness, and social equity, the commodification of mental functions, and the impact of enhancement practices on values such as hard work, as well as on personal identity and authenticity. Similar issues are raised by *brain stimulation and brain prostheses* (Farah 2009: 79–80), though other literature suggests that enhancement and privacy are by far not the central ethical issues in connection with stimulation and prostheses (Glannon 2014; Grübler and Hildt 2014).

With regard to *brain imaging*, the main "specifically neuroethical issue" is said to be "privacy of thought." This most "obvious concern" here relates to actual and potential uses of imaging technologies to detect lying and unconscious motivation and to "brainotype" mental and personality traits. Although the well-documented problem of considering brain scans "as more accurate and objective than in fact they are" is mentioned (Farah 2009: 76), the neuroethical treatment of brain imaging illustrates precisely that problem. It does so, however, not in connection with matters widely recognized as needing remedy (e.g., for fMRI: Whelan and Garavan 2014 on inflated predictions, Eklund et al. 2016 on inflated false-positive rates).

Rather, neuroethical practice illustrates beliefs about brain scans because (as mentioned) it embodies the assumption that imaging the brain "provides" or "reveals information about the mind" (Farah 2010: 4; 2009: 75)—in other words, that fMRI constitutes the "primary methodology to investigate mental states neuroscientifically" (Levy 2008: 7). As Fins (2011: 896) notes, neuroethics was made possible by technology and is to such an extent dependent on it that it can without exaggeration be said to be "essentially an ethics of technology." Such dependence is most radical with respect to neuroimaging (Vidal 2015). As the statements just quoted illustrate, even though neuroethics scholars such as Martha Farah, Eric Racine, and Adina Roskies, all quoted here, have helped bring to light myths and misunderstandings about neuroimaging, neuroethics *must assume* that brain imaging is a window into the mind and therefore into the constitutive features of personhood; its relevance depends on that postulate.

We do not wish to discuss to what extent or in what sense such claims about neuroimaging may be justified, but to underline that they constitute a *petitio principii* conveyed as a factual statement. Yet such a statement raises numerous philosophical and empirical issues that are not even hinted at, to begin with the confusion of "mental privacy" and "brain privacy" (Gilead 2015). Such issues should be at the heart of ethical reflection; if they were, however, then the treatment of brain imaging would no longer pertain to neuroethics as we know it. An illustration of this is Farah's (2014) thoughtful reply to criticisms of functional neuroimaging, which does not deal with the assumption in question—from which, however, everything else follows. De Vries (2007: S68) noted that the "uncritical way" in which some neuroethicists respond to neuroimaging research "is surprising to a social scientist." But we argue that it should not be, since what De Vries perceives as lack of critical perspective constitutes in this regard one of neuroethics' conditions of existence.

Finally, *brain and persons* emerges as a fourth category of specifically neuroethical issues because knowledge of brain function is "forcing us to reexamine our understanding of ourselves as moral agents and spiritual beings" (Farah 2009: 80). Perhaps some individuals feel forced to reexamine their self-understanding in light of neuroscientific information. Yet a reexamination of personhood is not an inherent and necessary consequence of such information. Brain science is to such an extent superfluous for questioning views about human nature that the "amoral deterministic viewpoint" which, it is asserted, "will probably gain a stronger hold on our intuitions" (Farah 2009: 80), has existed for centuries independently of knowledge about the brain and has never depended on empirical information. We are told that the problem with our "intuitive understanding of persons," which supposedly "includes the ideas that they have an essence that persists over time, that they are categorically either alive or dead, and that they have a nonmaterial dimension such as a spirit or soul," is that "none of these fits with the idea that a person is his or her brain" and that "as neuroscience reveals progressively more about the physical mechanisms of personality, character, and even sense of spirituality, there is little about a human being left to attribute to an immaterial soul" (Farah 2009: 80–81).

Yet believing that persons are their brains is obviously not the sole alternative to believing in an immaterial soul as the foundation of personhood. For example, although there are significant platonic currents in the history of Christian thought, the Christian tradition has emphasized that persons are intrinsically corporeal, that they require a full body, and that disembodied souls cannot be persons.[6] It posits a duality but nothing akin to "Cartesian dualism." History, moreover, amply demonstrates that materialism requires neither knowing anything about the brain nor the level of confusion displayed in the quoted argument. This confusion, which boils down to identifying the fact that we cannot be without our brains (or perhaps their functional equivalent) with the persuasion that we are our brains, inspired the late Stephan Chorover (2005: 2081), founding member of the Department of Brain and Cognitive Sciences at MIT, to see in neuroethics "an instance of anthropomorphism, in which characteristics properly belonging to the human social sphere are misattributed to brains."

5.5 A Concluding Reflection

Statements that from outside neuroethics seem to reflect an ignorance of history, neglect of the human sciences, and confusion of categories, epitomize the discipline's foundation: "the idea that a person is his or her brain." As Hoyer (2010: 94–95) points out, such a neuroanthropological view of the human (*Menschenbild*) is not a distinctive feature of neuroethics, but a cornerstone of the neuroscientific "thought style."[7] While it is not always clear whether it represents a methodological or an ontological position, unclarity bolsters its performative efficacy. Thus, we argue, giving up this position would entail giving up neuroethical exceptionalism. Statements such as "the brain is the origin of the mind" or "imaging the brain reveals information about the mind" can be interpreted in various ways, from emergentism to the strongest forms of reductionism. Within neuroethics, however, they are primarily performative, and leaving them vague is the best way of making them compatible with a range of interpretive options. That is why, as noted above, discussing the neuroethical project in terms of reductionistic neuroessentialism is off the mark: neuroethicists' claims about the mind-brain are not meant to explore philosophical issues or to clarify the relationship between mind and brain, but to sustain the neuroethical project. This observation does not imply that there are not many valuable neuroethical contributions; but it does provide insight into the discipline's ideological foundation and its function in the global dynamics of science—a foundation and a function that can in turn be embraced and furthered or criticized and resisted.

[6] The literature on this topic is vast. See Vergote (1991) for a concise statement and, most recently, Baschet (2016).

[7] Hoyer uses the term *Denkstil* introduced by the Polish biologist and philosopher of science Ludwik Fleck in his groundbreaking1935 book, *The Genesis and Development of a Scientific Fact*. "Neuroanthropological" does not refer to the field of neuroanthropology that has developed since the 1990s, but to a view of the human, such as the one defended by the German philosopher Thomas Metzinger (2009).

References

Baschet J (2016) Corps et âme. Une histoire de la personne au Moyen Âge. Flammarion, Paris

Baylis F (2015) Neuroethics and identity. In: Clausen J, Levy N (eds) Handbook of neuroethics. Springer, New York, pp 367–372

Brosnan C (2011) The sociology of neuroethics: expectational discourses and the rise of a new discipline. Sociol Compass 5(4):287–297

Buford C (2005) Neuroscience and metaphysics. Am J Bioeth 5(2):34–36

Buniak L, Darragh M, Giordano J (2014) A four-part working bibliography of neuroethics: part 1: overview and reviews—defining and describing the field and its practices. Philos Ethics Humanit Med 9:9. www.peh-med.com/content/9/1/9

Changeux J-P, Evers K, Simmenauer B (2014) Human brain project: first ethics and society report. www.humanbrainproject.eu/ec-deliverables

Changeux J-P, Evers K, Simmenauer B (2016) Human brain project: second ethics and society report. www.humanbrainproject.eu/ec-deliverables

Chorover SL (2005) Who needs neuroethics? Lancet 365:2081–2082

Clausen J, Levy N (2015) What is neuroethics? In: Clausen J, Levy N (eds) Handbook of neuroethics. Springer, New York, pp v–vii

Conrad EC, De Vries R (2011) Field of dreams: a social history of neuroethics. In: Pickersgill M, Van Keulen I (eds) Sociological reflections on the neurosciences. Bingley, Emerald, pp 299–324

De Vries R (2007) Who will guard the guardians of neuroscience? Firing the neuroethical imagination. EMBO Rep 8:S65–S69

Doucet H (2005) Imagining a neuroethics which would go further than genethics. Am J Bioeth 5(2):29–31

Eklund A, Nichols TE, Knutsson H (2016) Cluster failure: why fMRI inferences for spatial extent have inflated false-positive rates. PNAS 113(28):7900–7905

Evers K (2005) Neuroethics: a philosophical challenge. Am J Bioeth 5(2):31–33

Farah MJ (2009) Neuroethics. In: Caplan AL, Fiester A, Ravitsky V (eds) The Penn Center guide to bioethics. Springer, New York, pp 71–83

Farah MJ (2010) Neuroethics: an overview. In: Farah MJ (ed) Neuroethics: an introduction with readings. MIT Press, Cambridge, pp 1–9

Farah MJ (2011) Neuroscience and neuroethics in the 21st century. In: Illes J, Sahakian BJ (eds) Oxford handbook of neuroethics. Oxford University Press, New York, pp 761–781

Farah MJ (2014) Brain images, babies, and bathwater: critiquing critiques of functional neuroimaging. In interpreting neuroimages: an introduction to the technology and its limits. Hast Center Rep 45(2):S19–S30

Farah MJ, Smith ME, Gawuga C, Lindsell D, Foster D (2010) Brain imaging and brain privacy: a realistic concern? J Cogn Neurosci 21(1):119–127

Ferrari F, Villa RF (2016) The neurobiology of depression: an integrated overview from biological theories to clinical evidence. Mol Neurobiol

Fins JJ (2008) A leg to stand on: Sir William Osler and Wilder Penfield's "neuroethics". Am J Bioeth 8(1):37–46

Fins JJ (2011) Neuroethics and the lure of technology. In: Illes J, Sahakian BJ (eds) Oxford handbook of neuroethics. Oxford University Press, New York, pp 895–907

Fischbach R, Mindes J (2011) Why neuroethicists are needed. In: Illes J, Sahakian BJ (eds) Oxford handbook of neuroethics. Oxford University Press, New York, pp 343–376

Fitzpatrick SM (2012) Functional brain imaging: neuro-turn or wrong turn? In: Littlefield MM, Johnson JM (eds) The neuroscientific turn: transdisciplinarity in the age of the brain. University of Michigan Press, Ann Arbor, pp 180–198

Gilead A (2015) Can brain imaging breach our mental privacy? Rev Philos Psychol 6:275–291

Giordano JJ (2010) Introduction. Neuroethics: coming of age and facing the future. In: Giordano J, Gordijn B (eds) Scientific and philosophical perspectives in neuroethics. Cambridge University Press, New York, pp xxv–xxix

Glannon W (2011) Brain, body, and mind: neuroethics with a human face. Oxford University Press, New York
Glannon W (2014) Ethical issues with brain-computer interfaces. Front Syst Neurosci 8:art. 136
Grübler G, Hildt E (eds) (2014) Brain-computer interfaces in their ethical, social and cultural contexts. Springer, New York
Hagner M, Borck C (2001) Mindful practices: on the neurosciences in the twentieth century. Sci Context 14:507–510
Hall ZW (2002a) Excerpt of talk at the conference Neuroethics: Mapping the Field. http://dana.org/Cerebrum/2002/Neuroethics__Mapping_the_Field/
Hall ZW (2002b) Talk on "the future of neuroethics". In: Marcus SJ (ed) Neuroethics: mapping the field. Dana Press, Washington, pp 289–297
Hoyer A (2010) Neurotechnologie, Philosophie und Hirnforschung: ZurEntstehung und Institutionalisierung der Neuroethik. Unpublished MA Thesis, Johann Wolfgang Goethe-Universität Frankfurt am Main
Illes J (2010) Empowering brain science with neuroethics. Lancet 376:1294–1295
Illes J, Racine E (2005) Imaging or imagining? A neuroethics challenge informed by genetics. Am J Bioeth 5(2):5–18
Illes J, Lombera S, Rosenberg J, Arnow B (2008) In the mind's eye: provider and patient attitudes on functional brain imaging. J Psychiatr Res 43(2): 107–114
Leefmann J, Levallois C, Hildt E (2016) Neuroethics 1995–2012. A bibliometric analysis of the guiding themes of an emerging research field. Front Hum Neurosci 10:art. 336
Leshner AI (2005) It's time to go public with neuroethics. Am J Bioeth 5(2):1–2
Leshner AI (2011) Foreword. In: Illes J, Sahakian BJ (eds) Oxford handbook of neuroethics. Oxford University Press, New York, pp v–xi
Levy N (2007) Neuroethics: challenges for the 21st century. Cambridge University Press, New York
Levy N (2008) Introducing *neuroethics*. Neuroethics 1:1–8
Littlefield MM, Johnson JM (eds) (2012) The neuroscientific turn: transdisciplinarity in the age of the brain. University of Michigan Press, Ann Arbor
Lombera S, Illes J (2009) The international dimensions of neuroethics. Dev World Bioeth 9:57–64
Lu S (2016) Tailoring treatment by scanning the brain. Monit Psychol 47(8):54–57
Metzinger T (2009) The ego tunnel: the science of the mind and the myth of the self. Basic Books, New York
Mulkay M (1993) Rhetorics of hope and fear in the great embryo debate. Soc Stud Sci 23(4):721–742
O'Connor C, Joffe H (2013) How has neuroscience affected lay understandings of personhood? A review of the evidence. Public Underst Sci 22(3):254–268
Parens E, Johnson J (2007) Does it make sense to speak of neuroethics? EMBO Rep 8:S61–S64
Racine E (2010) Pragmatic neuroethics: improving treatment and understanding of the mind-brain. MIT Press, Cambridge
Racine E, Illes J (2008) Neuroethics. In: Singer PA, Viens AM (eds) The Cambridge textbook of bioethics. Cambridge University Press, New York, pp 495–504
Racine E, Bar-Ilan O, Illes J (2005) fMRI in the public eye. Nat Rev Neurosci 6:159–164
Rose N (2014) The human brain project: social and ethical challenges. Neuron 82:1212–1215
Roskies A (2002) Neuroethics for the new millennium. Neuron 35:21–23
Roskies A (2016) Neuroethics. In Zalta EN (ed) The Stanford encyclopedia of philosophy (Spring 2016). http://plato.stanford.edu/archives/spr2016/entries/neuroethics/
Safire W (2002) Introduction: visions for a new field of "neuroethics". In: Marcus SJ (ed) Neuroethics: mapping the field. Dana Press, Washington, pp 3–9
Schick A (2005) Neuroexceptionalism? Am J Bioeth 5(2):36–38
Shook JR, Giordano J (2014) A principled and cosmopolitan neuroethics: considerations for international relevance. Philos Ethics Humanit Med 9:1. www.peh-med.com/content/9/1/1
Singh I, Rose N (2006) Neuro-forum: an introduction. BioSocieties 1:97–102

Stahnisch F (2011) The essential tension: on ethical and historical conundrums in the trajectories of deep brain stimulation. In: Fangerau H, Fegert, JM, Trapp T (eds) Implanted minds: the neuroethics of intracerebral stem cell transplantation and deep brain stimulation. Bielefeld, transcript Verlag, pp 151–181

Tovino SA (2007) Functional neuroimaging information: a case for neuroexceptionalism? Fla State Univ Law Rev 34:415–289

Vergote A (1991) The body as understood in contemporary thought and biblical categories. Philos Today 35:93–105

Vidal F (2015) Historical and ethical perspectives of modern neuroimaging. In: Clausen J, Levy N (eds) Handbook of neuroethics. Springer, New York, pp 535–550

Vidal F, Ortega F (2012) Are there neural correlates of depression? In: Choudhury S, Slaby J (eds) Critical neuroscience: a handbook of the social and cultural contexts of neuroscience. Wiley-Blackwell, Oxford, pp 345–366

Vidal F, Ortega F (2017) Being brains: making the cerebral subject. Fordham University Press, New York

Wachbroit R (2008) The prospects for neuro-exceptionalism: transparent lies, naked minds. Am J Bioeth 8(1):3–8

Whelan R, Garavan H (2014) When optimism hurts: inflated predictions in psychiatric neuroimaging. Biol Psychiatry 75:746–748

Wilfond BS, Ravitsky V, Magnusson WG (2005) On the proliferation of bioethics sub-disciplines: do we really need "genethics" and "neuroethics"? Am J Bioeth 5(2):20–21

Part II

Focus, Theories, and Methodologies in Neuroethics

Section Introduction: Focus, Theories, and Methodologies in Neuroethics

6

Eric Racine and John Aspler

Abstract

In this introduction to the section on the focus, theories, and methodologies in neuroethics, we present part of the ongoing debate surrounding the disciplinary status of neuroethics (e.g., is it a field, is it a discipline), which reflect varied expectations about the ability or even the requirement for neuroethics to offer new approaches to academic inquiry. Accordingly, the following four chapters offer different perspectives on these theoretical and methodological issues.

Academic disciplines are usually defined by a focus and a set of theories and methodologies. For example, in the late nineteenth century, Durkheim defined the object of sociology as the study of social facts, global properties of institutions, and societies (and not the psychology of individuals) (Durkeim 1894). This galvanized the development of sociological theories and methods and, eventually, led to faculty positions and departments in sociology that sustained this new emerging discipline. Psychology is also a field marked by early tensions between distinct traditions of scholarship that defined the object and methods of psychology in sometimes strongly opposing ways. Indeed, between Wundt's psychophysical and

E. Racine (✉)
IRCM, Montréal, QC, Canada

Université de Montréal, Montréal, QC, Canada

McGill University, Montréal, QC, Canada
e-mail: eric.racine@ircm.qc.ca

J. Aspler
IRCM, Montréal, QC, Canada

McGill University, Montréal, QC, Canada
e-mail: John.Aspler@ircm.qc.ca

E. Racine, J. Aspler (eds.), *Debates About Neuroethics*, Advances in Neuroethics, DOI 10.1007/978-3-319-54651-3_6

experimental approach, and the introspective tradition that includes psychoanalysis, there are huge epistemological ruptures. The disciplines of sociology and psychology are now well established, even if debates about their focus, theories, and methodologies persist. These debates are often considered signs of their vitality. In contrast, neuroethics is a long way from becoming an established discipline. Indeed, such disciplinary aspirations might not even be appropriate, since the traditional model of the academic discipline may not align well with the scholarly and practical goals of neuroethics. For the sake of comparison, even after almost five decades of active scholarship, bioethics is not yet considered a discipline from both scholarly and institutional standpoints.

However, some scholars, in particular those who are more optimistic about the potential contribution of the neuroscience of ethics, suggest that the theoretical insights generated by neuroscience, as well as the behavioral and brain sciences more broadly, could form a new foundation for ethics, perhaps even establishing a new discipline (Changeux 1996; Gazzaniga 2005). Adopting a more cautious interpretation, others have suggested that the neuroscience of ethics can at least make an important contribution to how we understand moral thinking and moral behavior, but that this contribution has to be folded into a much broader set of studies in empirical ethics (Racine 2005; Churchland 2002). Still others argue that much of the neuroscience of ethics should remain the province of social psychology or of neuroscience itself (Farah and Wolpe 2004). However, this latter position neglects the potential enrichment that neuroscience could bring to ethics (Racine et al. 2017). Finally, there also exist questions about whether there is room within neuroethics for dissenting opinions about the value of the neuroscience of ethics at all—especially given external critiques (e.g., from critical neuroscience (Choudhury and Slaby 2012)). Only the future can tell whether the knowledge generated by neuroscience will have transformative effects on ethics and society to the point of establishing a discipline. But the future is notoriously hard to predict, notably in this case regarding whether the gaps between the humanities and social sciences, and the biological sciences, will give way to evermore creative and engaging scholarship that helps establish, in the words of Potter—one of the founders of bioethics—"bridges to the future" (Potter 1970, 1971).

In this spirit, the second section of the book surveys different theories and methodologies deployed in neuroethics, while critical contributions assess theoretical blind spots and the methodological standards relied on thus far. Chapter 7, *Theoretical Framing of Neuroethics: The Need for a Conceptual Approach,* is presented by Arleen Salles and Michele Farisco of the Centre for Research Ethics & Bioethics at Uppsala University, under the leadership of the philosopher and head of the Ethics and Society Subproject of the Human Brain Project, Kathinka Evers. They provide an overview of dominant theoretical perspectives in neuroethics, before arguing in favor of a model based on informed materialism, which they claim could provide a novel theoretical foundation for neuroethics. In Chap. 8, *Neuroethics: A Renewed View of Morality? Intentions and the Moral Point of View,* Bernard Baertschi, a long-time professor of philosophy at the University of Geneva and a neuroethics pioneer in the French-speaking world, explores the neuroscience of intentional actions and rejects interpretations of neuroscience evidence that call for radical shifts in the attribution of responsibility. In Chap. 9, *Is it Time to Abandon the Strong Interpretation of the Dual Process Model in Neuroethics?*, Veljko Dubljević, an emerging force in neuroethics located at North Carolina State University, rejects the dual process theory of moral judgment given mounting scientific evidence undermining this foundational neuroethics theory. Finally, in Chap. 10, *Neuroethics and Policy at the National Security Interface: A Test Case for Neuroethics Theory and Methodology*, Nicholas Evans, an expert in the ethics of dual-use research, and reputed American bioethicist Jonathan Moreno explore the significant issues that neuroscience raises with respect to armed conflict and national security. They draw conclusions about the methodological development needed to allow neuroethics to address broader social issues, such as the military application of neurotechnologies. The four chapters in this second section hint at the numerous theoretical and methodological questions to be addressed in future neuroethics scholarship.

References

Changeux J-P (1996) Le point de vue d'un neurobiologiste sur les fondements de l'éthique. In: Huber G (ed) Cerveau et psychisme humains: quelle éthique? John Libbey Eurotext, Paris

Choudhury S, Slaby J (eds) (2012) Critical neuroscience: a handbook of the social and cultural contexts of neuroscience. Willey-Blackwell, Toronto

Churchland PS (2002) Brain-wise: studies in neurophilosophy. MIT Press, Cambridge

Durkeim É (1894) Les règles de la méthode sociologique. Bibliothèque Paul-Émile-Boulet de l'Université du Québec à Chicoutimi, Chicoutimi

Farah MJ, Wolpe PR (2004) Monitoring and manipulating brain function: new neuroscience technologies and their ethical implications. Hastings Cent Rep 34:35–45

Gazzaniga MS (2005) The ethical brain. Dana Press, New York

Potter VR (1970) Bioethics the science of survival. Perspect Biol Med 14:127–153

Potter VR (1971) Bioethics: bridge to the future. Prentice-Hall, Englewood Cliffs

Racine E (2005) Pourquoi et comment doit-on tenir compte des neurosciences en éthique? Esquisse d'une approche neurophilosophique émergentiste et interdisciplinaire. Laval théologique et philosophique 61:77–105

Racine E, Dubljevic V, Jox RJ, Baertschi B, Christensen JF, Farisco M et al (2017) Can neuroscience contribute to practical ethics? A critical review and discussion of the methodological and translational challenges of the neuroscience of ethics. Bioethics

Theoretical Framing of Neuroethics: The Need for a Conceptual Approach

7

Kathinka Evers, Arleen Salles, and Michele Farisco

Abstract

There are different dominant perspectives, theories, and methodologies within neuroethics, each importantly shaping the identification, understanding, and discussion of the relevant ethical, social, philosophical and scientific issues. In this chapter, we first provide a brief overview of current neuroethical approaches calling attention to a common tendency to underestimate the role and value of conceptual analysis. Against that background, we present and develop the theoretical framework of fundamental neuroethics. Next, we suggest that neuroethics should be built on the sound scientific and philosophical foundations of informed materialism. Finally, we apply the proposed theoretical framework to the neuroethical discussion of brain simulation.

The twenty-first century has seen neuroscience develop rapidly and a new academic field emerge, neuroethics, investigating the neurobiological basis of morality as well as the ethical, social, and legal issues raised by neuroscientific research. There are different dominant perspectives, theories, and methodologies within neuroethics, each importantly shaping the identification, understanding, and discussion of the relevant ethical issues.

K. Evers (✉)
Centre for Research Ethics and Bioethics (CRB), Uppsala University, Uppsala, Sweden
e-mail: Kathinka.Evers@crb.uu.se

A. Salles
Centre for Research Ethics and Bioethics (CRB), Uppsala University, Uppsala, Sweden

Neuroethics Program CIF, Buenos Aires, Argentina

M. Farisco
Centre for Research Ethics and Bioethics (CRB), Uppsala University, Uppsala, Sweden

Biogem Genetic Research Centre, Ariano Irpino (AV), Italy

E. Racine, J. Aspler (eds.), *Debates About Neuroethics*, Advances in Neuroethics,
DOI 10.1007/978-3-319-54651-3_7

In this chapter, first we provide a brief overview of current neuroethical approaches calling attention to a common tendency to underestimate the role and value of conceptual analysis, focusing more on practical rather than on theoretical issues in the philosophical contributions.

Against that background, we present and develop the theoretical framework of *fundamental neuroethics*, which we argue provides the adequate methodology and theoretical foundations required in order to properly address the ontological, epistemological, and ethical challenges created by potential advances in brain science. Next, we suggest that neuroethics should be built on the sound scientific and philosophical foundations of informed materialism, a perspective that depicts the brain as an autonomously active, plastic, and projective organ evolved in sociocultural-biological symbiosis and that adopts an evolutionary view of consciousness as an irreducible part of biological reality, an evolved feature of the brain, and a suitable object of scientific study. Finally, we apply the proposed theoretical framework to the neuroethical discussion of brain simulation, conceived as a prospective new tool to describe, explain, and predict cerebral activities through their reproduction on a computer.

7.1 Introduction

Moral philosophers today confront an interesting situation: the natural sciences, in particular neuroscience, have entered "their" domains, purporting to shed scientific light on the phenomenon of moral thought. Neuroscientists are not merely suggesting areas for interesting applications of ethical reasoning or calling for assistance in solving problems arising from scientific discoveries, as other scientists have long done. Neuroscience purports to *offer scientific explanations* of important aspects of moral thought and judgement. And while the understanding of morality as a social phenomenon is primarily a matter of understanding cultural and social mechanisms, it is becoming increasingly apparent that empirical knowledge of the brain is relevant in this context. Progress in neuroscience, notably, in the dynamic functions of neural networks, can deepen our understanding of decision-making, choice, acquisition of character and temperament, and the development of moral dispositions.

This bold move has been received with varying levels of enthusiasm, ranging from optimistic faith in finding clarity on—and even possible solutions to—age-old philosophical problems (Greene 2008; Churchland 2008) to calls for caution (Legrenzi and Umiltà 2011; Choudhury and Slaby 2012: De Vos and Pluth 2016) and even rejection (Satel and Lilienfeld 2013). Moral conceptions and metaphysical notions (such as identity and free will) are often thought to stubbornly resist scientific understanding, and there is no agreement on whether neuroscience's incursion into the domains of ethics and philosophy is a cause for celebration or for scandal.

The possibilities opened by neuroscientific advances and the uncertainty and disagreement regarding what they mean and what their implications are have inspired several new academic lines of inquiry. One is "neuroethics", which is an interface between the empirical brain sciences, philosophy of mind, moral philosophy, ethics, psychology, and the social sciences. Though used early on to refer to ethical issues raised by neurologists in practice (Pontius 1973; Cranford 1989), the term now generally refers to the study of the myriad questions (both practical and conceptual) that arise when scientific findings about the brain are carried into philosophical analyses, medical practice, legal interpretations, and health and social policy. The topics the field[1] addresses are not new; some were raised already during the French Enlightenment, notably by Diderot (1966) who stated in his *Éléments de Physiologie*: "*C'est qu'il est bien difficile de faire de la bonne métaphysique et de la bonne morale sans être anatomiste, naturaliste, physiologiste et médecin…*" (suggesting that in order to be a good metaphysician or moral philosopher, one also needs knowledge of anatomy, physiology, and medicine). Other topics, such as the ethical quandaries arising from advances in neuroscientific research, have long been addressed by ethics committees throughout the world, though not necessarily under the label of neuroethics.[2]

'Neuroethics' is a very young and rapidly developing area of research generally characterized by its naturalistic orientation—supporting the relevance of scientific methods of explanation and justification—and its interdisciplinary nature (Farah 2010; Evers 2009; Racine 2010). As it grows into an established field of practice, two main issues internal to its development and distinctiveness are raised.

The first relates to its methodology. By virtue of its interdisciplinary character, neuroethics can be seen at times as the convergence of neuroscience and philosophy, of neuroscience and moral psychology, or of neuroscience and bioethics, depending on which perspective one wishes to emphasize. That the field draws from the insights and methodologies of the empirical sciences and of philosophy is key in providing a deeper understanding of the issues at stake. But at the same time, such interdisciplinarity raises questions about how to integrate methods and about which aims will provide the field with more explanatory depth.

The scientific method, which centrally involves repeatable observations and experiments, is usually assumed to guarantee objectivity and factual information, and yet, there is cause for caution and even scepticism regarding claims about neuroscientific objectivity in the production and interpretation of "facts" about the brain (Evers 2009). Neuroscientific findings are theoretically limited; they come with uncertainties often ignored and assumptions that can be questioned and are often shaped by extra-epistemic factors (i.e., societal, political, and economic expectations that affect which topics are chosen and how they are addressed). Finally, the rhetorical power of neuroscience typically outreaches its empirical findings and sometimes extends its discourses into other epistemological domains (Rose 2006; Frazzetto and Anker 2009; Racine 2010; Choudhury and Slaby 2012).

In contrast, the core of philosophical methodology is the definition of concepts and the identification of their essential components in search of conceptual clarity. But while in the past it was thought possible to discuss topics such as moral agency in the absence of biological and psychological knowledge, at present, a philosophical discussion of these topics that does not take empirical considerations into account is increasingly considered inadequate. This raises a number of issues that deserve attention, such as whether and how the integration of the scientific and

[1] For the purposes of this paper, we understand the term "field" to denote an area of research that may not yet be a discipline in a clearly defined sense.

[2] Cf., e.g., opinions by the *Comité Consultatif National d'Éthique*, France in the 1980s.

philosophical methodologies is possible, which has more authority and epistemological priority in a discussion of the issues, and whether they can be integrated to conform to a specific neuroethical methodology that distinguishes the field from other similar ones.

The second internal issue raised by the development of neuroethics has to do with the field's ontological commitments when addressing the ethical questions engendered by neuroscience. Indeed, the determination of whether neuroscientific findings are relevant to our understanding of the self, consciousness, and intentionality, for example, and of the usefulness of neuroethics itself, requires an understanding of the brain, its structural architecture, its development, and functions (Evers 2005; Evers 2009). But such understanding is not merely an issue of empirical interpretation, i.e., it does not concern the brain as it appears to us; it is one that requires philosophical interpretation as well, i.e., it concerns the brain as it is in itself (Northoff 2014; Wagner and Northoff 2015). The question here is about what is a scientifically and philosophically appropriate understanding of the brain that can be used to build a constructive neuroethics. Thus, it seems clear that both epistemological and ontological considerations have an impact on the viability of the neuroethical enterprise.

Consequently, in this first section, we explore these epistemological and ontological concerns. We begin by providing a brief overview of what we consider are dominant neuroethical perspectives and methodologies, calling attention to a common framing assumption regarding the role and value of conceptual analysis in neuroethics. Against that background, we then propose a conceptual framework that we call *fundamental neuroethics*, which provides the adequate methodology and theoretical foundations required in order to properly address the practical implications of potential advances in brain science—and their ontological, epistemological, and ethical impact on our understanding of human beings and their capacity for moral judgement.

7.2 Neuroethics: Methods and Topics

For the sake of simplicity, a distinction can be made among three different (not fully independent and often overlapping) methodological approaches within neuroethics: (a) neuro-bioethics, (b) empirical neuroethics, and (c) conceptual neuroethics.

7.2.1 Neuro-Bioethics

What we call "neuro-bioethics" corresponds to what some scholars label the "ethics of neuroscience" (Roskies 2002).[3] It typically addresses a number of issues that can be grouped into four major areas:

[3] James Giordano uses the term to refer to neuroethics in general, on the grounds that this term "grounds an understanding and use of neuroscience to the methodology of ethics and the interdisciplinarity and practicality of bioethics" (Giordano 2011: 18).

1. Ethical issues raised by neuroscientific research—e.g., informed consent, the handling of incidental findings, and privacy.
2. Ethical issues that emerge in psychiatric and psychological care and in clinical neurology—e.g., the extent to which progress in neuroscience meets the needs of patients with disorders of consciousness (i.e., coma, vegetative state, and minimally conscious state).
3. Policy considerations regarding some of the possibilities opened up by neuroscientific advances (e.g., cognitive neuro-enhancement and non-clinical uses of neuroimaging).
4. Public communication, media representation, and cultural and societal understanding of neuroscience's impact.

Although unique in that these discussions focus on the scientific study and manipulation of the brain, they mirror bioethical debates (Levy 2008). Their main goal is practical: to solve concrete moral problems.

It is well known that much bioethical theorizing has been characterized by the application of a limited set of concepts, such as rights, obligations, duties, and social good, and fundamental principles, such as autonomy, beneficence, justice, and utility, to the relevant moral problems. These bioethical concepts and principles (some of which have been much criticized for remaining largely untouched by empirical and contextual considerations) are justified by different ethical approaches. The advantages and disadvantages of these kinds of approaches (which generally resolve practical issues in a deductive fashion, i.e., starting from general principles or rules to assess particular issues) have been widely discussed (Clouser and Gert 1990; Clouser 1995; Beauchamp 1995). Unlike bioethics, neuro-bioethics generally makes no explicit mention of the method used when addressing the issues, nor is there a general consensus regarding which is the best methodology or whether there is one best methodology for all cases. However, a deductive approach seems to be present in the discussion of some particular neuroethical issues (e.g., cognitive enhancement), where often the debate seems to revolve around a few principles or rules and the extent to which the practice in question is or is not compatible with them (Farah et al. 2004).

Partly because of the alleged tendency of the deductive approach to promote a rather abstract ethical analysis of cases (Little 1996; Toulmin 1982), a number of alternative inductive approaches, such as casuistry, have been proposed within bioethics. Recently, some bioethicists and neuroethicists have been calling for a pragmatic methodology that puts an emphasis on understanding and enabling the interpersonal process of moral problem-solving in specific cases in an interdisciplinary fashion (Fins 2008; Racine 2010). In the pragmatic model, moral decision-making is dynamic (insofar as it is crucially concerned with interaction between stakeholders) and contextual. Moral judgement is not detached from interpretive complexities and specific circumstances; instead, it is nurtured by biological, physical, and historical considerations and requires responsiveness to the concrete clinical or research context.

Considering the above, it is evident that the role that philosophy plays within neuro-bioethical methodology varies. In the deductivist model, practical problems are identified and solved in terms of certain commonly accepted ethical frameworks grounded in moral theory and philosophy. In this model, philosophy plays a pivotal role. In the case of the pragmatic model, the role of philosophy is different. Indeed, the pragmatic model tries to incorporate insights from multiple sources and disciplines. Some pragmatist commentators suggest that while philosophy "tempers assertions and helps prevent ideological distortions" (Fins 2008: 44), it must take a back seat to scientific and clinical considerations. Other pragmatist neuroethicists call for recognizing the potential contributions of philosophy, while considering it one among the many other fields that can help in neuroethical decision-making (Racine 2010: 87).

7.2.2 Empirical Neuroethics

Following Northoff, we apply the term "empirical neuroethics" to the type of neuroethics research that seeks to use neuroscientific data to inform both practical and theoretical issues without focusing on translational concerns, such as *how* neuroscientific findings can be so used (Northoff 2009; Wagner and Northoff 2015).[4] Empirical neuroethics takes as a starting point the view that relevant knowledge about human beings, who they are, how they think and judge morally, and how they act can be achieved by looking at the empirical data on the workings of the nervous system and the brain. Accordingly, in our view, empirical neuroethics not only encompasses issues typically assumed to belong in the "ethics of neuroscience" (e.g., the social implications of a neuroscientific understanding of brain function) but also, and more fundamentally, those issues typically subsumed under what is known as the "neuroscience of ethics" (Roskies 2002). The reason for this is that, despite the fact that the focus of the neuroscience of ethics is the basis and meaning of concepts, it does not address the issue of *how* neuroscientific results can illuminate fundamental philosophical questions; justify a change in some beliefs we hold about autonomy (Levy 2008), for example, or personhood (Farah and Helberlein 2007); and possibly even refine and enhance the moral tools ethicists use (Roskies 2002; Levy 2011; Greene 2008).

The lack of attention to the issue of *how* is not a minor concern. The fact is that although the "neuroscience of ethics" is typically considered descriptive, it has long been prescriptive: implicit assumptions about brain facts, their value, and their normative weight underlie the claim that neuroscientific findings will lead us to revise particular metaphysical and ethical notions (Shook and Giordano 2014; Northoff 2009). Indeed, prescriptivity is not problematic in itself. However, what is problematic is that (1) there is no explanation of how to interpret those brain facts that function prescriptively and (2) that, unless one supposes that brain facts and normative

[4] Thus, we understand the term differently from Judy Illes who uses it to refer to empirical research in the field on neuroethics (Illes 2007).

concepts correspond one to one, it is not clear why such facts are prescriptive. Thus, a shortcoming of empirical neuroethics is that it is unable to account for one of the most challenging tasks for neuroethics: the determination of how to use biological data to have either explanatory or normative relevance.

7.2.3 Conceptual Neuroethics

The main difference between empirical and conceptual neuroethics is that whereas the former claims that observations about the brain are relevant to ethical and metaphysical concepts without explaining how, the latter does actually concern itself with the methodological issue of *how* this happens. For a conceptual approach, the link between descriptive considerations derived from observations about the brain and normative considerations is not necessarily self-evident: understanding it requires some kind of conceptual interpretation of neuroscientific findings. Thus, conceptual approaches call for developing a methodological modus operandi for fruitfully linking scientific and philosophical interpretations without necessarily giving primacy to either science or philosophy.

Neuroethics as Metaethics: One such conceptual approach proposes that neuroethics be understood as a type of metaethics, clarifying the operative underlying presuppositions and commitments involved in moral deliberation and action (Shook and Giordano 2014). In a number of articles, James Giordano and colleagues have insisted on the neuroethical need to develop a critical assessment of neuroscience, addressing "the validity of the tools and techniques utilized to develop neuroscientific and neuroethical theories" (Avram and Giordano 2014). While the need to engage in a critical neuroscience is not new (Choudhury and Slaby 2012), the authors suggest that attention be paid to *how* neuroscientific concepts are shaped in order to determine the usefulness of neuroscience in addressing fundamental human issues.

Fundamental Neuroethics: Developed by Kathinka Evers, this conceptual approach also posits that knowledge of the brain's structural and functional architecture and its evolution can deepen our understanding of personal identity, consciousness, and intentionality, including the development of moral thought and judgement. However, in contrast with other approaches, fundamental neuroethics focuses in particular on the issue of *how* natural science can deepen our understanding of moral thought and on the role of conceptual analysis.

Any attempt to address the issue of *how* necessarily leads to a discussion of whether human mental activity, including moral thought as a subset of thought in general, can be understood in biological terms. Indeed, this is not a new query: a version of the classical mind-body problem has been discussed for millennia and also in quite modern terms from the French Enlightenment until today. What is comparatively new is the realization of the *extent* to which ancient philosophical problems—such as whether humans possess a free will, the meaning of personal responsibility, how to understand the self, and the relationship between emotions and cognition, or between emotions and memory—vigorously emerge in the rapidly advancing neurosciences. While this has been acknowledged by other neuroethical approaches, fundamental neuroethics pays special attention to methodological and conceptual concerns typically left unexplained.

As the name suggests, the term "fundamental neuroethics" refers to basic research (in contrast to neuro-bioethics) that combines philosophical and scientific theoretical perspectives. From a methodological perspective, fundamental neuroethics argues that only a philosophical/conceptual level of interpretation of the scientific evidence allows for a legitimate connection between scientific evidence and philosophical concepts and issues without assuming that the empirical and the conceptual correspond one to one. Philosophical analysis also plays a key role in stifling unrealistic expectations regarding neuroscientific advances (Evers 2007). Empirical methodology and scientific interpretation, while providing important information, are by themselves insufficient to generate an adequate conceptual understanding of data, including data about the brain. Thus, philosophical analysis fosters understanding of the meaning and use of the main scientific concepts and thus complements scientific interpretation of theories and data.

Integration of scientific and philosophical methodologies is necessary because, as we will see in the next section, whether neuroscience can play a productive role in addressing fundamental philosophical issues depends on how the relevant neuroscientific concepts are interpreted. One such concept is the brain: fundamental neuroethics provides a philosophical account of the brain and how its functional architecture grounds human moral behaviour. Insofar as it does, this conceptual approach intends to offer both neuro-bioethics and empirical neuroethics the theoretical foundations needed to examine the ethical problems raised by neuroscience. Thus, a fundamental neuroethics perspective is key whether the focus is on the clinic, on neuroscientific research, on the neurobiology of moral judgement, or on the foundation for a new metaethics (Evers 2007; Evers 2009).

7.3 Informed Materialism

From the perspective of fundamental neuroethics, the relevance of neuroscience to understanding moral thought and judgement and the moral implications of neuroscientific research depends on which theoretical model of the brain is used to understand complex human behaviour.

The insight that the mind and brain are not dual ontological entities is today common scientific knowledge. However, this does not mean that objective knowledge about neuronal activities in the brain can tell us all there is to know about the mind and specifically about consciousness. In the absence of subjective accounts, no objective third-person description can fully capture the first-person view from its own perspective: it needs to be experienced first-hand to be known, and to live the experience of another is a logical impossibility (Evers and Sigman 2013). The field of neuroethics must therefore acknowledge both the embodiment of mind and of consciousness and the irreducible character of the subjective perspective. Accordingly, a naïve reductionist account, which postulates that we only have to wait for neuroscience to become sufficiently sophisticated to be able to explain all there is to know about consciousness (Crick 1994; Bickle 2003), is not acceptable because it is unable to account for the subjective view (Evers 2007, 2009). Moreover, insofar as the brain's subjective experience mobilizes emotions (themselves essential features of moral judgement),[5] neuroethics must also take into account the role of emotions and values. Naïve reductionism, failing to consider this, lacks explanatory power and relevance to understanding moral thought and judgement.

One of us (KE) has suggested an understanding of the brain that is epistemologically and morally relevant to provide an adequate theoretical framework for neuroethics: informed materialism (IM) (Evers 2007, 2009). IM has high explanatory value with regard to both the evolution and nature of moral judgement, notably with regard to subjective evaluations, and the personal and public responsibility to ensure that knowledge of the brain is put to practical use and yields societal benefits.

The concept of IM was originally used in chemistry (Bachelard 1953, "matérialism instruit"), but has since been used in neuroscience to oppose both dualism, which posits the existence of a mind or soul independent of the material body, and naïve reductionism, which excludes the subjective perspective from scientific study (Changeux 2004). IM is founded on the notion that all elementary processes of the brain are based on physicochemical mechanisms, and it adopts a view of consciousness as an evolved biological function of neuronal activities.

Departing from the view of the brain as a rigid automatic device whose operations are strictly determined, the informed materialist framework depicts the brain as an autonomously active, plastic, projective, and creatively narrative organ evolved in sociocultural-biological symbiosis. The three core ideas of informed materialism are that the mind cannot be understood independently of biology, that consciousness is an irreducible part of biological reality, and that the brain is an emotional, selectional system in which values are included as necessary constraints. No human being transcends its biological nature and, biologically speaking, no creature with a brain is born value-free.[6]

Evaluation becomes a fundamental feature of the brain thus conceived. A system cannot learn or remember without any values, or without the capacity to evaluate stimuli, preferring some stimuli to others. This classical idea in learning theory was expressed in neuronal terms by Dehaene and Changeux (1989, 1991) and by Edelman (1992) in his account of primary consciousness. These different neuroscientific accounts emphasize that learning is change of behaviour that results from cerebral categorizations of stimuli in terms of positive or negative values and value systems and emotions, conceived as essential to the selectional workings of the brain.

---

[5] Since an evaluation is essentially an expression not only of thought but of preference that is the result of emotion, cf. below.

[6] A "value" is here understood as something that is taken into account in decision-making and that influences a choice, selection, or decision, which can occur on many levels, (i.e., nonconscious as well as conscious), as a basic biological function or as a feature of advanced moral reasoning.

The human brain is both intrinsically active and in constant social interaction: it produces electrical and chemical activity, not only in response to external stimuli but also spontaneously—independently of the external world. This intrinsically active and motivated neural system is genetically predisposed to explore the world and to classify what it finds. Considering how our brains acquire knowledge of ourselves and the world, IM acknowledges that adequate understanding of our subjective experience must take into account self-reflective information, physiological observations, and physical measurements (Dehaene et al. 2003; Changeux 2004; Evers 2009).

7.3.1 Emotions

According to IM, sensitivity to reward signals, whether as basic biological functions or as higher-level cognitive/emotional functions, plays an important role in producing a number of mental activities; some researchers propose that this marks the dawn of consciousness (Denton 2005), motivation, planning, and volition and increases an organism's capacity to exert control over itself (LeDoux 1998). Current hypotheses about knowledge acquisition posit that reward signals select spontaneously arising pre-representations that are stabilized as representations after confirmation by both external experience and internal evaluation processes (Dehaene and Changeux 1991). Such pre-representations are stabilized through "cognitive games" where the brain tests the pre-representations in relation to context and previously established representations, discarding those that do not fit, and stabilizing those that do as permanent features of a developing cognitive apparatus (Dehaene et al. 1998; Changeux 2004). The anticipation of a reward creates a delay between the elaboration of tacit plans and the actual interaction with the world, which is an additional component of cognitive learning and a building block for the neural basis of morality (Schultz 2006; Schultz et al. 1997).

Herein lies the seed of morality: without emotions or preferences, there could be no morality because morality presupposes the capacity for preferential selection. The evolutionary connection between cerebral emotionality and consciousness is therefore highly relevant to neuroethics. Different forms of morality are developed systems of values that make human beings function in their environments. Thus, from an ethical point of view, knowing that neuronal processes of evaluation (assessing pre-representations, stimuli, anticipating reward, etc.) and their emotionality are fundamental characteristics of our brains is highly relevant.

A non-emotional system would be purely passive and incapable of self-organization, evaluation, or anticipation and hence of learning; without values (here understood as sensitivity to reward signals), a system cannot learn or remember (Edelman 1992). The complex behaviour of humans would have scarcely evolved without the basic feature of consciousness, and consciousness could not have evolved without emotions. From an IM perspective, dynamic processes of evaluation and the emotional systems involved in their generation are basic properties of our brains: we are neurobiologically predisposed to develop complex and diverse systems of moral (and other) values that enable us to establish appropriate relationships in our social, cultural, and physical environments. Evolution and natural selection have created an essentially evaluative cerebral architecture with a propensity for performing moral choices. Morality is a biologically conditioned, social phenomenon that has evolved genetically and epigenetically through the communication and interaction of self-aware organisms.[7]

7.3.2 Neuronal Epigenesis and Epigenetic Proaction

IM further emphasizes the plasticity of the brain and its epigenetic development in response to learning and experience. Genetic control over the brain's development is important, but by no means absolute: the epigenetic model of neuronal development postulates that the connections between neurons are not pre-specified in the genes, but that learning and experience influence the brain's development within the boundaries of a 'genetic envelope' (Changeux et al. 1973).

The formation of synapses takes place both prenatally and postnatally; it is far from complete at birth, and the postnatal development of the human brain lasts considerably longer than in any other mammalian species. The most intense development occurs during the first 2 years, but continues long after puberty; in fact, the highest executive functions determined by the frontal lobe are not fully mature until around the age of 25 (Lagercrantz 2005).

Throughout the exceptionally long period of postnatal development, an intense synaptogenesis steadily occurs in the human cerebral cortex, which persists, though to a smaller extent, in the adult. As a consequence of steady interactions with the physical, social, and cultural environment, an active epigenetic selection of neuronal networks results in the internalization of common cultural and ethical rules. This gives rise to the ability of human beings to be "epigenetically proactive": to influence the cultural imprints that are stored in our brains and to develop new overarching ethical rules to benefit and constructively interact with the ever-developing neuronal architecture of the human brain (Evers 2009, 2016; Evers and Changeux 2016).

The fundamental idea of epigenetic proaction is that it is possible to understand and influence human nature and the genesis of human norms in light of current knowledge of the brain. Being epigenetically proactive also means adapting and creating social structures, in both the short and the long term, to constructively interact with the ever-developing neuronal architecture of our brains.

This means that the sociocultural structures of the group and the neuronal functions of the individual develop symbiotically. The architecture of our brains contributes to our social behaviour and to the types of societies that we create,

[7] In this context, we are focusing on human morality, but without excluding that relevantly similar social phenomena could also develop in other species.

which in turn influence the development of each individual brain. The neuronal features that result from learning and experience can thus be passed on through generations (Evers and Changeux 2016).

This normative relevance of neuroscience should not be understood in the naïve sense of describing facts that entail or constitute norms. It is commonly accepted that normative judgements, moral or other, cannot logically be derived from descriptive statements—of committing what in philosophy is known as "the naturalistic fallacy" (Moore 1903). This fallacy has different versions; the one that is relevant here consists of deriving an 'ought to be' from an 'is', a conceptual mistake that IM does not make.

Mere factual descriptions of the brain's functional architecture are not tantamount to making recommendations or assertions of norms; nevertheless, there are important empirical, and notably causal, connections between biological facts, norms, and moral values. Norms are brain constructs that are biologically and culturally embedded in and constrained by sociocultural structures. A major responsibility of fundamental neuroethics is to decipher this network of causal connections between the neurobiological, sociocultural, and contingent historical perspectives which gave rise to moral norms at a given moment in human history and to evaluate their 'universal' character as pre-specified in our genome and shared by the human species. The 'fallacy' of the naturalistic approach thus becomes inverted into a responsibility. Trying to understand and build up human norms in light of what we today know about the brain and its development is not only a great scientific challenge but also raises equally important social and ethical challenges. Today, research collaborations between neuroscience, genetics, and social science provide rich and multifaceted knowledge about human beings and an increasingly integrated view of us as biological organisms interacting in complex natural and cultural environments in constant evolution.

Neuroscientific knowledge can deepen our understanding of who we are and of how we function as neurobiological and social creatures. It can help explain the mechanisms of normative judgement and how morality has evolved. An informed materialist view of the brain aims to provide a scientifically adequate and philosophically fruitful theoretical framework and philosophy for neuroethics.

7.4 Neuroethics of Brain Simulation

The theoretical framework of fundamental neuroethics and its commitment to informed materialism is not just productive in the discussion of how to understand human beings, their capacity for moral judgement, and normativity in general. It is also crucial when examining several particular issues emerging from contemporary neuroscience and related technologies. In this last section, we illustrate the use of a fundamental neuroethics framework by focusing on a timely issue: the increasing use of computer brain models and simulations for exploring, understanding, and explaining the brain, specifically the moral brain (i.e., the brain as the subject of moral judgement).

7.4.1 Why Does Neuroscience Need Brain Models and Simulations?

In the last few years, neuroscience has been moving away from a hypothesis-led approach grounded in empirical observation and classification to a data-led simulation-modelling approach grounded in the computational analysis and manipulation of big data repositories (Frackowiak and Markram 2015). This move defines a paradigm shift in neuroscience, i.e., from an empirical paradigm to a computational predictive paradigm. Proponents of this new approach consider this necessary in order to (a) overcome existing fragmentation in current neuroscientific methodology (i.e., a large amount of heterogeneous data emerging from different technologies with distinct spatial and temporal scales, such as EEG compared to fMRI) and (b) reach a more comprehensive and integrated understanding of the brain.

This novel approach, known as 'predictive biology', is notably exemplified by the European Human Brain Project and its goal of developing highly sophisticated computer models and simulations to achieve an integrated understanding of the brain (Markram 2013). Its proponents argue the following: since the brain is a complex, multi-scale, and multilevel system (i.e., organized into different spatio-temporal scales), and it is currently impossible to identify and describe the discrete cause-effect relationships critical to the brain system's functioning, progress in the development of a comprehensive brain description requires a different strategy. Within a Big Data Science framework, it is proposed that we start from already available data to infer general principles of organization (or fundamental rules) of the brain. The resulting abstract and formal models of the brain would allow the description and reconstruction of all the real brains (Frackowiak and Markram 2015). In this proposal, the validation of the resulting statistical model is crucial: the computer model of the brain obtained should be constantly revised according to incoming empirical data. Thus, in the end, a brain simulation is a statistical, dynamic prediction of the real brain.

As we explain below, the neuro-bioethical discussion about predictive computational neuroscience typically revolves around epistemological and practical considerations (i.e., on the one hand, the intrinsic limitations of existing empirical methodology and, on the other hand, the need for new therapeutic and clinical applications) (Christen et al. 2016). We believe that the development of predictive computational neuroscience raises further issues that call for a fundamental neuroethics approach.

7.4.2 Neuro-Bioethics of Brain Modelling and Simulation

The need for a neuroethical assessment of the new neuroscientific paradigm shift arises, notably, though not exclusively, from the ethical justification suggested by its proponents, i.e., the demand for new ethical research and clinical tools. Proponents of the computational paradigm point out that huge investments in research have resulted in only a few new effective drugs and treatments (Markram 2013). They also note that while the technical limitations of the existing neuroscientific paradigm depend on technological development and will be increasingly reduced, its intrinsic shortcomings seem unavoidable.

However, this reasoning seems insufficient for showing that the computational predictive paradigm is more effective and therefore morally justified. After all, other paradigm shifts in neuroscience have been suggested as new potential frameworks for improving both research and treatment options (Mayer et al. 2014; Brembs 2015). Thus, proponents of brain models and simulations provide additional ethical arguments to support the need for a computational paradigm in neuroscience. They argue that computational simulations that replace or limit the involvement of human participants in experimental procedures offer several ethically relevant advantages, among them (Markram 2013):

- They place no limit on what we can record, which means that it is possible to obtain a potentially unlimited amount of data from a simulation.
- They place no limit on the number of manipulations that can be performed.
- They enhance the replicability of experiments.
- They could help unveil the relative correlation between different space and time scales within the brain.
- They may allow simulation of brain diseases which would have important diagnostic and prognostic implications.

Thus, ultimately the new computational paradigm in neuroscience appears to be justified on the basis of two main practical reasons: fewer risks for patients resulting from their reduced involvement in experimental trials and increased effectiveness of treatments that would follow from the better diagnostic and prognostic tools potentially resulting from the new paradigm.[8]

Although from a neuro-bioethical perspective the use of brain models and simulations does not appear to raise particular practical concerns (especially since such use would limit or avoid potential risks resulting from more traditional procedures and might give us new opportunities for improving actual clinical protocols and for developing new treatments for very challenging conditions still lacking effective treatment)—brain models and simulations do raise additional neuroethical issues. In what follows, we sketch a conceptual fundamental neuroethical analysis of brain models and simulations, with particular reference to their relevance for understanding moral judgement. We consider the use of a fundamental neuroethics approach critical in order to assess the potential benefits and limitations of brain models and simulations (Farisco et al. 2016).

[8] Of course, both these arguments rest on a fundamental epistemological premise – that the new paradigm is methodologically well grounded and actually able to advance our present knowledge. We are analysing the epistemological issues elsewhere.

7.4.3 Fundamental Neuroethics of Brain Modelling and Simulation

The possibility of developing brain models and simulations raises important and often unexplored conceptual issues (Farisco et al. In preparation). In the first place, brain models and simulations are intrinsically limited by methodological constraints, including the selection of relevant data to include in the model/simulation, the selection of the specific level to model/simulate, and the lack of a complete knowledge of how different brain levels connect to each other. These limitations need to be taken into account when considering the impact of neurotechnology in general and the need to ensure realistic expectations and to achieve balanced interpretations of its results. Over-interpretations, and the promotion of unrealistic hope and hype, are to be avoided, and the distinction between experimental results and possible ethical speculations must be clearly established (Wagner and Northoff 2015).

A different and particularly relevant issue raised by brain models and simulations is whether and to what extent it can provide new insight, not only about human cognition generally but also about our normative thought, including how we make moral judgements.[9] Will it help us understand which values we may be neurobiologically predisposed to develop and how moral judgements relate to brain functions? The relationship between moral judgement and brain functions can be theoretically understood in many different ways, including as reciprocal independence, as deep mutual connection, or as relative or absolute dependence of the first on the latter. Regardless, given the specific interpretation of their mutual relationship, the already explained IM perspective implies that knowledge of the brain is highly relevant to understanding moral judgement. Thus, greater knowledge of how different cerebral areas work should enhance our understanding of how moral judgement is carried out, at least in part. This means that if brain models and simulations really offer a better understanding of brain functioning, it is ipso facto relevant for neuroscientific perspectives in explanations of moral judgement.

However, to say that brain models and simulations might be relevant for analysing moral judgement does not tell us how explanatory they can be. In this sense, the computational predictive paradigm might not be much different from neuroscience's current empirical paradigm: its explanatory ability might be equally limited for technical and conceptual/methodological reasons (i.e., both use proxy elements for studying brain features, be it in computational or in vitro/vivo models). For this reason, theoretically, the issues raised by brain models and simulations may not differ from those raised by traditional empirical neuroscience applied to moral judgement.

Moreover, brain models and simulations' relevance in explaining moral judgement might be limited not only for epistemological and methodological features intrinsic to the technology itself but also because of the characteristics of the brain.

[9] In this context, rather than referring to morality as a general, abstract concept, it seems useful to refer to moral judgement as the particular instantiation of morality. Even though moral judgement and morality can clearly overlap, the latter cannot be reduced to the former.

We have seen that to be properly described, the brain needs an interdisciplinary methodology. Such interdisciplinarity entails, among other things, the need to engage with other approaches for deciding the best experimental design and for properly interpreting the emerging data. Interdisciplinarity is also critical for transferring findings to real-life situations.

If so, the issue becomes to what extent can brain models and simulations enrich our understanding of human morality in general? This is an important question because such knowledge has potentially significant ethical, social, and political implications. *Scientia potentia est* (*Science is power*): it is possible to speculate on different scenarios that could result from better knowledge of moral judgement. For instance, an improved understanding of how we make moral decisions can be the basis for more appropriate political and educational *praxis*. However, the discussion of whether brain models and simulations can contribute to a better explanation of morality and/or moral judgement requires a careful examination of the main concepts: morality and moral judgement.

As we have seen, within a naturalistic framework, morality can be roughly defined as the evaluative interaction with the world, which at the individual level is mediated by emotional and rational dimensions and at the super-individual level is institutionalized in different shared values and norms expressed in ethical codes of conduct. Accordingly, moral judgement can be broadly defined as a special type of interaction with the world mediated by the brain.

A brain-based simulation of moral judgement is, in principle, possible through the simulation of the relevant cerebral elements. However, according to IM, these elements are interactive (i.e., intrinsically related to external factors affecting their characteristics). This means that an adequate simulation of the cerebral functions and structures critical for moral judgement should also include extra-brain elements, namely, the interaction of the brain with the rest of the body (e.g., gut and endocrine system) and with the external environment (e.g., social and cultural world). Thus, from the perspective of IM, developing brain models and simulations for investigating moral judgement, although possible in principle, is premature and might be too problematic in practice for a number of reasons, including the conceptual ones we presented. Ultimately, the success of brain models and simulations depends on two things: (a) their epistemological soundness (i.e., whether brain models and simulations give us a reliable knowledge of the brain, especially of the brain areas critical for moral judgement) and (b) the extent to which they include the relevant extra-cerebral elements.[10]

Conclusion

Neuroethics is still a very young research area in need of further theoretical and methodological assessments. So far, we suggest that there has been a tendency in neuroethics to underestimate the role and value of conceptual analysis, focusing

[10] Actually this second premise follows from the first, because brain models and simulations are not epistemologically sound if they do not acknowledge the dynamic and interactive nature of the brain, including the relevant external elements in its simulation.

more on practical rather than on theoretical issues. However, this is slowly changing with the appearance of more theoretical philosophical approaches. Here, we have presented (a) one such approach and argued for the need to develop conceptual neuroethics further and (b) a specific understanding of the brain's functional architecture: informed materialism.

Under the heading of conceptual neuroethics, we argued that the fundamental neuroethics framework is a type of conceptual approach that focuses on *how* natural science can deepen our understanding of moral thought. Fundamental neuroethics combines perspectives from philosophy and neuroscience, stressing that conceptual investigation is a necessary prerequisite for addressing practical concerns about the impact of neuroscience. Thus, fundamental neuroethics importantly contributes to forming the ground upon which we can develop other kinds of neuroethics discourse (e.g., neuro-bioethics and empirical neuroethics.)[11]

Finally, on the basis of our theoretical assessment in the first two parts of the chapter, we focused on a particular example in the form of brain models and simulations in part three. In part three, we first sketched a neuro-bioethical analysis of brain models and simulations, so that we could then propose a further conceptual assessment of the issues from the perspective of fundamental neuroethics. A main conclusion of that analysis is that brain models and simulations may be relevant yet limited for contributing to explanations of the mechanisms of moral judgement. In particular, they would also need to include extra-cerebral elements, both internal and external to the body.

The case of brain models and simulations illustrates how an appropriate neuroethical assessment of the issues emerging from contemporary neuroscience and related applications needs to involve a conceptual fundamental approach in order to achieve the conceptual clarity that may help provide adequate interpretations of new ideas and findings.

Acknowledgments This project has received funding from the European Union's Horizon 2020 research and innovation programme under grant agreement 720270(HBP SGA1).

References

Avram M, Giordano J (2014) Neuroethics: some things old, some things new, some things borrowed...and to do. AJOB Neurosci 5(4):1–3

Bachelard G (1953) Le matérialisme rationnel. Presses Universitaires de France, Paris

Beauchamp T (1995) Principlism and its alleged competitors. Kennedy Inst Ethics J 5(3):181–199

Bickle J (2003) Philosophy and neuroscience: a ruthlessly reductive account. Kluwer Academic Publishers, Dordrecht

[11] We are aware that our conceptual assessment of neuroethics raises questions that require further treatment: For example, how can conceptual neuroethics concretely contribute to the development of other forms of neuroethics? Is the conceptual work internal or external to empirical scientific knowledge? How should empirical data be included in philosophical theories? These are issues we are exploring at present.

Brembs B (2015) Watching a paradigm shift in neuroscience. http://bjoern.brembs.net/2015/03/watching-a-paradigm-shift-in-neuroscience/?utm_content=bufferbb4ac&utm_medium=social&utm_source=twitter.com&utm_campaign=buffer. Accessed 18 May 2016
Changeux JP (2004) The physiology of truth. Belknap Press, Cambridge
Changeux JP, Courrège P, Danchin A (1973) A theory of the epigenesis of neuronal networks by selective stabilization of synapses. Proc Natl Acad Sci U S A 70:2974–2978
Choudhury S, Slaby J (eds) (2012) Critical neuroscience. A handbook of the social and cultural contexts of neuroscience. Wiley, Chichester
Christen M, Biller-Andorno N, Bringedalb B, Grimesc K, Savulescu J, Walter H (2016) Ethical challenges of simulation-driven big neuroscience. Am J Bioeth Neurosci 7(1):5–17
Clouser KD, Gert B (1990) A critique of principlism. J Med Philos 15(2):219–236
Clouser KD (1995) Common morality as an alternative to principlism. Kennedy Inst Ethics J 5(3):219–237
Crick F (1994) The astonishing hypothesis: the scientific search for the soul. Scribner, New York
Cranford RE (1989) The neurologist as ethics consultant and as a member of the institutional ethics committee. Neurol Clin 7:697–713
Churchland P (2008) The impact of neuroscience on philosophy. Neuron 60(3):409–411
De Vos J, Pluth E (eds) (2016) Neuroscience and critique. Routledge, New York
Dehaene S, Kerszberg M, Changeux JP (1998) A neuronal model of a global workspace in effortful cognitive tasks. Proc Natl Acad Sci U S A 95:14529–14534
Dehaene S, Changeux JP (1989) A simple model of prefrontal cortex function in delayed-response tasks. J Cognitive Neurosc 1:244–261
Dehaene S, Changeux JP (1991) The Wisconsin card sorting test: theoretical analysis and modeling in a neuronal network. Cereb Cortex 1:62–79
Dehaene S, Sergent C, Changeux JP (2003) A neuronal network model linking subjective reports and objective physiological data during conscious perception. Proc Natl Acad Sci U S A 100:8520–8525
Denton DA (2005) Les émotions primordiales et l'éveil de la conscience. Éditions Flammarion, Paris
Diderot D (1966) Eléments de Physiologie. Quote from Oeuvres complètes, Jules Assérat & dMaurice Tourneux (ed), 20 vol, Paris 1875–1877 (II-322)
Edelman GM (1992) Bright air, brilliant fire: on the matter of the mind. Basic Books, New York
Evers K (2005) Neuroethics: a philosophical challenge. Am J Bioeth 5(2):31–33
Evers K (2007) Toward a philosophy for neuroethics. EMBO Rep, special issue on neuroscience and society 8:48–51
Evers K (2009) Quand la matière s'éveille. Éditions Odile Jacob, Paris
Evers K (2016) Can we be epigenetically proactive? In: Metzinger T, Windt JM (eds) Open mind: philosophy and the mind sciences in the 21st century. MIT Press, Cambridge, pp 497–518
Evers K, Changeux JP (2016) Proactive epigenesis and ethical innovation: a neuronal hypothesis for the genesis of ethical rules. EMBO Rep
Evers K, Sigman M (2013) Possibilities and limits of mind-reading: a neurophilosophical perspective. Conscious Cogn 22:887–897
Farah MJ, Illes J, Cook-Deegan R, Gardner H, Kandel E, King P et al (2004) Neurocognitive enhancement: what can we do and what should we do? Nat Rev Neurosci 5:421–425
Farah MJ, Helberlein AS (2007) Personhood and neuroscience: naturalizing or nihilating? Am J Bioeth 7(1):37–48
Farah MJ (2010) Neuroethics: an introduction with readings, 1st edn. The MIT Press, Cambridge
Farisco M, Evers K, Salles A (2016) Big science, brain simulation and neuroethics. Am J Bioeth Neurosci 7:28–30
Farisco M, Hellgren Kotaleski J, Evers K (In preparation) Mirroring the conscious brain? A conceptual investigation of brain models and simulations
Fins J (2008) A leg to stand on: Sir William Osler and Wilder Penfield's "neuroethics". Am J Bioeth 8:37–46
Frazzetto G, Anker S (2009) Neuroculture. Nat Rev Neurosci 10:815–821

Frackowiak R, Markram H (2015) The future of human cerebral cartography: a novel approach. Philos Trans R Soc Lond 370
Giordano J (2011) Neuroethics: interacting "traditions" as a viable meta-ethics. AJOB Neuroscience 2(2):17–19
Greene J (2008) The secret joke of Kant's soul. In: Sinnott Armstrong W (ed) Moral psychology, vol 3. MIT Press, Cambridge, pp 35–80
Illes J (2007) Empirical neuroethics. Can brain imaging visualize human thought? Why is neuroethics interested in such possibility? EMBO Report 8:S57–S60
Ledoux J (1998) The emotional brain, Phoenix, UK
Legrenzi P, Umiltà C (2011) Neuromania. On the limits of brain science. Oxford University Press, Oxford
Lagercrantz H (2005) I barnets hjärna, Bonnier Fakta, Stockholm
Levy N (2008) Introducing neuroethics. Neuroethics 1:1–8
Levy N (2011) Neuroethics: a new way of doing ethics. AJOB Neurosci 2(2):3–9.
Little M (1996) Why a feminist approach to bioethics? Kennedy Inst Ethics J 6(1):1–18
Markram H (2013) Seven challenges for neuroscience. Funct Neurol 28:145–151
Mayer EA, Knight R, Mazmanian SK, Cryan JF, Tillisch K (2014) Gut microbes and the brain: paradigm shift in neuroscience. J Neurosci 34:15490–15496
Moore GE (1903) Principia ethica. Cambridge University Press, Cambridge
Northoff G (2009) What is neuroethics? Empirical and theoretical neuroethics. Curr Opin Psychiatry 22:565–569
Northoff G (2014) Minding the brain: a guide to philosophy and neuroscience. Palgrave MacMillan, Croydon
Pontius AA (1973) Neuro ethics of "walking" in the newborn. Percept Mot Skills 37(1):235–245
Racine E (2010) Pragmatic neuroethics. The MIT Press, Cambridge
Rose S (2006) The future of the brain. Oxford University Press, New York
Roskies A (2002) Neuroethics for the new millenium. Neuron 35:21–23
Satel S, Lilienfeld SO (2013) Brainwashed: the seductive appeal of mindless neuroscience. Basic Books, New York
Schultz W (2006) Behavioral theories and the neurophysiology of reward. Annu Rev Psychol 57:87–115
Schultz W, Dayan P, Montague PR (1997) A neural substrate of prediction and reward. Science 275:1593–1599
Shook J, Giordano J (2014) A principled and cosmopolitan neuroethics: considerations for international relevance. Philos Ethics Humanit Med 9(1):1–13
Toulmin S (1982) How medicine saved the life of ethics. Perspect Biol Med 25(4):736–750
Wagner NF, Northoff G (2015) A fallacious jar? The peculiar relation between descriptive premises and normative conclusions in neuroethics. Theor Med Bioeth 36:215–235

8 Neuroethics: A Renewed View of Morality? Intentions and the Moral Point of View

Bernard Baertschi

Abstract

In the traditional view of morality, intentions play a central role: they define what a typical action consists of and allow for the assignment of both blame and praise. Actions are intentional bodily movements, and if actions are morally assessed, it is first and foremost because they are intentional. Recently, several psychologists have investigated the neural basis of these mental phenomena. Although many studies confirm the traditional view, others point in the opposite direction: intentions play only a subordinate role in morality. For Joshua Knobe, intentionality is not central but depends on ascriptions of responsibility, far from grounding them. For Joshua Greene, moral judgement is based on intentions only when we rely on alarm emotions. If these studies are found to be convincing, it would oblige us to modify our view of morality: responsibility would be linked with outcomes rather than with intentions. On the legal level, the doctrine of *mens rea* would also be modified, and perhaps even abandoned. Neuroethics would then be a field that purports to offer a renewed view of morality. However, I think that a careful examination of the data and their interpretation shows that this conclusion is mistaken: intentions remain at the centre of morality even if it is not easily noticed in some situations, especially when side effects are involved.

B. Baertschi
Institute for Ethics, History, and the Humanities, University of Geneva, Geneva, Switzerland
e-mail: bernard.baertschi@unige.ch

E. Racine, J. Aspler (eds.), *Debates About Neuroethics*, Advances in Neuroethics,
DOI 10.1007/978-3-319-54651-3_8

8.1 Introduction

In the traditional view of morality, intentions play a central role: they define what a typical action consists of and allow for the assignment of both blame and praise. Actions are *intentional* bodily movements, and if actions are morally assessed, it is first and foremost *because* they are intentional. For example, accidental harm does not arouse anger, but intentional harm does, and only the latter causes moral reproach. With the progress of neuroscience and brain imaging, several psychologists have investigated the neural basis of these mental phenomena. Many studies confirm the traditional view, but others point in the opposite direction: intentions might only play a restricted and subordinate role in morality. This constitutes a challenge to the traditional view that has been raised in particular by Joshua Knobe and Joshua Greene, two influential neuropsychologists. If these studies are found to be convincing, it would oblige us to modify our view of morality, perhaps profoundly. Neuroethics would then be a field that purports to offer a renewed view of morality. However, are these studies convincing? This is the question I try to answer in this paper.

It demands first that I carefully spell out what the traditional view of morality says about intentions. Sections 8.2 and 8.3 are devoted to this task. In the first section, I argue that intentions are an essential part of what constitutes an action: strictly speaking, an unintentional action is a contradictory expression. Actions are morally judged, but intentions are also crucial to assign responsibility, as I show in Sect. 8.3: in ordinary circumstances, we blame and praise what is done intentionally, not what happens accidentally. In Sect. 8.4, I present several empirical studies confirming the traditional view: intentions are really at the centre of morality, and when people judge that they are not, it is often because they are afflicted by some psychiatric or neurological conditions. In the two following sections, I examine challenges based on experiments to this traditional view of morality. Joshua Knobe (Sect. 8.5) has claimed that intentionality does not ground attribution of responsibility: on the contrary, we see an action as intentional only if we already consider its agent as responsible (this phenomenon is called the *Knobe effect*). For Joshua Greene (Sect. 8.6), our morality does not represent a seamless whole and is formed of two parts: the first (System 1) is intuitive and assigns a central role to intentions, but the second (System 2) does not – it is rational and places most of the moral weight on outcomes (i.e., the important factor is not whether we behave intentionally, but that some good is done).

Knobe and Greene's arguments place the traditional view of morality in jeopardy, and they suggest it is in need of revision. But are their arguments convincing? In Sect. 8.7, I argue that they are not, since a thorough reinterpretation of the studies they rely on suggests that their evidence has only limited validity: it concerns not all actions – only actions with side effects. Consequently, it is not the intentionality of the actions themselves that is concerned but only the intentional character of the side effects (i.e., an action can have several effects, and not all of them are willed – some are even unforeseen). In the last section, I add some precisions in discussing the *doctrine of double effect*, a view that has been proposed in traditional morality to manage actions with side effects.

8.2 Intention, the Criterion of Action

"I had intended to visit my mother after lunch, so I could not attend your meeting". Such an utterance is commonplace, and it is not difficult to imagine a context where it would be completely appropriate. An intention is a mental act, akin to a decision.[1] In ordinary language, an intention is often less strong than a decision: I am disposed to reconsider what I intend to do, much more than what I have decided to do. But both can be described as mental acts aiming at a goal – here, a visit to my mother.[2]

"I went to a meeting, when I happened to run into my mother. It was not intentional because I did not know that my mother was in town". This encounter with my mother is fortuitous and, consequently, not intentional. Of course, when I stumbled upon her, I was fulfilling an intention – to attend a meeting – but I did not aim to see my mother there.

Intentions are about something (i.e., we intend to do something). *To be about something* is what philosophers call "intentional", and Daniel Dennett warns us: "This *aboutness* that, for example, sentences, pictures, beliefs, and (no doubt) some brain states exhibit, is known in philosophical jargon as *intentionality*, an unfortunate choice as a technical term, since outsiders routinely confuse it with the everyday idea of doing something intentionally" (Dennett 2013: 62). It is particularly misleading for actions, because they are intentional in both senses: they are about something and accompanied or shaped by an intention.

What interests me in this paper is the dual property of typical actions: to be about something while being directed by an intention.[3] This dual property in fact covers two properties, since they can be separated in some pathological conditions like *anarchic hand syndrome*, where the patient observes his hand moving purposely (it aims at an object), but without having had any intention to move it: he discovers his hand's "action" (Marcel 2003). However, what I am interested in here are "normal" actions, and not what would be better named "pseudo-actions". Such normal or typical actions must also be recognised as *mine*; this sense of ownership is crucial (Forest 2014: 99–105).[4]

[1] Mental acts like decisions have been extensively studied by neuropsychologists since Benjamin Libet in the debate concerning free will (Fried et al. 2011). My subject focuses on another important point in action theory, unrelated to the free will debate.

[2] Here, I follow Franz Brentano, who said: "Each mental act is primarily directed to an object" (Marek 2013). For a mental act, to be directed to an object and to have a goal are synonymous.

[3] Rigato and col. notice: "What philosophers call 'intentional,' neuroscientists call 'goal-directed'" (2014: 181). However, everything that is goal-directed is not intentional in the relevant sense, but this terminological difference is not important for my argument.

[4] There exist other kinds of nontypical actions, like impulsive actions, which seem to be intentional only in the sense of aboutness, actions performed under coercion or actions made while sleeping, during an episode of REM sleep behaviour disorder (Maoz and Yaffe 2015; Cerri 2016). I will not investigate them.

This dual property is constitutive of actions, which Donald Davidson noticed.[5] To see this, let us imagine four situations where a roofer causes the death of a pedestrian.

- John is working on a roof. Suddenly, he trips and falls on a pedestrian. The pedestrian softens the fall, saving John's life, but is killed.
- Andrew is working on a roof. When he passes a tool to a workmate, he loses his balance and falls on a pedestrian. The pedestrian softens the fall, saving Andrew's life, but is killed.
- Paul is working on a roof. Suddenly, he has a vision: God orders him to kill a nearby pedestrian. Paul throws himself off the roof and lands on the pedestrian, who is killed.
- Peter is working on a roof. He sees his worst enemy walking down the street. Peter throws himself off the roof and lands on his enemy, who is killed.

Intention – and responsibility, as we soon will see – is understood differently in each of the four cases. Only Paul and Peter fall intentionally; Andrew's fall is a non-intentional effect of a former intention: to pass a tool, and what John does contains nothing intentional at all. We are even tempted to say that John does nothing and that things happen to him against his will: John does not perform an action. Notice that I use here the expressions "to perform *X* with an intention to do it" and "to perform *X* intentionally" as synonymous; I will generally follow this use, in agreement with what has been named the *Simple View,* that is, the thesis "that anyone who *A*'s intentionally intends to *A*" (McCann 1991: 205).

John's case allows a distinction to be drawn between an *action* and an *event.* What distinguishes them is the presence of an intention in the former. Every action is intentional, which means, as Elisabeth Anscombe suggests, "intentional under some description that we give (or could give) of it" (Anscombe 1963: 29).[6] What John does cannot be described as intentional, even if we try hard. We can also say that the goal of an action should be identical with the content of the corresponding intention: the goal of Peter's action and the content of his intention are that his enemy will be dead. This analysis is meaningless for John.

8.3 Intention, the Prime Bearer of Responsibility

What happens in the four cases above is bad, given that a man dies as a result. Therefore, questions of ethics are relevant. More precisely, when someone is harmed, the question of who bears responsibility arises, and to answer that question,

[5] "I follow a useful philosophical practice in calling anything an agent does intentionally an action" (Davidson 2002: 5).

[6] Every bodily movement can be described in different manners. Consequently, an intentional action can be described without any reference to the intention, but of course, it does not deprive it of its intentional character: a bodily movement is intentional – it is an action – if there exists a description of it mentioning an intention; otherwise, it is (a part of) an event.

it is necessary to take intentions into account. John and Andrew do not kill the pedestrian intentionally, because they do not aim at the death of the pedestrian. Therefore, they are not morally responsible for the pedestrian's death. They are causally responsible for it indeed, but not morally.[7] Paul acts intentionally (he wants to kill the pedestrian), but without being morally responsible for his actions because of a cognitive impairment (i.e., he suffers from a hallucination): intentionality is not the only condition that needs be met when determining someone's responsibility. Only Peter is fully responsible and, in this case, blameworthy.

In order to blame or to praise someone, we have to consider intentions, because we are first and foremost responsible for what we aim at consciously and willingly, and moral responsibility is a prerequisite for blame and praise. Sometimes, we hear that consequentialists disagree, because they only take outcomes into account. But this is false, as John Stuart Mill adamantly stated: "There is no point which utilitarian thinkers (and Bentham pre-eminently) have taken more pains to illustrate than this. The morality of the action depends entirely upon the intention – that is, upon what the agent *wills to do*" (Mill 1991: 150, n. 1), but he adds that we should not confuse intention with motive, that is, we should not confuse what we intend to do with its cause. If I rescue a famous drowning person, my intention is to save her, but my motive can be that I want to be rewarded. I will insist on this distinction in Sect. 8.5.

In criminal matters – but not in civil law – the lesson is the same: the doctrine of *mens rea* in common law is an expression of it. A defendant cannot be culpable if he did not have a bad intention when he acted. Michael Treadway and colleagues showed in a study that graphic descriptions of harmful acts amplify amygdala activity and willingness to punish, but only if the act is intentional. They conclude that "Justice […] requires that punishment takes into account not only the negative emotions elicited by harm, but also an evaluation of the transgressor's intent" and comment: "An actor's mental state – whether it is purposeful, knowing, reckless, negligent or blameless – can markedly affect how severely he or she is punished for the harm committed" (Treadway et al. 2014: 1270).[8]

The centrality of intention in morality has a developmental signature: studies have shown that children are very soon aware of the distinction between what is intentional and what is not. Chris Frith, recalling how important it is to discriminate between intentional and non-intentional movement, adds that very young children can do just that:

> "It is important for us to make a distinction between deliberate actions and accidents. If my arm movement accidentally spills the wine on you then everyone is very compassionate about my embarrassment. But if, with much the same arm movement, I do it deliberately, the action is meant, and taken, as a severe insult. Infants as young as 9 months can distinguish between deliberate and accidental actions made by other people, for example, whether the toy was withheld deliberately or dropped accidentally" (Frith 2010: 13).

---

[7] I leave here the question of negligence or carelessness aside.

[8] Notice that *lex talionis* (an eye for an eye, a tooth for a tooth) and many ancient laws focus on action's effects rather than on intentions.

Here, Frith uses the expression "deliberate". As we can already see, many words express the same idea: willed, voluntary, intentional and deliberate. We can also add: knowingly, purposely and probably others.[9] Frith contrasts deliberate with accidental, but it can also be an antonym for impulsive, opening another semantic repertoire. Depending on the context of an utterance, these expressions are strictly synonymous or not. We will see in Sect. 8.5 that this has some importance, when I examine and discuss the *Knobe effect*.

8.4 Empirical Confirmations

The central place of intention in our traditional view of morality has been highlighted and buttressed by several experimental studies. I have already mentioned some of them, and I will now examine three others more thoroughly.

The first has to do with the *Ultimatum Game*, a well-known economic experiment assessing our sense of fairness. Someone (*B*) receives a sum of money from *A*, but *B* can keep it only if he transfers a part to a third person (*C*) and if *C* accepts the gift. If *C* refuses, the sum is entirely given back to *A*. Usually, *C* refuses when *B* gives him less than 30% of the sum.

Berna Güroglu and colleagues have introduced a modification to the game, in order to test the influence of intentions. *B* is instructed to give 20% of the sum. However, *C* is told that *B* can choose between three different offers: to give 20%, 50% or 80% of the sum, so that *C* believes that *B* makes the 20% offer freely. Consequently, *C* refuses very often (75% of the time). Later, *C* is informed that *B* has had no choice and did not intend to give the lesser sum; consequently, C's refusal drops (only 30%). The authors of the study conclude that the refusal is, in large part, a consequence of the perception of *B*'s intention. The moral quality of the intention counts more than the fairness of the result (the offer): "Information that highly influences fairness judgments is intentionality, that is, perceptions of fairness are influenced by the intentions of the interaction partner. A seemingly unfair act might evoke less negative affect if one believes that it was not done intentionally" (Güroglu et al. 2010: 414). These findings are robust, since they have been confirmed by other studies (Güney and Newell 2013).

The second experimental study concerns patients with ventromedial prefontal cortex (VMPFC) damage, who are individuals that suffer from a brain lesion responsible for a serious neuropsychiatric condition: acquired sociopathy. It was conducted by Liane Young et al. (2010). They presented four scenarios to their subjects (VMPFC patients and typical people), where the intention of the agent and the outcome of the action vary in an ordered manner. Grace (the agent in the four scenarios) has two opposite intentions, one bad (to poison a friend) and one neutral (to offer him sugar), and her action has two opposite results, one bad (her friend is poisoned) and one neutral (her friend is fine). When combined, we obtain four possibilities:

[9] Legal systems consider some of these expressions to be synonyms (Zangrossi et al. 2015: 2).

1. Grace thinks the powder is sugar. It is sugar. Her friend is fine.
2. Grace thinks the powder is sugar. It is toxic. Her friend dies.
3. Grace thinks the powder is toxic. It is sugar. Her friend is fine.
4. Grace thinks the powder is toxic. It is toxic. Her friend dies.

Asked to judge Grace's act, typical people evaluate 4 (the successful attempt to harm) as the worst scenario, followed by 3 (the failed attempt to harm). Scenario 2 (accidental harm) is considered unfortunate, but not so bad. On a scale where 1.0 means strictly forbidden and 7.0 means completely permissible, they put the successful attempt to harm (Scenario 4) at 1.1, the accidental harm (Scenario 2) at 3.5 and the failed attempt (Scenario 3) at 2.2. These participants believed that it is morally worse to have bad intentions than to have good ones that lead to bad consequences.

With VMPFC patients, situation 4 remains the worst scenario, but 2 and 3 are assessed differently: accidental harm is ranked at 3.1 and the failed attempt at 5.0. Accidental harm is then worse than the failed attempt to harm for these patients. Young and colleagues comment: "Notably, VMPFC participants also judged attempted harms as significantly more permissible than accidental harms" (Young et al. 2010: 848), because they consider the result to be of greater moral importance – when a bad intention does not succeed, it is not so serious (even if good intentions followed by good results are judged better, 6.0 on the scale).

Both groups of subjects (brain-damaged patients and typical people) evaluate actions and intentions. For typical people, the value of the action is significantly determined by the intention of the agent and the intentional character of the outcome. For VMPFC patients, the intention plays a less central role. Young and colleagues comment: "Patients with bilateral damage to the VMPFC were more likely to deliver utilitarian moral judgments" (Young et al. 2010: 845). But utilitarianism is not a neurological condition! Does it mean that utilitarians have a mistaken conception of morality? A utilitarian would answer that an empirical study shows what people think, not what they ought to think and that ordinary people are perhaps wrong when they place so much importance on intentions in the assessment of responsibility and wrongness. I will address this question later, but for the moment, I continue with my analysis of traditional morality, which is summarised in the following way by Young and colleagues: "A fundamental component of normal moral judgment is the ability to blame those who intend harm, even when they fail to cause harm. [...] The ability to blame for failed attempts not only features prominently in mature moral judgments but emerges quite early in development" (Young et al. 2010: 849). Moral education or maturation is nevertheless necessary to render normal human beings more charitable: bad outcomes tend to tip the balance unfavourably for children, but in the end, intentions win the game of responsibility assessment and consequently of blame and praise.

Chris Frith has suggested "that the cognitive basis for the feeling of responsibility is, first, a mechanism that binds intentions to outcomes" (Frith 2014: 139).[10] The

[10] See also Christensen and Gomila (2012: 1259), and Yoshie and Haggard (2013).

study by Young and colleagues confirms this claim, even if people with different capacities bind them differently.

A third study also supports this conclusion. This study has been conducted with Asperger's patients by Joseph Moran and colleagues. The participants read vignettes combining intention and outcome, as in Grace's scenarios (*no harm*, neutral intention and neutral outcome; *accidental harm*, neutral intention and bad outcome; *attempted harm*, bad intention and neutral outcome; *intentional harm*, bad intention and bad outcome). The participants had to pass moral judgement on the four types of scenarios. Moran and colleagues observed that Asperger's patients do not place the same importance on intentions in morality when compared to typical adults. If "neurotypical adults weigh a person's intention more heavily than the outcome of their action when evaluating the moral permissibility of an action" (Moran et al. 2011: 2688), it is not the same for Asperger's patients. They rate accidental harm lower on the scale of permissibility (1.0–7.0): 3.5 for Asperger's patients and 4.8 for typical adults.[11] Like VMPFC patients, Asperger's patients favour outcomes, but not in the same manner: if the former exculpates bad intentions when the result is neutral (i.e., when no harm is done), the latter blames innocent intentions if the result is bad. It reminds us of what Young and colleagues said of children, as Moran acknowledges: "In several respects the pattern of results displayed by the adults [with Asperger's] mirrors that displayed by typically developing children" (Moran et al. 2011: 2690).

The picture of morality emerging from these studies is on par with our intuitions: morality is centred on intention or the intentional aspect of our deeds. Intention is necessary for distinguishing actions from events: actions are purposeful, whereas events occur independently of goals. Some actions, but not all, relate to morality, notably actions involving a harm. However, every harm is not wrong: in order to be wrong, a harm should be intended and not forgivable. Consequently, intentionality is also an important ground for moral responsibility, blame and praise. Children and some patients have difficulties with this, but not "neurotypical" adults.

However, two different objections have been directed against this view. The first contends that a certain phenomenon (the Knobe effect) casts some doubt on it, whereas the second claims that intentions play a major role only in one kind of moral judgements. I examine these objections in the next two sections.

8.5 The Knobe Effect

Joshua Knobe (2003) imagined two scenarios where an assignment of intentionality is made; he has a surprising result for our traditional conception of morality and responsibility. In each scenario, the vice president of a company proposes a new program to the chairman of the board which will have two effects:

[11] When the outcome is not bad, Moran and colleagues did not observe any significant difference between the moral judgements of Asperger's patients and typical adults (Moran et al. 2011: 2690).

- First scenario: the profits of the company will rise (first effect), but the environment will be severely harmed (side effect).
- Second scenario: the profits of the company will rise (first effect), but the environment will be benefited (side effect).

In both scenarios, the chairman's answer is the same: he does not care about the environment and only wants to increase the company's profits. Then, Knobe asks the participants if the chairman intentionally harmed/benefited the environment. With regard to traditional morality, it seems that the answer will be the same in both cases: the harm/benefit is not intentional, even if the harm is reckless. But the responses are different, as Neil Levy noted. The harm (first scenario) is judged to be intentional by 82% of the participants, and the benefit (second scenario) is assessed as non-intentional by 77% of the participants: "Surprisingly, altering the moral valence of the side effect dramatically alters subjects' perception of its intentionality: The majority of subjects now judged that helping the environment was not intentional" (Levy 2011: 7). Both answers are surprising, the first because the chairman's goal is not aiming at modifying the environment and the second because it is not consistent with the first answer.

However, there is an easy way out, emphasised by Levy: "If judgments of intentionality are sensitive to moral considerations, then it might be because people judge the intentionality of a side effect on the basis of its moral permissibility, rather than judging the permissibility of an action on the basis of the intentionality (or unintentionality) of the side effect" (Levy 2011: 7). Permissibility or responsibility comes first; intentions are only second.

If this interpretation is correct, then the traditional view of morality is in jeopardy: intentions are not so crucial. In the experiments examined in the previous section, intentions were central, at least for typically developing mature adults. Are people confused and inconsistent? Are they ambivalent? For many authors, we can explain these reactions by distinguishing two processes at work in our mind: Systems 1 and 2, the first being fast and intuitive, the second being slow and rational. A recent study suggests this reading: it was observed that participants who were good at the *Cognitive Reflection Task*, which measures a person's capacity to suppress spontaneous responses and to reflect on the task at hand, are less prone to the *Knobe effect*, "suggesting that the Knobe effect may arise from a System 1 process" (Ngo et al. 2015: 2).[12]

I will examine this interpretation in the next section. Here, I will offer two other comments on the Knobe effect, establishing that it is in fact not a challenge to the traditional conception of morality.

The first is semantic. As I have said, many expressions are used when we speak of an action as intentional. I have added that, depending on the context, some of them can be synonymous or not. In my opinion, the Knobe effect suffers from such ambiguities. In ordinary language, it is clear that responsibility for bad results is linked with intentionality. Think of this frequent reproach: "I am sure he did it on

[12] Ngo and colleagues have nevertheless not been able to confirm the results (Ngo et al. 2015: 5).

purpose"; a reproach that has no counterpart for good actions. However, it is not easy to apply this claim to the chairman, because his purpose is not specifically to harm the environment.[13] But think of: "I am sure he did it knowingly", or "consciously", or "willingly" or "deliberatively". The context authorises the application of such sentences to the chairman's behaviour. Moreover, the same context shows that "to have the intention to do *X*" and "to do *X* intentionally" are not always synonyms.[14] Our ordinary language is subtle, but sometimes too much so for philosophical precision, especially since intentions create complicated situations – in this respect, traditional morality is not a fully unified theory; but it is not a surprise.

The asymmetry between bad and good results – responsibility is usually only invoked in the case of the former – also suggests that "responsible" is very often used as a synonym for "blameworthy" or "culpable". As Vladimir Chituc and colleagues note: "Judgments in many domains are distorted by a motivation to blame" (2016: 22), even if, philosophically speaking, we are morally responsible for good results too. The Knobe effect is also linked with this semantic fact, which impacts other moral judgements too; it highlights a kind of imbalance in folk morality (Doris et al. 2007).

Another ambiguity consists of a frequent confusion between intentions and motives or reasons: Henry Sidgwick observes that "the distinction between 'motive' and 'intention' in ordinary language is not very precise" (Sidgwick 1981: 202). When we look for more accuracy, we notice that if both are sometimes identical (a goal can be a motive or a reason), it is not always the case. It is not surprising because they are conceptually different: motives and reasons are causes, preceding the actions, whereas intentions are mental acts embodied in actions and aiming at a goal. For instance, if I see someone drowning and I help him, my intention is to save his life, but my motive can be very different: it could be that I want to be at peace with my conscience or that I hope to be rewarded. In the stories imagined by Knobe, the reason the chairman has to begin the new program is precisely the goal of it: to increase the company's profits. The fact that his decision is motivated by a reason that appears to be morally suspect for many – his reason could be described as "increasing profits even at the expense of the environment" – and that this reason is also the goal aimed at could explain why the bad consequences were considered to have been intended (the good ones are not, because their value is at odds with the morally dubious motive).

Secondly, I observe that the Knobe effect impacts morality only at its periphery. It concerns side effects only, not primary effects: nobody ever doubted that the chairman's project to increase the company's profits was intentional, that he had the intention of increasing them and that he was making a genuine action. The problem focuses on the side effects – the Knobe effect is also named "the side-effect effect" – and, as another study has shown, it also focuses on the means: if the vice president suggests to the chairman that they shorten the worker's coffee break (a bad means)

[13] See nevertheless (Leslie et al. 2006: 425).

[14] Joshua Knobe (2004) acknowledges this. Some authors also emphasise that "intentionally" has several meanings; see, for example, Cova et al. (2012).

in order to increase productivity, the participants will likely consider the chairman to have acted intentionally in shortening the break – but not if he gives the workers a 1-hour nap break (a good means) for the same purpose (they will be in better health, and so will work more) (Cova and Naar 2012). Édouard Machery observes that "people take the costs that are incurred in order to reap some benefits to be intentionally incurred" (Machery 2008: 166).[15]

It is not difficult to see that both findings are linked to a psychological problem, known as the direction of intention. Take side effects first: when our action has several effects, it is often difficult to say which of those effects are willed and which are not. Classic debates about euthanasia (to kill someone in order to stop his suffering), abortion (to destroy a foetus in order to save the mother's life) and warfare (to bomb a strategic bridge where children are playing) are full of such difficulties[16]: can the agent confidently say that he does not aim at the bad effect, even if he knows for sure that his action will cause it? But if we are causally responsible for bad effects, and if we cannot claim that they are mere unintentional by-products, should we instead argue that we intended to bring them forth and so are fully morally responsible for them? Here, intentions and responsibilities are intermingled; it is not surprising that both notions contaminate each other and that the Knobe effect reflects this.

The case of means is a little different, because they are necessarily chosen. Consequently, we are not surprised "that means are generally judged more intentional than side effects" (Cova and Naar 2012: 837). But this judgement seems to be valid only for bad means and not for good ones. Why? Since the scenarios tested by Cova and Haar are inspired by the Knobe effect, the same mechanisms are probably at work. But here too, all the discussion takes place because we are faced with a genuine action, which is an intentional one.

In such complicated cases where effects and means are not easily evaluated, we seek simpler ways out, and one solution is to focus on anomalies and resort to characters: if we think that the person is greedy or benevolent, we tend to judge her acts accordingly. Grant Gillett notices: "In the normal course of events, human behaviour is, more or less, explicable on the basis of character or personal narrative" (Gillett 2008: 122). Peter Railton has imagined two scenarios where an olive tree owner sprays them against pests, knowing that his neighbour's goats will be harmed/benefited by the product, and he observes:

> "Intuition makes use of whatever evidence it can, and given our experience, someone who 'doesn't care at all' about whether he harms his neighbours is, happily, statistically rare, while someone who 'doesn't care at all' about whether he helps his neighbours is, perhaps regrettably, much more common. Statistical learning systems pay special attention to anomalies, since they carry more information than events that are more predictable" (Railton 2014: 853).

[15] However, Machery seems to understand "intentionally" as "deliberately", since he says: "Because [people] believe that costs are intentionally incurred, they judge that harming the environment is intentional" (Machery 2008: 177).

[16] They have been tackled by the *doctrine of double effect*. I will make some observations on this doctrine in the last section.

Such a disregard for the interests of others is rare and often revealing of a more general attitude: to act on bad intentions or at least on not good ones. Railton concludes: "What may matter in such intuitive social attributions of intent with respect to side effects is the fit of the action with the causal-attitudinal-intentional model of the agent we tacitly construct in light of his or her behaviour – for example, as 'anti-social' versus 'self-concerned' versus 'prosocial' – rather than the moral quality of the side effect itself" (Railton 2014: 854). A bad guy generally has bad intentions; consequently, the bad side effects of his actions will (probably) be intentional and considered by an observer to be so. Add in the asymmetry of good and bad, and you will arrive at the Knobe effect.

As we see, in the end, the Knobe effect does not constitute a challenge to the picture of traditional morality I have presented: it can take place within it, because it presupposes genuine actions, i.e., intentional actions. However, some authors link intentions with intuitions. Are intentions consequently tied to a System 1 process? And if this is the case, what impact does being tied to System 1 have?

8.6 A Dual Process Approach

Joshua Greene was the first to study moral dilemmas with the aid of neuroimaging in order to better understand moral judgement and moral decision-making (Greene et al. 2001). The dilemma that interested him most was *the trolley problem*, where subjects are asked if it is permissible to divert a threat (an out-of-control trolley) with the effect that only one person is killed instead of five. The responses are surprising, since the participants think it is permissible if the agent turns a switch, but forbidden if he has to push a fat man onto the rails to stop the trolley. However, in both cases one man dies and five are saved.

Greene also observes that different brain areas are mobilised in each case: rational ones (VMPFC) in the case of the switch, emotional ones (amygdala) in the fat man's version. In order to interpret these data, he turns to a dual process theory, inspired by Daniel Kahneman and Amos Tversky. When we make a moral judgement or a moral decision, we have two different resources: an intuitive, emotional and swift one (System 1) and a rational and slow one (System 2). Greene uses the metaphor of a camera to illustrate them: "The human brain is like a dual-mode camera with both *automatic settings* and a *manual mode*. A camera's automatic settings are optimised for typical photographic situations ('portrait', 'action', 'landscape'). [...] A dual-mode camera also has a manual mode that allows the user to adjust all of the camera's settings by hand" (Greene 2013: 133). In other words, in usual situations, we resort to System 1, since it is efficient and quick, but in unusual ones, when the situation is not clear, System 2 is more appropriate: rational deliberation is better here than intuitions and emotions.

Sometimes, especially in intricate situations such as moral dilemmas, we are fooled. In the *trolley problem*, it is better if five people survive instead of one; however, if we have no problem with turning the switch (i.e., we judge that it is morally permissible), why are we reluctant to push the fat man? Because System 1 enters the game and arouses emotions in order to prevent us from causing harm through personal physical force. Of course, such an emotion is usually very appropriate from a moral point of view: we should refrain from personally harming people, but in the fat man case, it seems to deliver a wrong answer.

In order to better understand what is at stake in System 1 judgments, Greene and colleagues have investigated some other versions of the *trolley problem*. They conclude that "harmful actions [are] judged to be less morally acceptable when the agent applied *personal* force to the victim. [… However] the personal force factor only affects moral judgments of intended harms, while the intention factor is enhanced in cases involving personal force. Put simply, something special happens when intention and personal force co-occur" (Greene et al. 2009: 369). This result is not surprising: harmful actions are morally problematic, especially when they are intended, since intention is at the centre of responsibility. However, it is not in this traditional sense that Greene sees the matter, since the intentional factor depends also on the personal force, and permissibility is lower when force increases, even if intention remains constant. System 1 appears then to have a certain complexity and a holistic function: it is not only an emotional immediate response (sometimes, Greene speaks of "alarm emotions") because the automatic setting

> "responds to harms that are *specifically intended*. Second, it responds more to harm caused *actively*, rather than passively. And, third, it responds more to harm caused directly by *personal force*, rather than more indirectly. It seems that these are not three separate criteria, employed in checklist fashion. Rather, they appear to be intertwined in the operation of our alarm gizmo, forming an organic whole" (Greene 2013: 246).

In Greene's dual process theory, intention counts, but only within System 1 processes, where it combines with personal force in an active pattern. System 2 focuses exclusively on outcomes: it is morally better if only one man dies. Moreover, from the point of view of ethics, System 1 and System 2 are not on the same footing, because when they conflict, Greene claims that we should rely on System 2, which is rational, and put System 1 on hold. Such a move has a price and might not be psychologically possible in some circumstances or for some of us, since we also read in *Moral Tribes*:

> "If you don't feel that it's wrong to push the man off the footbridge, there's something wrong with you. I, too, feel that it's wrong, and I doubt that I could actually bring myself to push, and I'm glad that I'm like this. What's more, in the real world, not pushing would almost certainly be the right decision. But if someone with the best of intentions were to muster the will to push the man off the footbridge, knowing for sure that it would save five lives, and knowing for sure that there was no better alternative, I would approve of this action, although I might be suspicious of the person who chose to perform it" (Greene 2013: 251).

If Greene were a supporter of virtue ethics, such ambivalence could be a genuine difficulty, but he is not, and for him, despite all psychological difficulties, pushing the fat man is the correct and ethical decision. Morality as it is and morality as it should be do not always coincide, as is well-known.

Consequently, for Greene, the traditional view of morality and responsibility is mistaken, because it is one-sided: the intentional component of an action has and should only have a secondary weight in morality. In usual situations, it is only one part in a shortcut allowing us to determine what is right and where responsibility lies; and in unusual or intricate ones, it is irrelevant, since we have to focus on outcomes, as utilitarianism and consequentialism are urging us to do (Greene 2008).

8.7 The Death of Traditional Morality?

If Greene's conclusions about the role and place of intentions are correct, this will result in an upheaval of our traditional view of morality. Even Mill would be thrown overboard because, as we saw, he also placed intentions at the heart of ethics. Morality as traditionally understood would come to an end and be replaced by something else, a new morality conceived perhaps as a kind of social management of misfortunes or a social control for bad behaviours – a move probably not unwelcome for Bentham and some other utilitarians. Nevertheless, at first glance, this seems to be absurd: John, Andrew, Paul and Peter, the protagonists of my four stories, ought then to be judged alike, as murderers, because they have caused the death of a human being. But what appears to be absurd for our traditional view of morality may well not be for rational morality even if we are psychologically unable to switch completely to this new stance – at least, it would take time. Greene has already argued for such a move concerning punishment in a paper written with Jonathan Cohen (Greene and Cohen 2004): retributivism should be replaced by consequentialism. Instead of punishing a criminal in a spirit of revenge or of debt paying, we ought to aim at social reintegration and rehabilitation.[17] Notice that, for Greene and some other authors, there is a link between intentions and retribution, and they contend that "our criminal justice system should change radically in the light of new neuroscience as it is imprudently concerned with an agent's intention" (Gkotsi and Gasser 2016: 63).

The question should nevertheless be frankly faced: is our traditional view mistaken when it makes intentions central to morality? Greene has given his reasons; can we find other ones?

The fact that children perceive intentions very early on can be an argument for linking them to System 1, which is a kind of "primitive" system. An experiment conducted by Jean Decety and Stephanie Cacioppo also shows that the perception of intention comes very early – as a kind of intuition – when we watch others' behaviour and that it is linked with emotional arousal: "We demonstrate for the first time how intention understanding […] and then affective processing occurs in very early stages of moral cognition processing […]. These results support the view that intentionality judgments both precede and guide moral cognition" (Decety and Cacioppo 2012: 3071). However, these data are inconclusive, because they also confirm the traditional view, for which intentions come first in order to guide moral assessment.

[17] Their argument presupposes hard determinism and is presented in the context of the free will debate.

Other studies cast doubts on Greene's interpretation. Michael Koenigs and colleagues have shown that sociopaths tend to push the fat man more frequently than typical subjects (Koenigs et al. 2007), and we already know that they assess intentions differently too. However, it seems difficult to say that they practise a better morality than we do. Greene nevertheless bites the bullet. After having reaffirmed his view under the name *The No Cognitive Miracles Principle* (NCMP), stating that "when we are dealing with unfamiliar moral problems, we ought to rely less on automatic settings (automatic emotional responses) and more on manual mode (conscious, controlled reasoning), lest we bank on cognitive miracles", he adds: "A corollary of the NCMP is that we should expect certain pathological individuals – VMPFC patients? Psychopaths? Alexithymics? – to make better decisions than healthy people in some cases" (Greene 2014: 715).[18] It remains nevertheless to be investigated if these better decisions are due to more rationality or less emotivity. As Jean Decety and Jason Cowel observe: "Are individuals who make utilitarian judgments in personal situations more rational and calculating, or are they simply colder and less averse to harming others?" (Decety and Cowel 2015: 10).

Let us return to intentions. For Greene, it is probably a mistake to say that psychopaths assess intentions differently: rather, they discard them for the benefit of outcomes. They put System 1 aside or at least rely less on it. Paradoxically, since they suffer from severe conditions, the judgements of these patients do not show the importance of intentions but their unimportance.

However, I think that all this reasoning is grounded in a misunderstanding: intentionality is not absent from System 2 but is as ubiquitous and as important there as in System 1. When someone judges that an agent should hit the switch or push the fat man, he presupposes that this agent has the intention of saving the life of the five persons. He does not act by accident, but deliberately, and this element is taken into account to determine the agent's responsibility. Consequently, Greene's argument cannot be generalised but is only valid (if it is) in a specific situation, when harms are caused as side effects – it was already the case with the Knobe effect.

More precisely, what is illustrated by the *trolley problem* is a situation where two effects are caused, a good one (to save five lives) and a bad one (the death of one person). Greene contends that in such situations, only the outcome should count and not the fact that the bad effect (the death of the victim) is willed as a means or permitted as a side effect of the good one. In contrast, traditional morality tends to suggest that in such situations we intend means, but not side effects, or at least not in the same manner: such effects are intentional or deliberately accepted, but not intended in the sense that they were the object of an intention and aimed at. Consequently, it is morally wrong to push the fat man but right to turn the switch. Greene objects to this thesis – and in my mind, he is partly right, as I will argue[19] – but this has nothing to do with the place and importance of intention in morality generally.

[18] The connection between utilitarianism and psychopathy is nevertheless weak and probably misguided; see Jaquet (2015).

[19] See also Baertschi (2013: chap. 1–2). Contrary to Greene, I conclude that we should refine our conception of intention and of its direction when several effects, good and bad ones, are present and not throw intentions overboard.

8.8 The *Doctrine of Double Effect* and Traditional Morality

Some classical moralists have tried to systematise such complex situations with the help of the *doctrine of double effect*. This doctrine consists of several rules saying that we should intend the good effect only, that the bad effect should not be the cause of the good one and that we have to take proportionality into account (e.g., a small bad effect is wiped out in relation to a larger good effect). This doctrine is not a general principle of morality and so should only be used in intricate situations where it is not easy to see the right option because a bad effect that cannot be avoided is involved (Goffi 2004: 238).[20] In brief, it has been conceived as a rational tool when our intuitions are confused or even muted, whereas for Greene it is "just an (imperfect) organising summary of the intuitive judgments. […]. It's our moral intuitions that justify the principle" (Greene 2013: 223). We can indeed object to the summary and be suspicious of a doctrine that fits with our intuitive judgements; but once more, it does not cast doubt on the intentional character of actions and the importance of intention for responsibility – Greene himself acknowledges that these questions are raised in the frame of an "action plan" (Greene 2013: 247), and what is planned is willed.

One of Greene's moral concerns is System 1's emotional blindness to distant and impersonal harms. He states: "When we harm people (including future people) by harming the environment, it's almost always as a side effect, often passive, and never through the direct application of personal force to another person. If harming the environment felt like pushing someone off a footbridge, our planet might be in much better shape" (Greene 2013: 253). Here, System 2 is necessary to avoid catastrophic effects, but is difficult to mobilise. I completely agree with him on this point: reflection is necessary in such situations; however, once again, it has no relevance for intentionality. It has to do with the relationships between agency, causality and responsibility: Am I responsible for the acts I intentionally perform and for voluntary omissions? Yes, of course, but I am not responsible for those I do or allow to happen accidentally or inadvertently. Am I responsible for *all the effects* of my deliberate acts and omissions? No, of course not, because these effects can be unforeseeable. And if they are foreseen, but not willed? Here the difficulties begin, which the *doctrine of double effect* tries to clarify. If intentions are at the centre, then responsibility cannot extend in the same manner to all the effects: the fact that they have been willed or not counts, and here, the doctrine has merits.

Utilitarianism provides another answer: we have to balance foreseeable good and bad effects and choose the best option. Accordingly, it objects to the *doctrine of double effect*, suggesting that it wrongly dismisses our responsibility for bad effects when they are not intended. Sidgwick made this proposal, linking responsibility to intentions: "For purposes of exact moral or jural discussion, it is best to include under the term "intention" all the consequences of an act that are foreseen as certain or probable (Sidgwick 1981: 202)". Utilitarianism keeps up Mill's lesson on the

[20] Notice that in the case of the Knobe effect, the *doctrine of double effect* considers that the chairman's project is not permissible: harming the environment is too high a cost.

moral importance of intentions, and it is necessary if it wants to remain a *moral* theory: our intentions always matter. However, morality is not alone in the struggle toward happiness, since, for many bad effects and undesirable states of affairs, we can instead turn to risk management, social security and political action.

Conclusion

Neuroethics, with the aid of neuropsychology, questions several traditional moral views. One of them is the central place given to intentions: for traditional morality, they characterise what actions are and ground the attribution of responsibility. Several empirical studies confirm this view, but others seem to refute it: the Knobe effect suggests that the intentional character of our actions is not central but derived from allocations of responsibility. For the dual process theory (at least Greene's version), intentions are important only when we rely on moral intuitions. I have argued that these two charges are not conclusive, mainly because they are at most valid only for actions' side effects. Consequently, traditional morality can stand firmly on its foundations, at least with regard to the place and role of intentions.

Traditional morality abides, of course, through many debates, and there exists pressure to modify parts of it. One place of disagreement concerns responsibility when an action has several effects. Here, utilitarianism and the *doctrine of double effect* are in tension, showing that this part of traditional morality is not on a firm ground. It is to their credit that Knobe and Greene's arguments have brought this difficulty to the forefront of neuroethics and to discussions about moral reasoning.

Acknowledgements I wish to thank Florian Cova, Ugo Gilbert Tremblay, Jean-Yves Goffi, François Jaquet, Pierre Le Coz and Yves Page for their helpful comments on a first version of this paper.

References

Anscombe E (1963) Intention. Blackwell, Oxford
Baertschi B (2013) L'éthique à l'écoute des neurosciences. Les Belles Lettres, Paris
Cerri M (2016) The ethical ghost in the brain: testing the relationship between consciousness and responsibility in the special case of REM sleep behavior disorder. In: Lavazza A (ed) Frontiers in neuroethics. Cambridge Scholar Publishing, Cambridge, pp 117–133
Chituc V, Henne P, Sinnott-Armstrong W, De Brigard F (2016) Blame, not ability, impacts moral "Ought" judgments for impossible actions: toward an empirical refutation of "Ought" implies "Can". Cognition 150:20–25
Christensen J-F, Gomila A (2012) Moral Dilemmas in cognitive neuroscience of moral decision-making: a principled review. Neurosci Behav Rev 36:1249–1264
Greene J, Cohen J (2004) For the law, neuroscience changes nothing and everything. Philos Trans R Soc Lond B 359:1775–1785
Cova F, Naar H (2012) Side-effect without side effects: the pervasive impact of moral considerations on judgments of intentionality. Philos Psychol 25(6):837–854
Cova F, Dupoux E, Jacob P (2012) On doing things intentionally. Mind Lang 24(4):378–409
Davidson D (2002) Essays on actions and events. Clarendon Press, Oxford

Decety J, Cacioppo S (2012) The speed of morality: a high density electrical neuroimaging study. J Neurophysiol 108:3068–3072
Decety J, Cowel J (2015) Empathy, justice, and moral behavior. AJOB Neurosci 6(3):3–14
Dennett D (2013) Intuition pumps and other tools for thinking. Penguin Books
Doris J, Knobe J, Woolfolk R (2007) Variantism about responsibility. Philos Perspect 21:183–214
Forest D (2014) Neuroscepticisme. Ithaque, Montreuil
Fried I, Mukamel R, Kreiman G (2011) Internally generated preactivation of single neurons in human medial frontal cortex predicts volition. Neuron 69:548–562
Frith C (2010) Consciousness, will and responsibility: what studying the brain tells us about the mind. ICR Monograph Series 58. www.thebrainandthemind.co.uk/. Accessed 11 June 2016
Frith C (2014) Action, agency and responsibility. Neuropsychologia 55:137–142
Gillett G (2008) Subjectivity and being somebody. Human identity and neuroethics. Imprint Academic, Exeter
Gkotsi G, Gasser J (2016) Neuroscience in forensic psychiatry: from responsibility to dangerousness. Int J Law Psychiatry 46:58–67
Goffi J-Y (2004) Le principe des actions à double effet. In: Goffi J-Y (ed) Hare et la philosophie morale. Vrin, Paris, pp 231–248
Greene J (2008) The secret joke of Kant's soul. In: Sinnott-Armstrong W (ed) Moral psychology. MIT Press, Cambridge, pp 35–79
Greene J (2013) Moral tribes. Atlantic Books, London
Greene J (2014) Beyond point-and-shoot morality: why cognitive (neuro)science matters for ethics. Ethics 124(4):695–726
Greene J, Sommerville B, Nystrom L, Darley J, Cohen J (2001) An fMRI investigation of emotional engagement in moral judgment. Science 293:2105–2108
Greene J, Cushman F, Stewart L, Lowenberg K, Nystrom L, Cohen J (2009) Pushing moral buttons: the interaction between personal force and intention in moral judgment. Cognition 111:364–371
Güney S, Newell B (2013) Fairness overrides reputation: the importance of fairness considerations in altruistic cooperation. Front Hum Neurosci 7(252):1–123
Güroglu B, van den Bos W, Rombouts S, Crone E (2010) Unfair? It depends: neural correlates of fairness in social context. Soc Cogn Affect Neurosci 5(4):414–423
Jaquet F (2015) Les conséquentialistes ne sont pas (des psychopathes), Implications philosophiques. http://www.implications-philosophiques.org/actualite/une/les-consequentialistes-ne-sont-pas-des-psychopathes/. Accessed 11 June 2016
Knobe J (2003) Intentional action in folk psychology: an experimental investigation. Philos Psychol 16:203–231
Knobe J (2004) Intention, intentional action and moral considerations. Analysis 64:181–187
Koenigs M, Young L, Adolphs R, Tranel D, Cushman F, Hauser M et al (2007) Damage to the prefrontal cortex increases utilitarian moral judgments. Nature 446:908–911
Leslie A, Knobe J, Cohen A (2006) Acting intentionally and the side-effect effect. Psychol Sci 17(5):421–427
Levy N (2011) Neuroethics: a new way of doing ethics. AJOB Neurosci 2(2):3–9
Machery É (2008) The folk concept of intentional action: philosophical and experimental issues. Mind Lang 23(2):165–189
Maoz U, Yaffe G (2015) What does recent neuroscience tell us about criminal responsibility? J Law Biosci 3(1):120–139
Marcel A (2003) The sense of agency: awareness and ownership of action. In: Roessler J, Eilan N (eds) Agency and self-awareness: issues in philosophy and psychology. OUP, Oxford, pp 48–93
Marek J (2013) Alexius Meinong. Stanford encyclopedia of philosophy. http://plato.stanford.edu/entries/meinong/. Accessed 29 Aug 2016
McCann H (1991) Settled objectives and rational constraints. Am Philos Q 28(1):25–36
Mill JS (1991) Utilitarianism. Oxford World's Classics, Oxford

Moran J, Young L, Saxe R, Lee SM, O'Young D, Mavros P et al (2011) Impaired theory of mind for moral judgment in high-functioning autism. Proc Natl Acad Sci U S A 108(7):2688–2692
Ngo L, Kelly M, Coutlee C, Carter M, Sinnott-Armstrong W, Huettel S (2015) Two distinct moral mechanisms for ascribing and denying intentionality. Sci Rep 5(17390)
Railton P (2014) The affective dog and its rational tale: intuition and attunement. Ethics 124:813–859
Rigato J, Murakami M, Mainen Z (2014) Spontaneous decisions and free will: empirical results and philosophical considerations. Cold Spring Harb Symp Quant Biol LXXIX:177–184
Sidgwick H (1981) The methods of ethics. Hackett PC, Cambridge
Treadway M, Buckholtz J, Martin J, Jan K, Asplund C, Ginther M et al (2014) Corticolimbic gating of emotion-driven punishment. Nat Neurosci 17(9):1270–1277
Yoshie M, Haggard P (2013) Negative emotional outcomes attenuate sense of agency over voluntary actions. Curr Biol 23:2028–2032
Young L, Bechara A, Tranel D, Damasio H, Hauser M, Damasio A (2010) Damage to ventromedial prefrontal cortex impairs judgment of harmful intent. Neuron 65:845–851
Zangrossi A, Agosta S, Cervesato G, Tessarotto F, Sartori G (2015) "I Didn't Want to Do It!" the detection of past intentions. Front Hum Neurosci 9(608). eCollection 2015

Is It Time to Abandon the Strong Interpretation of the Dual-Process Model in Neuroethics?

9

Veljko Dubljević

Abstract

Pioneering neuroimaging studies in the "neuroscience of ethics" (Greene et al. 2001) have led to the development of the dual-process model of moral judgment (Greene 2008). These studies have also provided the much-needed impetus for neuroethics itself, having inspired a range of empirical studies on moral judgment (see Christensen and Gomila 2012) and conceptual analyses in ethics informed by neuroscience (see Levy 2007; Racine 2010; Glannon 2007, 2011). However useful dual-process model has been in the past, mounting empirical counterevidence (Koenigs et al. 2007; Duke and Begue 2015), and the conceptual implications of fallibilism lead to a conclusion that this model should be shelved as neuroethics moves forward.

Fallibilism as a pragmatic attitude emanates from the logic and spirit of science: scientific inquiry and progress actually mean overthrowing previously established beliefs and tentatively establishing new beliefs based on science (Dewey 1929). When this is applied to the study of moral judgment, it yields a drastically different picture from the one painted by champions of the dual-system model in moral decision-making: moral beliefs are fallible (no matter how fast we might reach them), have the logical status of hypotheses, and do not provide absolute certainty. For example, utilitarian calculus, though deliberate and slow, is as open to biases as an approach using heuristics. The point is that fallibility needs to be recognized in all endeavors, whether they are guided by quick intuitive processes or by time-consuming and explicit reasoning.

V. Dubljević
Institut de recherches cliniques de Montréal (IRCM); and North Carolina State University, Raleigh, NC 27695, USA
e-mail: Veljko.Dubljevic@ircm.qc.ca

E. Racine, J. Aspler (eds.), *Debates About Neuroethics*, Advances in Neuroethics,
DOI 10.1007/978-3-319-54651-3_9

9.1 Introduction

Neuroethics, often defined as a field with two distinct branches – the "ethics of neuroscience" and the "neuroscience of ethics" (Roskies 2002) – is a groundbreaking approach to moral problems at the intersection of numerous other fields. On the one hand, it studies the social and ethical issues stemming from the uses of cutting-edge neuroscience and technology (i.e., the ethics of neuroscience). On the other hand, it relies on the use of neuroscientific tools, such as functional magnetic resonance imaging (fMRI), to elucidate patterns of moral decision-making and neural correlates of other processes important for ethics (i.e., the neuroscience of ethics). Pioneering studies in the latter category (Greene et al. 2001) led to the development of the dual-process model of moral judgment (Greene 2008), a key model in the history of neuroethics. These studies have also provided the much-needed impetus for the creation of the field itself. Indeed, they have inspired a range of empirical studies on moral judgment (for a review, see Christensen and Gomila 2012) and conceptual analyses in ethics informed by neuroscience (see Levy 2007; Racine 2010; Glannon 2007, 2011).

But what exactly is the dual-process model of moral judgment, and what does it explain and predict? In their initial fMRI study, Greene and colleagues (Greene et al. 2001) drew on Philippa Foot's work in moral philosophy (Foot 1967) to formulate a classification of dilemmas including the so-called impersonal, such as the original trolley dilemma[1] and the personal, such as the footbridge dilemma.[2] Proponents of the dual-process view on moral judgment take neuroimaging data to show that the patterns of responses and brain activation in trolley dilemmas favor a "utilitarian" view of morality based on abstract thinking and calculation, while responses and patterns of brain activation in the footbridge dilemma suggest that emotional reactions drive non-consequentialist answers. The purported upshot is that rational (driving utilitarian calculation) and emotional (driving aversion to personally causing injury) processes are competing for dominance. This conclusion has been met with severe criticism from many sides: some questioned the interpretation of data (e.g., Berker 2009; Dean 2010; Bluhm 2014), others criticized the methodological usefulness of the approach (Kahane and Shackel 2010; Rosas and Koenings 2014), and still others reported a link between the purportedly "rational" utilitarian preferences and distorted, disturbed, or antisocial cognition (see Bartels and Pizarro 2011 for Machiavellianism, Koenigs

[1] "A runaway trolley is headed for five people who will be killed if it proceeds on its present course. The only way to save these people is to hit a switch that will turn the trolley onto a sidetrack, where it will run over and kill one person instead of five. Is it okay to turn the trolley in order to save five people at the expense of one?" (Greene 2008: 41–42)

[2] "[…] a runaway trolley threatens to kill five people, but this time you are standing next to a large stranger on a footbridge spanning the tracks, in between the oncoming trolley and the five people. The only way to save the five people is to push this stranger off the bridge and onto the tracks below. He will die as a result, but his body will stop the trolley from reaching the others. Is it okay to save the five people by pushing this stranger to his death?" (Greene 2008: 42)

et al. 2012 for psychopathy, Crockett and Rini 2015 for decreased responsibility, Koenigs et al. 2007 for vmPFC damage, Mendez 2009 for frontotemporal dementia, and Duke and Begue 2015 for blood alcohol levels). Despite these important criticisms, the leading proponent of the approach, Joshua Greene, maintains that a dual-process model "which emphasizes both emotional intuition and controlled cognition, is supported by multiple fMRI studies using different behavioral paradigms, multiple behavioral studies of neurological patients, and a variety of behavioral studies using both experimental manipulations and individual difference measures" (Greene 2015).

Based on the results of his studies, and other corroborating evidence, Greene denies that any of the challenging views are a problem for his dual-process approach and paints a picture of moral judgment where neuroscience "proves" that the "cold" rational calculation of consequences competes with the "hot processing" of intentional harm. A normative prediction of the approach is that consequentialist judgment is more "correct" than other types of moral judgment (e.g., deontological), since it is based on reasoning, as opposed to emotional reactions. According to Greene "we have innate responses to personal violence that are powerful but rather primitive. That is, we might expect humans to have negative emotional responses to certain basic forms of interpersonal violence, where these responses evolved as a means of regulating the behavior of creatures who are capable of intentionally harming one another, but whose survival depends on cooperation and individual restraint" (Greene 2008: 43).

There is nothing wrong with being enthusiastic about what one's hypothesis can predict or one's theory can explain. However, Greene's enthusiasm for absolute foundations leads him to suppose "that consequentialist principles [...] provide the best available standard for public decision making and for determining which aspects of human nature it is reasonable to try to change and which ones we would be wise to leave alone" (Greene 2008: 77).

The basic issue with this view (apart from stubborn resistance to counterevidence) is that it disregards fallibilism and supposes that an absolute foundation for correct moral judgment can be found – once and for all, this time with the aid of neuroimaging. Greene follows many philosophers in this respect and, again, like many before him, thinks that reasoning, or in the current scientific parlance, effortful, deliberative consciously controlled "System 2" processes, is an absolute foundation for moral judgment. His argument rests on several assumptions: (1) that consequentialist judgment is based on reasoning and not on intuition, (2) that moral intuition is simply emotion, and (3) that reasoning is better or more correct than intuition when it comes to decision-making. If any of these constituent assumptions fail, so does the whole structure of the argument.

By examining recent relevant empirical evidence, I will argue that all of these assumptions ultimately fail and that the dual-process approach to moral judgment should be viewed as severely flawed, with overwhelmingly convincing objections. More specifically, I will argue that (1*) consequentialist, outcome-directed judgment is not always, or even most of the time, based on reasoning, that (2*) moral intuition is much more than just emotion and should not be equated with it, and that (3*) just as automatic, "System 1" processes can sometimes be wrong, so can deliberative, "System 2" processes. I will not claim that we should replace reasoning with emotion, but that certain intuitive processes (which are mostly based on prior acquisition of explicit and tacit knowledge) outperform consciously controlled processes in terms of accuracy and speed.

9.2 Is Consequentialist Judgment Based on Reasoning?

Let us start with claim 1. Even though Greene was right to invoke initial studies that seem to corroborate his hypothesis, this claim still remains controversial, with certain empirical findings appearing to remain at odds with the dual-process approach.[3] In particular, if utilitarian, outcome-based judgment is caused by abstract thinking (System 2), whereas non-consequentialist intent- or duty-based judgment is intuitive (System 1) and thus irrational, how come children ages 4–10 focus more on outcome than on intent (see Cushman 2013)? Given that abstract thought develops mostly after age 12, "fully rational" utilitarian judgments should not be observable in children. And yet they are not only observed but seem to dominate immature or dysfunctional adult moral cognition. The answer is that outcome-based moral judgment is intuitive as well, and if there is an interplay between System 1 and System 2, that might be the case for all types of moral judgment (i.e., intuitive reactions may also be corrected or overseen by effortful reasoning), not just in the "consequentialist" case. To make things more complicated, the difference in reaction times in answers to personal and impersonal dilemmas, used by Greene and colleagues to establish which judgments were intuitive and which were deliberative, was recently shown to have been misleadingly interpreted in the original study (McGuire et al. 2009). Additionally, more recent evidence points to a conclusion that reaction times are a poor measure of intuition vs. deliberation (see Evans et al. 2015).

Thus, the status of Greene's conclusions regarding the exact nature of specific moral judgments stemming from neuroscientific research in general, and neuroimaging in particular, is more tentative than definitive. Apart from the problems with reaction times, another reason for doubting that outcome-based judgment is based on reasoning is Greene's ample use of "reverse inference," a process of inferring the presence of particular mental states based on brain activation patterns in response to experimental procedures (see Poldrack 2011; Mathews and Bok 2015). Greene's interpretation relies on the data from the initial neuroimaging study, which detected activations in brain areas associated with cognition (in other unrelated studies) during "impersonal" moral judgment tasks. However, a reverse inference assumes that a particular pattern of activity represents a network for the specific task studied – when that same activity could represent broader functions or behaviors. Therefore,

[3] In this paragraph, I draw and substantially expand on Dubljević 2016.

the use of reverse inference has been criticized due to the lack of adequate background information, lack of specificity of many brain regions, lack of consistency in attributing functions to different brain regions, and lack of adequate operationalization of the subject matter (see Poldrack 2011; Schleim 2015; Mathews and Bok 2015; Christensen and Gomila 2012; Dubljević and Racine 2014). If the goal is generating hypotheses, reverse inference is a valid strategy, but reifying current research hypotheses as proven empirical facts is controversial at best and inappropriate at worst. Therefore, it is far from being "proven" by neuroscience that consequentialist judgment is based on "cold cognition," and Greene's normative prediction turns out to just be utilitarian wishful thinking.

Furthermore, recent research in cognitive psychology suggests that judgments traditionally considered to arise from reasoning, System 2, or Type 2 processes are accomplished intuitively. According to Bialek and colleagues, "[t]he role of Type 1 processes in the analysis of numerical base rates or in the evaluation of logical validity supports a model in which alternative responses are activated automatically and in parallel" (Bialek et al. 2014: 22).

Finally, Greene's "just-so" evolutionary story (e.g., in Greene 2008), as an explanation for "emotional buttons" and the calculation of consequences, seems rather naïve, especially given the fact that even relatively simple organisms dissociate between (and prefer) five items of value to one item of value. Claiming that "innate responses to personal violence" are "rather primitive" as opposed to the "abstract thinking" of numerical evaluations, which are exhibited by preverbal children and nonhuman animals, is somewhat puzzling. In sharp contrast to Greene's assumption, experts in the field of development and evolution of numerical cognition report that "the brain regions recruited during approximate number representations are shared by adult humans, nonhuman primates and young children who cannot yet count to 30, suggesting a *primitive basis in the brain*" (Geary et al. 2015: 226, emphasis added). Given the social nature of humans, and their long evolutionary history, it would be very surprising indeed if all types of moral judgment were not fast, automatic, and fundamental (or *primitive* in Greene's parlance).

9.3 Is Moral Intuition Based on Emotion?

How about the second claim, namely, that moral intuition boils down to emotion? Again, even though there are many influential authors that agree on this, reanalyses of the initial evidence for this claim have concluded that emotion has only a minor effect on moral judgment (see, e.g., May 2014) and that the initial reported effects are actually the result of artifacts in stimulus design, not of underlying neural systems (McGuire et al. 2009). Also, recent evidence shows that, even though they often co-occur and interconnect, moral intuition is both dissociable from emotional processes (Batson et al. 2009) and can occur prior to them (Decety and Cacioppo 2012; Yang et al. 2013; Yang et al. 2014), so it is safe to assert that results of more recent studies call this second assumption into question.

So what evidence does Greene have for associating moral intuition with emotion? In his work, apart from using reverse inference to analyze detected activations in brain areas associated with emotion (again in other unrelated studies) during "personal" moral judgment tasks, he draws substantially on work by Jonathan Haidt (e.g., Haidt 2001), in which moral judgment is equated with emotion and, more specifically, feelings of disgust, which purportedly have major effects in shaping moral judgment.[4] And yet, a closer look at the evidence provides a very different picture.

First, effects of disgust on moral judgment are very minor – a reanalysis of evidence has shown that only minor increases in blameworthiness are recorded for some populations (e.g., those with increased sensitivity to physical disgust) and conditions, whereas no effects are present in other populations (May 2014). An unrelated reanalysis of Greene's initial study found that results pointing to the conclusion that emotional processes drive moral judgment were themselves driven by poor design and analysis of items in certain dilemmas only and are not generalizable to other, let alone to all, moral dilemmas. The authors even conclude that "there is no reason to assume that emotionally salient moral decisions are processed in a qualitatively different way to those dilemmas that are not emotionally salient. Furthermore, there is no evidence here to support the theory that there are two competing moral systems at work" (McGuire et al. 2009: 580).

Second, recent work has successfully dissociated emotion from moral judgment.[5] For instance, in the "torture case" study (Batson et al. 2009; Batson 2011), American respondents were asked to rate the moral wrongness of specific examples of torture and their own emotional arousal. The experimental group is presented with a vignette in which an American soldier is tortured by militants, while a control group read a text in which a Sri Lankan soldier is tortured by Tamil rebels. Even though there was no significant difference in the intensity of moral judgment, the respondents were emotionally "riled up" only in the case of a member of their in-group being tortured.

Neuroimaging work also confirms this dissociation. For instance, Yang and colleagues recorded event-related potentials for disgust- and morality-based questions in a go/no-go task.[6] They report that moral questions are answered faster than disgust-cued questions, which suggests that respondents make moral judgment before they can process the emotional information (Yang et al. 2013). A separate, high-density EEG study by Decety and Cacioppo (Decety and Cacioppo 2012) has also shown that the cognitive apprehension of mental state information during moral judgment is in fact very fast and begins prior to the activation of brain regions

[4]This should not be understood as denying the fact – noted by Haidt – that humans usually reach moral judgments quickly and are not always aware of the processes that led to them.

[5]In this paragraph, I draw on Dubljević 2016.

[6]Go/no-go task is a psychological measure in which stimuli are presented in a continuous stream and participants perform a binary decision on each stimulus. One of the outcomes requires participants to make a motor response (go), whereas the other requires participants to withhold a response (no-go).

related to processing emotions. As Rottman and Young explain, "[t]he differentiation between intentional and accidental harms [...] occurs within less than one-tenth of a second after a stimulus is perceived, demonstrating that adults automatically and immediately integrate information about intent into their harm-based moral judgments" (Rottman and Young 2015: 135). However, it would be a mistake to conclude that this automatic processing is innate, instinctive, and/or emotional, as opposed to learned. Namely, as previously noted, young children focus more on outcomes than intentions when making moral judgments (see Cushman 2013), and so it can be concluded that rapid intuitive integration of intention is a hallmark of adult moral competence and a result of cognitive development and maturation.

To conclude, it is safe to assert that there is growing evidence that emotions in moral judgment are dissociable from and modulated by prior nonconscious cognitive processes. The purported impact of emotions on moral judgment is founded on poor item design, questionable reaction time data, reverse inference, and the use of fMRI – a technique that has good spatial resolution but limited temporal resolution. In contrast, EEG studies provide data suggesting that moral intuition is based on fast cognitive processes, such as distinguishing between intentional and unintentional harm. Therefore, in contrast to Greene's second assumption, moral intuition is much more than just emotion and should not be equated with it, even though it has to be noted that moral intuition is often followed by emotional reactions (Huebner et al. 2008). Indeed, it is crucial to recognize that what Greene's dual-process approach to moral judgment equates and lumps together as System 1 is actually comprised of very heterogeneous processes (Evans 2008; Glockner and Witteman 2010; Kauppinen 2015). This leads us to the final point (which is discussed below): namely, some intuitive (System 1) processes are likely to result in false beliefs or incorrect outputs, especially when triggered outside of the context in which they are adaptive, while other intuitive processes are more likely to produce true beliefs.

9.4 Are Deliberative Processes Less Error Prone Than Intuitive Processes?

Now, let us consider Greene's third assumption, namely, that reasoning is more reliable than intuition. In fact, evidence suggests that just as automatic processes can sometimes be wrong, so can deliberative processes. It also shows that some forms of automatic processing are much more reliable across a range of domains, such as language, chess, medical diagnosis, and managerial decisions. In the literature on intuitive decision-making and bounded rationality, intuition in general is described as a product of learning and experience and highly adaptive, whereas emotion-driven judgment is described as a product of more instinctive urges and less reliable in all forms of decision-making (cf. Simon 1987).

Intuition in general is adaptive in certain environments – where there are multiple cues that appear in parallel (as opposed to a linear sequence) or when many cues are actually "noise" and when relevant cues need to be rapidly identified (cf. Sadler-Smith and Sparrow 2008). Moral judgment and decision-making exhibit all of these characteristics – the moral agent has to dissociate quickly between relevant and irrelevant aspects of the situation and make a (potentially life-saving) decision in adequate, often very short, time. Therefore, the most promising approach is to consider moral intuition as a subset of general intuition and thus more likely to be based on adaptive leaps in cognition than on emotional reactions (Gigerenzer 2008, 2010). It has to be noted that intuition does not operate independently of conscious analysis (cf. Simon 1987). Simon notes that intuitive and reasoning processes are complementary components of effective decision-making systems. Intuition simply reduces the cognitive load by introducing ever larger steps in reasoning. These steps are based on the level of experience an agent has in the subject domain: experts make giant intuitive steps without making errors, whereas novices can rely only on tiny steps (cf. Simon 1987: 67f).

So does this mean that intuitive moral judgment is always correct? Not at all. The psychological research into intuitive judgment gives ample evidence of how our quick decisions can lead to errors. However, there are two possible interpretations of this evidence. The first interpretation is that intuitive judgments are faulty, whereas the second is that they just need to be treated as fallible. Greene's third assumption critically relies on the former interpretation. Actually, the whole dual-process model is one sided in an ongoing debate about nonconscious cognition, since it relies on evidence from only one approach in the study of fast, intuitive, or "heuristic" judgment: the "heuristics and biases" approach. This approach associates the use of heuristics with irrational decision-making, while the opposing "fast and frugal heuristics" approach defines heuristics in more favorable terms (see Kelman 2011). Both of these approaches agree on the fact that intuition in general and specific heuristic strategies in particular work well most of the time, but occasionally misfire, and hence give more credence to the *second* interpretation above.

The general gist of the overall evidence is that intuition or heuristics should be properly viewed as cognitive strategies that ignore certain information to make decisions faster, more frugally, and/or more accurately compared to more complex methods (Gigerenzer and Gaissmaier 2011). An important issue with the assessment of the error proneness of heuristics is that novices are more likely to demonstrate context-free application of heuristics, which results in errors, from which they can learn, whereas experts are more sensitive to context and tend to avoid the misapplication of heuristics (Kahneman and Klein 2009).

Now, whether "professional experts" exist for morality (comparable to chess experts) and whether such an expert would have more reliable moral intuitions are separate issues that I do not have the space to analyze in this essay. Suffice it to say that we do generally make a distinction between the moral judgment of children, or adolescents, and adults, and we also recognize intuitive moral competence. This intuitive moral competence has been compared with intuitive linguistic competence (see Rini 2014), which basically means that a person can quickly and intuitively form and evaluate sentences without much effortful grammatical analysis. Analogously, after a long period of cognitive maturing and learning, competent human adults are able to quickly and intuitively make moral decisions and judgments (see also Mikhail 2011) for which they are held responsible by society at large.

And so we come to the conclusion that fallibilism is the most appropriate interpretation of current evidence. Errors in intuitive judgment certainly exist, but this does not mean that intuition is faulty. Indeed, fallibilism as a pragmatic attitude emanates from the logic and spirit of science: scientific inquiry and progress mean overthrowing previously established beliefs and tentatively establishing new beliefs based on current evidence (Dewey 1929). When this is applied to normative judgments, it yields a drastically different picture from the one painted by champions of the dual-process model in moral decision-making: moral beliefs are fallible (no matter how fast or slow we might reach them), have the logical status of hypotheses, and do not provide absolute certainty.

For example, utilitarian calculus, even if it might be deliberate and slow, is as open to biases as an approach using intuition or heuristics. The biases are of a different kind, certainly, but nevertheless can act as systematic distortions. Having sufficient time to deliberate on the moral decision does not guarantee that the decision will be moral and not tainted with self-interest. Perhaps an example might help put things into perspective. During times of disaster, people often help each other without much thought. This is something that we learn and expect from fellow humans on a regular basis, and disasters commonly do not allow much time for effortful reasoning. However, in certain conditions, there is sufficient time for deliberation, and this solitary deliberation may not always convey "correct" moral judgment. For instance, in the Titanic disaster, the crew members had several hours "to deliberate on the consequences of saving [their] own life, or sacrificing [themselves] for the lives of other more vulnerable passengers. [...] [T]he crew's survival rate was 18% higher than that of the passengers [...] [pointing to a conclusion that] the members of the crew took advantage of better access to information and lifeboats, thus behaving in line with their self-interest" (Suter 2011: 454).

This conclusion based on a historical example is corroborated by more recent evidence. For instance, an aptly named recent study (*Spontaneous Giving and Calculated Greed*) used priming of intuitive or deliberative decision-making in economic games. The results show that cooperative behavior is the default, intuitive response for many people, whereas deliberative processes often "taint" the decision-making with self-centeredness, selfishness, and greed (Rand et al. 2012). Indeed, as authors of a recent review of evidence regarding cooperation and morality conclude: "[c]ooperation with unrelated individuals is a hallmark of humankind despite the temptation to behave selfishly. [...] [S]ocial heuristics form [...] advantageous behavioral strategies as intuitive defaults, and thus mechanisms that promote cooperation may lead us to cooperate automatically [...]. Such "irrational" cooperation is central to the success of human civilization and forms a key component of morality" (Jordan et al. 2015: 95–96).

This puts the simplistic assumption "deliberation is good, intuition is bad" into question and points to the need to recognize fallibility in all types of cognition, whether they are guided by quick intuitive processes or slow reasoning. The upshot might be that systematic distortions of intuitive or heuristic-based moral judgment and decision-making are connected to overly hasty generalizations favoring certain groups at the expense of others, whereas the key bias affecting conscious thought in morality is the temptation to behave selfishly. The best way to deal with both kinds of biases might be to have public scrutiny of moral decisions and to favor humility and fallibility instead of certainty and foundationalism in moral matters.

Concluding Remarks

The dual-process model of moral judgment, stemming from pioneering studies in the neuroscience of ethics, served the purpose of generating academic and public interest for issues at the intersection of ethics and neuroscience and thus helped establish neuroethics as a field. However, since neuroethics is characterized by due consideration of available empirical data from behavioral and brain sciences, and this evidence points to the conclusion that the dual-process model of moral judgment is inconsistent and empirically unsupported and in fact has been falsified; it is time to relegate this model, at least in its strong interpretation, to the past – to the gallery of outdated beliefs that were previously established but ultimately overthrown by evidence. The evidence suggests that a content-based intuitionist model of moral judgment is more likely. Still, this analysis did not go into sufficient detail to prove or even consider the relevant assumptions of available alternative models: the moral foundations theory (Haidt and Graham 2007), the universal moral grammar (Mikhail 2007, 2011), and the ADC approach (Dubljević and Racine 2014).

This chapter has been limited to reviewing the three assumptions that the dual-process model relies on, namely, that consequentialist judgment is based on reasoning and not on intuition, that moral intuition is nothing more than emotion, and that reasoning is more reliable than intuition, and the evidence shows that these assumptions are false. Based on this analysis, it is uncontroversial to conclude that it is far from proven "that consequentialist principles [...] provide the best available standard for public decision making" (Greene 2008). This does not mean that consequentialist principles are faulty or misguided. On the contrary, the take-home message should be one of fallibilism – principles from different moral theories offer valuable guidelines which are nevertheless open to error. They are useful for public scrutiny of moral judgment and decision-making as they shine light onto the biases inherent in both reasoning and intuition. However, they are not and cannot be unmistakable foundations that are not open to theoretical revision or empirical falsification.

References

Bartels DM, Pizarro D (2011) The mismeasure of morals: antisocial personality traits predict utilitarian responses to moral dilemmas. Cognition 121:154–161

Batson CD (2011) What's wrong with morality? Emot Rev 3(3):230–236

Batson CD, Chao MC, Givens JM (2009) Pursuing moral outrage: anger at torture. J Exp Soc Psychol 45:155–160

Berker S (2009) The normative insignificance of neuroscience. Philos Public Aff 37(4):293–329

Bialek M, Terbeck S, Handley S (2014) Cognitive psychological support for the ADC model of moral judgment. AJOB Neurosci 5(4):21–23

Bluhm R (2014) No need for alarm: a critical analysis of Greene's dual-process theory of moral decision-making. Neuroethics 7(3):299–316
Christensen JF, Gomila A (2012) Moral dilemmas in cognitive neuroscience of moral decision-making: a principled review. Neurosci Biobehav Rev 36:1249–1264
Cushman FA (2013) Action, outcome and value: a dual-system framework for morality. Personal Soc Psychol Rev 17(3):273–292
Dean, R (2010) Does neuroscience undermine deontological moral theory? Neuroethics 3. 43–60
Decety J, Cacioppo S (2012) The speed of morality: a high density electrical neuroimaging study. J Neurophysiol 108:3068–3072
Dewey J (1929) The quest for certainty: a study of the relation of knowledge and action. Milton, Balch & Company, New York
Dubljević V (2016) An interesting time for the study of moral judgment and cognition. iCog Blog. http://icog.group.shef.ac.uk/an-interesting-time-for-the-study-of-moral-judgment-and-cognition/. Accessed 12 Jan 2016
Dubljević V, Racine E (2014) The ADC of moral judgment: opening the black box of moral intuitions with heuristics about agents, deeds and consequences. AJOB Neurosci 5(4):3–20
Duke AA, Begue L (2015) The drunk utilitarian: blood alcohol concentration predicts utilitarian responses in moral dilemmas. Cognition 134:121–127
Evans JSBT (2008) Dual processing accounts of reasoning, judgment and social cognition. Annu Rev Psychol 59:255–278
Evans AM, Dillon KD, Rand DG (2015) Fast but not intuitive, slow but not reflective: decision conflict drives reaction times in social dilemmas. J Exp Psychol 144(5):951–966
Foot P (1967) The problem of abortion and the doctrine of double effect. The Oxford Review 5:5–15
Geary DC, Berch DB, Mann Koepke K (2015) Evolutionary origins and early development of number processing. Elsevier/Academic Press, Amsterdam
Gigerenzer G (2008) Moral intuition = fast and frugal heuristics? In: Sinnott-Armstrong W (ed) Moral psychology, vol 2: the cognitive science of morality—intuition and diversity. MIT Press, Cambridge, pp 1–26
Gigerenzer G (2010) Moral satisficing: rethinking moral behavior as bounded rationality. Top Cogn Sci 2(3):528–554
Gigerenzer G, Gaissmaier W (2011) Heuristic decision making. Annu Rev Psychol 62:451–482
Glannon W (2007) Bioethics and the brain. Oxford University Press, New York
Glannon W (2011) Brain, body and mind: neuroethics with a human face. Oxford University Press, New York
Glockner A, Witteman C (2010) Beyond dual-process models: a categorisation of processes underlying intuitive judgment and decision making. Think Reason 16(1):1–25
Greene JD (2015) The cognitive neuroscience of moral judgment and decision making. In: Decety J, Wheatley T (eds) The moral brain: a multidisciplinary perspective. MIT Press, Cambridge, pp 197–220
Greene JD (2008) The secret joke of Kant's soul. In: Sinnott-Armstrong W (ed) Moral psychology, vol. 3: the neuroscience of morality. MIT Press, Cambridge, pp 35–79
Greene JD, Sommerville RB, Nystrom LE, Darley JM, Cohen JD (2001) An fMRI investigation of emotional engagement in moral judgment. Science 293:2105–2108
Haidt J (2001) The emotional dog and its rational tail: a social intuitionist approach to moral judgment. Psychol Rev 108(4):814–834
Haidt J, Graham J (2007) When morality opposes justice: conservatives have moral intuitions that liberals may not recognize. Soc Justice Res 20(1):98–116
Huebner B, Dwyer S, Hauser M (2008) The role of emotion in moral psychology. Trends Cogn Sci 13(1):1–6
Jordan J, Peysakhovich A, Rand D (2015) Why we cooperate. In: Decety J, Wheatley T (eds) The moral brain: a multidisciplinary perspective. MIT Press, Cambridge, pp 123–142
Kauppinen A (2015) Moral intuition in philosophy and psychology. In: Clausen J, Levy N (eds) Handbook of neuroethics. Springer, Dordrecht, pp 169–183

Kahane G, Shackel N (2010) Methodological issues in the neuroscience of moral judgement. Mind Lang 25(5):561–582
Kahneman D, Klein G (2009) Conditions for intuitive experts: a failure to disagree. Am Psychol 64(6):515–526
Kelman M (2011) The heuristics debate. Oxford University Press, Oxford
Koenigs M, Young L, Adolphs R, Tranel D, Cushman F, Hauser M et al (2007) Damage to the prefrontal cortex increases utilitarian moral judgements. Nature 446:908–911
Koenigs M, Kruepke M, Zeier J, Newman JP (2012) Utilitarian moral judgment in psychopathy. Soc Cogn Affect Neurosci 7(6):708–714
Levy N (2007) Neuroethics: challenges for the 21st century. Cambridge University Press, Cambridge
Mathews DJH, Bok H (2015) Brain research on morality and cognition. In: Clausen J, Levy N (eds) Handbook of neuroethics. Springer, Dordrecht, pp 1151–1166
McGuire J, Langdon R, Coltheart M, Mackenzie C (2009) A reanalysis of the personal/impersonal distinction in moral psychology research. J Exp Soc Psychol 45:577–580
Mendez MF (2009) The neurobiology of moral behavior: review and neuropsychiatric implications. CNS Spectr 14(11):608–620
Mikhail J (2007) Universal moral grammar: theory, evidence and the future. Trends Cogn Sci 11(4):143–152
Mikhail J (2011) Elements of moral cognition. Cambridge University Press, New York
Poldrack RA (2011) Inferring mental states from neuroimaging data: from reverse-inference to large-scale decoding. Neuron 72:692–697
Prinz J (2015) Is the moral brain ever dispassionate? In: Decety J, Wheatley T (eds) The moral brain: a multidisciplinary perspective. MIT Press, Cambridge, pp 51–67
Racine E (2010) Pragmatic neuroethics: improving treatment and understanding of the mind-brain. MIT Press, Cambridge
Rand DG, Greene JD, Nowak MA (2012) Spontaneous giving and calculated greed. Nature 489:427–430
Rini RA (2014) Analogies, moral intuitions, and the expertise defense. Rev Philos Psychol 5:169–181
Rosas A, Koenings M (2014) Beyond 'utilitarianism': maximizing the clinical impact of moral judgment research. Soc Neurosci 9(6):661–667
Roskies A (2002) Neuroethics for the new millennium. Neuron 35(1):21–23
Rottman J, Young L (2015) Mechanisms of moral development. In: Decety J, Wheatley T (eds) The moral brain: a multidisciplinary perspective. MIT Press, Cambridge, pp 123–142
Sadler-Smith E, Sparrow P (2008) Intuition in organizational decision making. In: Hodgkinson GP, Starbuck WH (eds) The Oxford handbook of organizational decision making. Oxford Scholarship Online, Oxford, pp 305–324
Schleim S (2015) The half-life of a moral dilemma task: a case study in experimental (neuro-) philosophy. In: Clausen J, Levy N (eds) Handbook of neuroethics. Springer, Dordrecht, pp 185–199
Simon H (1987) Making management decisions: the role of intuitions and emotion. Acad Manag Exec 1(1):57–64
Suter RS, Hertwig R (2011) Time and moral judgment. Cognition 119: 454–458
Yang Q, Li A, Xiao X, Zhang Y, Tian X (2014) Dissociation between morality and disgust: an event-related potential study. Int J Psychophysiol 94:84–91
Yang Q, Yan L, Luo J, Li A, Zhang Y, Tian X et al (2013) Temporal dynamics of disgust and morality: an event-related potential study. PLoS One 8:e65094

Neuroethics and Policy at the National Security Interface: A Test Case for Neuroethics Theory and Methodology

10

Nicholas G. Evans and Jonathan D. Moreno

Abstract

Cutting-edge research in neuroscience has been frequently sponsored or conducted by elements of national security. This elicits three important but underexplored methodologically relevant issues in neuroethics: (1) the use of force through neurotechnologies, (2) the use of neurotechnologies by or on members of the national security establishment, and (3) priority setting and the ethics of pursuing basic scientific research in support of national security. We describe these issues and then give an account for how neuroethics theory will need to accommodate new sources of normative insight in order to provide answers.

Cutting-edge research in neuroscience has been frequently sponsored or conducted by elements of national security. This elicits three important but underexplored methodologically relevant issues in neuroethics: (1) the use of force through neurotechnologies, (2) the use of neurotechnologies by or on members of the national security establishment, and (3) priority setting and the ethics of pursuing basic scientific research in support of national security. We describe these issues and then give an account for how neuroethics theory will need to accommodate new sources of normative insight in order to provide answers.

N.G. Evans (✉)
Department of Philosophy, University of Massachusetts Lowell, Lowell, MA, USA
e-mail: nicholas_evans@uml.edu

J.D. Moreno
David and Lyn Silfen University Professor, University of Pennsylvania,
Philadelphia, PA, USA
e-mail: morenojd@mail.med.upenn.edu

E. Racine, J. Aspler (eds.), *Debates About Neuroethics*, Advances in Neuroethics,
DOI 10.1007/978-3-319-54651-3_10

10.1 Introduction

Modern medicine and public health have a long historical connection to national security. By "national security," we mean securing a nation-state and its citizens as both valuable for its own sake and as a means to increase and maintain the well-being of those citizens, as well as the institutions whose stated goals are accomplishing that task (Herington 2012). Statistical methods for tracking disease were pioneered by figures such as Florence Nightingale in the Crimean War, as a method of demonstrating ways to reduce fatalities in military hospitals (Winkel 2009). Public health efforts to stem the transmission of infectious disease emerged in no small part due to the effects that diseases like influenza and syphilis had on troop deployments and force strength (Watterson and Kamradt-Scott 2016). Modern radiotherapy to treat cancers owes its existence to the Manhattan Project and its successors, without which the majority of our knowledge about the effects of radiation on biological matter would not exist (Evans 2013).

Cognitive neuroscience and its progenitors have a similar relationship to national security. The use of pharmaceuticals to enhance soldier strength, resilience, or alertness dates back to antiquity: the Greeks took opium mixed with wine to calm their nerves both before and after battle, while inhabitants of the North Asian steppe would consume dried, psychoactive toadstools to enhance stamina and inure themselves against pain (Kamienski 2016). Cognitive science research on human nervous system responses to biochemical agents at times played a role, sometimes unwittingly, in the manufacture and use of chemical and biological weapons (Dando 2015). The use of both pharmacological and behavioral interventions in interrogation is long-standing; famous episodes include the US Central Intelligence Agency's experiments with lysergic acid (LSD) (Wheelis 2012) and the recent revelations of psychologist complicity in torturing captives during the US occupations of Afghanistan and Iraq (Risen 2015).

It is curious, then, that neuroethics—both the study of how neuroscience informs ethical thought and the ethics of neuroscience research and practice (Roskies 2002)—has had relatively little engagement with issues of national security. On the one hand, this isn't surprising: despite its history, bioethics has often ignored the ethical applications and implications of medicine and biology for national security. This has only changed in the last decade, with a rise in concern about "dual-use" microbiology research that could be used by aspiring bioterrorists (Evans 2015; Selgelid 2016) and with the reframing of infectious disease outbreaks such as the Ebola virus, as a national security threat (Evans 2016). Given the length of time it has taken to make modest links between national security and bioethics, it stands to reason that neuroethics is less developed on account of the youth of the discipline.

On the other hand, advances in psychology and cognitive science have been tied to military concerns in a stronger way than, say, those of genetics. War has been understood to be as much of a mental as a physical burden for thousands of years, and the use of both active interventions on soldiers (e.g., pharmacological and medical) and passive measures on civilian populations (e.g., propaganda) has been extensive. From Sun Tzu's determination that governing the "moral law of a community" is the first factor determining the outcome of battle (Tzu 2015), to attempts to "win hearts and minds" in defeating counterinsurgencies, the idea that people's minds are key features in a national security strategy is long-standing (Dixon 2009).

In this chapter, we examine national security as a challenge for neuroethics theory and methodology. After a brief survey of the interface of cognitive neuroscience and national security, we argue that national security presents a set of interesting and importantly *different* cases with which neuroethics must grapple. We identify three important areas for neuroethics to focus in dealing with national security issues: (1) the use of force through neurotechnologies, (2) the use of neurotechnologies by or on members of the national security establishment, and (3) priority setting and the ethics of pursuing basic scientific research in support of national security. We highlight the challenges these pose for neuroethics as a discipline and offer recommendations on how the discipline should proceed in tackling these important issues.

10.2 Contemporary Neuroscience and National Security

The changing nature of conflict and of societies has informed recent interest in cognitive neuroscience as a tool in promoting national security. Modern forms of global violent extremism and nontraditional conflicts, a public increasingly disillusioned with the prospect of committing conventional forces to armed conflicts, and strained budgets have led to searches for technological means of maintaining a nation's national security while limiting the resort to conventional warfare. Neuroscience along with information technology and the life sciences are seen as important avenues to contributing to national defense.

Four applications of the neurosciences are particularly attractive for use by security forces: (1) social behavior prediction, (2) behavior modification, (3) extension of capacities, and (4) neuroweapons. These research areas overlap at various points and are often able to be used together in practice.

10.2.1 Social Behavior Prediction

Social behavior prediction makes use of neurological, neuropsychological and neurobiological, and machine learning research to develop algorithmic methods for predicting human behavior. Typically, the focus of this research is to determine how groups of individuals will behave under particular stimuli or how a given stimuli will affect members of a target population. For example, the Defense Advanced Research Projects Agency (DARPA) pursued, from 2011 to 2015, a project titled *Narrative Networks (N2)*:

> Through an improved basic understanding of narrative effects, tools are being developed to detect brain activity associated with narrative influence and to emulate this activity in the context of larger environmental factors with models of narrative influence on individual and group behavior (Miranda et al. 2015: 61).

The N2 project utilized a range of neuroscientific techniques, including functional magnetic resonance imaging (fMRI), psychological testing, and brain-computer interfaces to develop an account of how different narratives (i.e., stories) influence neurological states.

In terms of behavior prediction, the N2 project provides an insight into determining how different storytelling techniques can trigger neurological activity that in turn generates particular kinds of behavior. In principle, the study of narratives allows for the analysis of media to determine how, for example, insurgent-made videos might incite violence and predict different triggers that may serve to radicalize vulnerable members of a population. When added to other sources of information, narrative analysis can help intelligence communities predict future sites of violence by monitoring the kind of narratives being used by suspects or non-state groups (Giordano and Wurzman 2011).

Other neurotechnologies such as electroencephalogram (EEG) and MRI could be used during interrogation to detect deception, despite concerns about the current validity of this technology for lie detection purposes (Tennison and Moreno 2012). These same technologies could be used during risk assessments for imprisoned or detained individuals to determine whether they present a risk for future criminal or terrorist acts (Nadelhoffer et al. 2012).

10.2.2 Behavior Modification

The N2 program also identified how narratives can *modify* behavior. In soldier training, the use of noninvasive brain-computer interfaces could allow narratives to be shaped, in real time, to the neurological responses of warfighters. This efficiently transmits information in ways that are tailored to an individual's neurology (Canli et al. 2007; Miranda et al. 2015). It could also be used in therapeutic contexts for individuals suffering from post-traumatic stress disorder (PTSD) or the so-called moral injury, a psychological trauma arising from engaging in acts or omissions in war that violate or disrupt an agent's personal morality, such as killing, maiming, or mistreating civilians (Drescher et al. 2011).

The use of narratives is also purported to be useful in modifying the behavior of insurgents or aspiring terrorists (Casebeer and Russell 2005). By using narratives that are designed to prime individuals toward de-escalating violence, terrorist acts could be averted through persuasion rather than prevented through the use of force. This would represent a significant advance in degree, if not kind, from current methods of debriefing and converting enemy combatants.

Future applications involve the use of electromagnetic interventions to enhance human cognitive and psychological capabilities. Noninvasive brain stimulation has shown some promise in modifying human cognition. Transcranial magnetic stimulation (TMS) and transcranial direct current stimulation (tDCS), involving the application of a magnetic field or electric current, respectively, to a person's skull, have been shown to enhance learning, memory, and mood (Tennison and Moreno 2012). Some have argued this could be useful in clinical or pedagogical settings in both civilian and military contexts (Sehm and Ragert 2013; Tennison and Moreno 2012). Others have posited that in the future, electrodes fitted to the inside of helmets could be used in the field to enhance warfighters by enabling them to quickly and calmly process battlefield information both through their senses and surveillance delivered electronically from nearby unmanned aerial vehicles (Canli et al. 2007).

The power and specificity of behavioral augmentation using pharmacological agents are, moreover, increasing. For example, recent studies have demonstrated that administering propanol shortly after a traumatic event can reduce the likelihood a person will develop PTSD; the same drug can reduce PTSD symptoms if administered after recalling the memory that has led to the development of PTSD (Tennison and Moreno 2012). As modern neuroscience progresses, we should expect the power of these interventions to increase, as well as the range of behaviors that can be modified.

Together, the novel areas of narrative, electromagnetic stimulation techniques, and pharmaceuticals could be combined to modify the behavior of terrorists or insurgents directly (Rose 2011). This recalls earlier attempts by the US intelligence services to develop "mind control drugs," most infamously the MK ULTRA project by the Central Intelligence Agency, albeit with greater potential for success, given early positive results. The aim here would be to replace invasive and harmful "enhanced interrogation" (i.e., torture) methods with forms of behavior modification that encourage cooperation with security forces.

10.2.3 Capacity Extension

The use of pharmaceutical or other medical interventions to enhance warfighter capacities is likewise being explored by defense organizations. Modafinil, which allows people to retain their wakefulness, memory, and executive functions under conditions of sleep deprivation (Repantis et al. 2010), is a particularly promising drug for use among fighter pilots and sentries in armed conflict zones (Moreno 2012). Modafinil straddles the fuzzy border between behavioral modification and human enhancement.

It is important to note here that the difference between behavioral modification and capacity extension is, in part, one of the strategic priorities for national security organizations. The conceptual distinction between therapies and enhancement is long debated (Anderson 1989) and interventions such as tDCS feature in the literature on so-called cognitive enhancement (e.g., Bostrom and Sandberg 2009). In terms of how national security institutions function, in the USA and elsewhere, there may be a distinction between science and technology used to treat soldiers post-deployment (say through the Veteran's Administration) and science and technology used to train or modify soldiers pre-deployment. These technologies are also treated as distinct from the perspective of the emerging literature on neuroethics and national security (Canli et al. 2007; Giordano and Wurzman 2011), and thus we leave them as separate categories with the understanding that there are numerous points at which capacity enhancement and behavior modification intersect.

The renaissance of drones in armed conflict over the last decade could be enhanced further through the use of brain-computer interfaces (BCIs; sometimes also referred to as "brain-machine interfaces or BMIs). Soldiers would be able to fly sophisticated aircraft from the comfort of home soil or from secure deployment on aircraft carriers, without risking their lives (Evans 2011). This capacity extension would extend to the use of BCIs to inform soldiers of emerging threats, provide analysts with superior integration with their data, and allow for command and control systems to be directly in touch with soldiers.

The future possibilities for the use of brain-computer interfaces to enhance soldiers by connecting them directly to their machines are potentially limitless. That is, given the development of a robust platform to translate machine code into electrical impulses readable by the brain, there is no limit in principle to the kind of device to which warfighters—or anyone else—could be connected (Evans 2011; Miranda et al. 2015). In-field combatants could have direct access to databases for retrieval of strategic information. The relationship between humans and machines could also work the other way, with implanted medical devices that detect, via a BCI, subtle changes in a warfighter's body, counteracting adverse effects or enhancing capacities to suit operational requirements.

10.2.4 Neuroweapons

The use of neuroscience to develop weapons goes back to the development of acetylcholinesterase inhibitory nerve gases used in the Second World War and later against the Kurds in 1988 (Rose 2011). Since then, a plethora of neurobiological agents, both biological and chemical, have been created with both lethal and nonlethal applications. Hypnotics such as zolpidem can cause anterograde amnesia in clinical doses, but can also cause hallucinations in high doses (Giordano and Wurzman 2011). The Moscow theater siege in 2002 was broken with a chemical agent later revealed to be a fentanyl derivative, highlighting the potential for weaponized pharmaceuticals to be used in law enforcement in addition to armed conflict (Crowley and Dando 2014).

The use of chemical and biological neuroweapons in armed conflict is prohibited under international humanitarian law. However, nonlethal or incapacitating chemical agents, including those that affect the cognition, may be stockpiled for law enforcement purposes (Crowley and Dando 2014). Into the gray area between armed conflict and law enforcement, such as long-term occupation or reconstruction efforts such as those undertaken in Afghanistan and Iraq, there is the potential for the use of nonlethal chemical agents in national defense scenarios. Recent US military policy, for example, contends that these agents may nonetheless be employed as "defensive means of warfare" for such tasks as controlling riots, dispersing civilians who are being used as shields, and conducting rescue missions (Evans and Moreno 2016). The possibility of "defensive" uses of nonlethal chemical agents has spurred interest in developing novel agents for riot control and crowd dispersal, as well as for subduing large groups that conceal insurgents

(Gross 2010). Agents under consideration for use as nonlethal weapons include opioids, benzodiazepines, alpha2-adrenoreceptor agonists, and neuroleptic anesthetics (Dando 2015; National Research Council National Academy of Engineering 2014).

10.3 Unique Challenges

The above developments present a set of ethical issues: (1) the use of force through neurotechnologies, (2) the use of neurotechnologies by or on members of the national security establishment, and (3) priority setting and the ethics of pursuing basic scientific research in support of national security. These issues, however, present a challenge for neuroethics methods and practice.

10.3.1 The Use of Force

On a conceptual level, the justifications for acts that pertain to national security are typically different from those in other spheres of endeavor. In particular, they are acts that are deliberately *liberty violating* or *coercive.* Whereas one might have concerns, for example, that the use of neurotechnology in legal proceedings might violate an individual's rights to privacy (Fisher 2010), in intelligence collection, it is precisely the privacy of individuals that is being compromised for some other end. Likewise, while death might be an unfortunate result of a surgical procedure, the death of enemy combatants is an expected, if not sought-after result in armed conflict. National security is typically justified based on the need to protect sovereignty and citizens, and the use of force is permissible if (and usually only if) it is necessary to safeguard the rights of citizens (Rawls 2001: 89–98).

Different security institutions may have coercive or liberty-violating actions proscribed on different normative bases; the use of force in law enforcement is not the same, on a normative basis, as the use of force in armed conflict. In the case of military organizations, the ethics of contemporary military actions are described by principles such as those found in just war theory: *jus ad bellum*, the ethics[1] of declaring or going to war; *jus in bello*, the ethics in war; and *jus post bellum*, the ethics/law of war's end (Allhoff et al. 2013). Within war, for example, *jus in bello* describes limits on the use of force to those acts that are necessary to advance or resolve the conflict, are proportionate to the goals sought in a particular action, and respect the rights of noncombatants such as civilians, POWs, and wounded soldiers (Walzer 2015).

In the case of law enforcement, the use of force on citizens should require considerably higher justifications, as the purpose of law enforcement is to protect the

[1] *Jus* (pronounced "y-us") is literally the Latin for "law," but *jus ad bellum/in bello/post bellum* are conventionally used to refer to the normative ethical foundations for armed conflict, in addition to customary or formal International Humanitarian Law that governs contemporary armed conflict.

human rights of citizens. Moreover, the invasion of privacy by law enforcement is bounded by the property and civil rights of citizens. It typically requires a warrant issued by an authority such as a judge in order to violate a citizen's rights (Miller et al. 2006).

The use of force by intelligence agencies is controversial, but the limits of their actions are often divided between foreign intelligence and domestic intelligence services. In each case, obligations may differ depending on the strength of obligations one owes to one's fellow citizens, compared to foreign nationals. Moreover, the ethics of intelligence collection arguably operates under similar principles to those of armed conflict: necessity and proportionality and respecting the rights of civilians (Bellaby 2012).

A central task for neuroethics is to determine how novel neurotechnologies ought to be used in these different theaters. In some cases, this may be straightforward. In keeping with the extended cognition thesis put forward by Levy, among others (Levy 2007), some technologies may be permissible or impermissible insofar as they perform the same functions as non-neurotechnologies. Neuroscience-informed social behavior prediction may thus be normatively the same as other behavior prediction algorithms (e.g., the Pentagon's aborted Total Information Awareness project) that accomplish similar goals.

Even if neurotechnology is similar to other technologies on the face of things, the details matter. Social behavior prediction, like medical diagnostic tests, has what an epidemiologist would call a "response operator characteristic." Put simply, strategies for predicting behavior will sometimes return false positives or false negatives. These are not merely technical features of a technology, but have ethical implications. Justifying the use of a test to detect whether someone is at risk of radicalization will, in part, depend on the risk of falsely identifying, or failing to identify, a person or group as potential suspects, or carrying the potential to radicalize and commit terrorist acts.

The degree to which it is permissible to use social behavior prediction grounded in neuroscience will depend in part on the ability of the science to produce useful, reliable information. Moreover, the justification for these discoveries, in practice, may depend on how useful and reliable they are relative to other kinds of intelligence collection. The act of determining if these neuroscientific developments really are the same as non-neuroscience-informed technologies, and if intelligence services are justified in using them, is a task for neuroethics. To date, little has been done to evaluate these new technologies and set lower limits on how accurate these technologies would have to be in order for their utility to outweigh other considerations, such as the rights of citizens.

Other technologies will need to be examined in terms of their effects, where these effects might challenge existing paradigms or conceptual tools. Consider neuroweapons, in particular those that use chemical or biological agents to affect the central nervous system or human cognition. Chemical and biological weapons have different legal and normative statuses in international law. The norm against the development of biological weapons is set forth in the Biological and Toxin Weapons Convention (BTWC):

Each State Party to this Convention undertakes never in any circumstances to develop, produce, stockpile or otherwise acquire, or retain:

1. Microbial or other biological agents or toxins whatever their origin or method of production, of types, and in quantities that have no justification for prophylactic, protective, or other peaceful purposes
2. Weapons, equipment, or means of delivery designed to use such agents or toxins for hostile purposes or in armed conflict (United Nations n.d.)

The first article of the BTWC is paralleled by that of the Chemical Weapons Convention (CWC):

Each State Party to this Convention undertakes never under any circumstances:

(a) To develop, produce, otherwise acquire, stockpile or retain chemical weapons, or transfer, directly or indirectly, chemical weapons to anyone
(b) To use chemical weapons
(c) To engage in any military preparations to use chemical weapons
(d) To assist, encourage, or induce, in any way, anyone to engage in any activity prohibited to a State Party under this Convention (Organisation for the Prohibition of Chemical Weapons 1993)

Where weapons, in the case of the CWC, are defined as any kind or quantity of chemical agent not intended for (a) industrial, agricultural, research, medical, pharmaceutical, or other peaceful purposes; (b) protective purposes, namely, those purposes directly related to protection against toxic chemicals and to protection against chemical weapons; (c) military purposes not connected with the use of chemical weapons and not dependent on the use of the toxic properties of chemicals as a method of warfare; or (d) law enforcement including domestic riot control purposes. The BTWC does not specify a positive definition of "weapon," but in subsequent review conferences, the parties to the Convention agreed to the broad interpretation "that the Convention prohibits the development, production, stockpiling, other acquisition or retention of microbial or other biological agents or toxins harmful to plants and animals, as well as humans, of types and in quantities that have no justification for prophylactic, protective or other peaceful purposes" (United Nations 2012).

The two disarmament conventions are bracketed by the Protocol for the Prohibition of the Use in War of Asphyxiating, Poisonous or other Gases, and of Bacteriological Methods of Warfare, otherwise known as the Geneva Protocol. The Geneva Protocol concerns the use of agents in wartime and is regarded by the International Committee of the Red Cross, among others, as a part of customary international law. The BTWC and CWC, in contrast, are in its scope, concerning the production and stockpiling of biological and toxin weapons in addition to their use, implicitly in the case of the BTWC and explicitly in the case of the CWC.

Though the use of chemical weapons in armed conflict is prohibited under the CWC, they are permitted for use in law enforcement, including riot control (Crowley and Dando 2014). One implication for this is what kinds of chemical agent might be properly used in law enforcement. A chemical agent that kills indiscriminately is unlikely to be justified under the ethics of law enforcement and thus wouldn't be permitted under the CWC's law enforcement provision. But chemical calmatives for crowd control might be permitted for law enforcement, for example, depending on the strength of their effect and any adverse reactions that occur as a result of prolonged exposure. Parsing between what is and is not permitted for development, much less use, in law enforcement thus has international implications, but also depends on complex questions about what kinds of force law enforcement is permitted to use.

Biological weapons are prohibited for use in armed conflict, and even the development of biological agents in "quantities that have no justification for prophylactic, protective, or other peaceful purpose" is prohibited by treaty. But what constitutes an appropriate protective purpose is less than clear. Moreover, both chemical and biological agents that target the nervous system have been viewed as having high utility in counterinsurgency scenarios (Gross 2010). How counterinsurgency falls in between armed conflict and law enforcement is an open topic in military ethics, which struggles to deal with the just use of "force short of war," such as the use of drone strikes or special forces strikes that fall short of the level of hostilities associated with armed conflict, but are clearly outside the bounds of law enforcement (Ford 2013). Similarly, neuroscience might contribute to a "creeping militarization" of biology and thus undermine the norm against biological weapons (Dando 2015).

A problem for military neuroethics is what kinds of effects constitute an appropriate use of force in different scenarios. In particular, conventional understandings of the use of force are tailored to kinetic weapons, such as firearms or explosives. While changes in mental states no doubt accompany the use of kinetic force, weapons that explicitly target mental states or cognition have an ambiguous status in the ethics of armed conflict and policing.

The neuroethics literature that deals with the etiology and phenomenology of pain and its relation to other mental states could be useful in parsing the kinds of harm, or violation, that count in determining the permissibility of military, police, or intelligence actions. Some headway has been made in this: for example, Justo and Erazun have argued against so-called lie detectors that utilize fMRI to determine whether a detainee is telling the truth, since it may in principle be more invasive or present a greater violation of a person's autonomy than the use of torture. They argue that while torture violates a person's autonomy and human rights, so-called lie-detectors undermine a detainee's ability to resist their captors by remaining silent or lying, thus presenting an additional violation (Justo and Erazun 2007).

By and large there is no consensus on what kinds of pain and suffering are permissible in the use of force for national security purposes. One position might be that if killing is permitted under certain conditions, then anything short of killing is also permissible under those conditions. But in general, there is a counter-presumption against weapons that simply mutilate, rather than kill, as in the case of white phosphorus or other "nonlethal" chemical irritants. An answer to the question of why certain acts might be less permissible than killing surely has a cognitive component, and it is here that neuroethics might be mobilized to make important comment on armed conflict.

10.3.2 Neuroscience and Warfighters

Ethical concerns surrounding weapons that affect mental states dovetail into human enhancement. In particular, ethical concerns arise when we consider warfighters being enhanced through neuroscience and neurotechnologies. A civilian seeking to use enhancement technologies might do so through voluntary means, including but not limited to seeking medical care; warfighters may not have the same kind of choices.

This is not an uncommon concern for the military and novel scientific technologies. As Tennison and Moreno note:

> According to the Uniform Code of Military Justice, soldiers are required to accept medical interventions that make them fit for duty. Experimental treatments are a harder case, but the US government has shown a tendency to defer to commanders in a combat situation if they think some treatment is likely to do more good than harm, even if unproven (Tennison and Moreno 2012).

In a more general sense, the military is often at the forefront of technological development, and soldiers frequently "field test" new technologies adapted to war zones, such as the rapid adoption of mine-resistant ambush-protected (MRAP) vehicles introduced to the US Army during the occupations of Afghanistan and Iraq (Eisler 2007). Warfighters, historically, are not only compliant with orders to adopt new technologies, but enthusiastically adopt anything they believe will give them an edge in battle (Kamienski 2016).

But therapies that render someone "fit for service" are not clearly those of human enhancement. While there has been considerable debate about the distinction between therapy and enhancement (Anderson 1989), there is a separate question about how enhanced we should require soldiers to be to do their job and the kinds of risks we should require them to undertake in order to meet that threshold. If we hold that soldiers are required to undertake *any* risk—the so-called unlimited liability thesis—we might think that enhancement of any kind is warranted. On the other hand, we might think that the risks imposed on soldiers must be weighed against the operational advantage a nation state has when forced into armed conflict. In this case, the value of enhancement technologies should be proportionate to the objectives of armed conflict.

It is not clear, however, that human enhancements are the same as being deployed with new armor or weapons, particularly in the case of the enhancement of warfighters' neurocognitive abilities. Invasive medical or pharmacological interventions may have a potentially devastating impact on a warfighter's cognition if something goes wrong, either in the installation or in the field: what happens if a warfighter runs out of pills in the field, and does that carry ethical weight? It is also unclear what ought to happen after a tour of duty: whether a warfighter should be required to surrender her enhancements (or whether the military is permitted to require this surrender), and what implications this might have for warfighters returning from service.

Answering "what should happen to enhanced soldiers after they return home" requires balancing two considerations. First, the degree to which human neuroenhancements might constitute a reframing of an individual's agency: how modifying a person's cognition through, say, expanding their ability to process information, the range of sensory inputs they can comprehend, or their emotional states under stress change the warfighter's sense of self. This question of significance is both absolute and relative to other cognitive changes beyond a warfighter's service (e.g., growing old or infirm) or noninvasive changes that might simulate cognitive enhancement (e.g., using night vision goggles, rather than having one's sight medically modified). In either counterexample, it is arguably not the case that growing old or surrendering one's equipment are widely held to be morally salient features of a warfighter's life. The degree to which we should treat enhancements as different thus remains an open question.

In the balance lie what burdens military personnel must assume in pursuit of maintaining a nation's security, and what support a society reciprocally owes them as they deploy to and return from combat. Given that veterans often suffer serious psychosocial challenges reintegrating into society post-deployment (Sayer et al. 2010), how we weigh these obligations against the kinds of changes neurocognitive enhancement might bring to warfighters will determine everything from the interventions we could reasonably expect them to adopt, to the kinds of care and support they are entitled to on their return. It may also inform the kinds of neurocognitive advances we think are permissible or most valuable to pursue and develop.

10.3.3 Post-ELSI Ethics

A feature of the study of the ethical, legal, and social implications (ELSI) of emerging science and technology, historically, is that it is exclusively concerned with the *downstream* ethical, social, and legal implications of scientific research. It is not equipped—intentionally so—to deal with questions of whether or not certain scientific projects ought to be pursued (Kitcher 2003).

This is a huge problem for any inquiry, including neuroethical inquiry that deals with national security. The above challenges have implications for the development of certain technologies in addition to their use. In the case of neuroweapons, some developments may run contrary to major international arms control treaties. In the case of warfighters, the health of service personnel may lie in the balance. There is at least a *prima facie* case for thinking about this as we consider our priorities vis-à-vis enhancement.

The quantity of investment in neuroscience that passes through the military motivates the conclusion that neuroethics has a particularly strong need for this kind of analysis, relative to other areas of inquiry. A full third of the US Brain Research through Advancing Innovative Neurotechnologies (BRAIN) Project's 2016 funding, for example, is allocated to DARPA (Office of Science and Technology Policy 2015). While national security institutions certainly fund other biomedical research, these funding pools are comparatively small: even biodefense research, which has significant investment from defense agencies, is dominated by civilian agencies (Boddie et al. 2015). What makes this a distinct challenge for neuroscience is the strong but largely unexplored connection that already exists between neuroscience and national security.

A core methodological challenge for neuroethics is thus developing the tools to render normative analysis on proposed projects in the neurosciences. In particular, applying normative tools to scientific questions—asking not only what science is valuable *qua* scientific knowledge but *qua* human values—is an important and necessary development in neuroethics. This is not an insignificant challenge. Thus far, contributions to applied ethics that seek to identify in advance valuable lines of scientific inquiry from an ethical perspective are relatively rare (see, e.g., Brown and Evans 2016). This new set of tools, to examine and critique projects as or before they occur, would need to infer possible technological developments from particular experimental designs and chart trajectories of research programs and their implementation by national security organizations, as well as both the ease and consequences of their misuse.

10.4 Neuroethics Theory and Methodology

These three cases provide a series of methodological and theoretical challenges for neuroethics. In the case of neuroweapons, neuroethics needs to incorporate the conceptual frames of military ethics and the policy landscape of international humanitarian law into its existing conceptual frameworks in order to make progress. In the case of enhancement and warfighters, the unique challenges warfighters face as a specific population need to be considered. And in the case of post-ELSI ethics, neuroethics needs to grow beyond other fields of ethical inquiry and critically analyze the trajectory of neuroscience, in addition to its products.

Detailing how to respond to these challenges is beyond the purview of our work, but here we provide some insights into how neuroethicists might go about answering these challenges.

First, neuroethics needs to broaden its scope to accommodate scholarship on the ethics of armed conflict and policing. These are necessary to answer questions about the ethics of using neurotechnologies—whether initially designed for civilian use or purpose built for supporting national security institutions—to violate the rights of others. This literature falls outside of the historical antecedents of neuroethics, which we take to be *inter alia* biomedical ethics and philosophy of mind.

Second, neuroethics must engage with the literature on dual-use research. Given that major neuroscience basic research efforts such as the US BRAIN Project involve significant investment in military or defense-related organizations, such as DARPA, the technologies developed are likely to have the capacity to either help or harm humanity. There is an existing literature on dual-use in biomedical ethics (see, e.g., Selgelid 2016), but so far contributions that focus on neuroscience have been sparse (Bartolucci and Dando 2013; Dando 2013; Moreno 2011, 2013; Resnik 2007; Tennison and Moreno 2012). Providing an analysis of dual use from the perspective of neuroethics could incorporate progress from biomedical ethics while offering a novel perspective on dual-use.

As neuroscience develops into a tool for reflecting on policy, neuroethics at the intersection of neuroscience and national security will inevitably have to deal with international arms treaties and International Humanitarian Law. This is a nontrivial addition to any discipline, as these topics are themselves major components of law, philosophy, and social science. This is a ripe area for development; however, the BTWC is presently a stagnant treaty, mired even this year in a lack of consensus between state parties to the Convention, struggling to make progress on issues such as whether and how to periodically review advances in science and technology for their capacity to benefit or run contrary to the Convention (Dando 2013). The introduction of a new breed of activist scholars in the form of neuroethics would be a new venue for neuroethicists to conduct their work, but also a welcome addition to the academic observers and nongovernmental organizations that work for change around major arms control treaties.

Finally, neuroethics has an opportunity to branch out into the post-ELSI world ahead of allied disciplines such as bioethics. To date, there has been little normative analysis of research projects before they are initiated. Here, neuroethics enters a largely underdeveloped territory for *any* applied ethical discipline. However, the literature on risk analysis in the realm of biosecurity may provide a fruitful link to analyzing neuroscience research looking forward (Haas 2002), as would recent attempts to provide forward-looking risk assessments of potentially dangerous virology experiments (Gryphon Scientific 2016; Selgelid 2016).

Whether or not neuroethics is or ought to be a distinct discipline (Parens and Johnston 2007), our discussion of the national security implications of neuroscience demonstrates that neuroethics needs to pay attention to a wide range of existing ethical disciplines in order to tackle ethical issues presented by emerging neuroscience. This is a tall order for any discipline. In our view, the best way to incorporate this knowledge is through capacity building: bringing new scholars into the discipline of neuroethics whose expertise can be brought to bear on topics of interest. In the case of national security, international relations, science and technology studies, and security and peace studies, to name a few, provide the best chance to bridge the knowledge gaps that would hamper a full analysis of some of these issues.

Conclusion

In this chapter, we highlighted some of the implications of modern neuroscience for national security: behavior prediction, behavior modification, capacity extension, and neuroweapons. We then identified three areas in which neuroethics is ideally suited to address these implications—neuroweapons and armed conflict, neuroscience and warfighters, and post-ELSI ethics—and the challenges the discipline faces to provide satisfactory answers.

As a developing discipline, neuroethics has an opportunity to tackle these challenges. It is, moreover, a socially valuable area of research. Institutions that are designed to promote national security such as the military, law enforcement, and intelligence agencies can also diminish security when they are ineffective, vulnerable, or corrupt. Given the rush of interest in neuroscience by the national security establishment, we advocate the development of the tools and theory of neuroethics to attack these issues head on.

Acknowledgment This work was funded by the Greenwall Foundation Making a Difference Grant "Dual-Use Neurotechnologies and International Governance Arrangements: A Mentored Project."

References

Allhoff F, Evans NG, Henschke A (2013) Not just wars: expansions and alternatives to the just war tradition. In: Allhoff F, Henschke A (eds) The Routledge handbook of ethics and war. Routledge, New York, pp 1–8

Anderson WF (1989) Human gene therapy: why draw a line? J Med Philos 14(6):681–693

Bartolucci V, Dando M (2013) What does neuroethics have to say about the problem of dual use? In: Rappert B, Selgelid MJ (eds) On the dual uses of science and ethics. ANU E-Press, Canberra, pp 29–44

Bellaby R (2012) What's the harm? The ethics of intelligence collection. Intell Nat Secur 27(1): 93–117

Boddie C, Sell TK, Watson M (2015) Federal funding for health security in FY 2016. Health Secur 13(3):186–206

Bostrom N, Sandberg A (2009) Cognitive enhancement: methods, ethics, regulatory challenges. Sci Eng Ethics 15(3):311–341

Brown R, Evans NG (2016) The social value of candidate HIV cures: actualism versus possibilism. J Med Ethics

Canli T, Brandon S, Casebeer W, Crowley PJ, DuRousseau D, Greely HT et al (2007) Neuroethics and national security. Am J Bioeth 7(5):3–13

Casebeer WD, Russell JA (2005) Storytelling and terrorism: towards a comprehensive 'counter-narrative strategy'. Strategic Insights 4(3):1–16

Crowley M, Dando MR (2014) The incapacitating chemical agents loophole. http://thebulletin.org/incapacitating-chemical-agents-loophole7750. Accessed 4 Mar 2016

Dando MR (2013) Educational implications of the potential for hostile applications of advances in neuroscience. In: Allhoff F, Henschke A (eds) The Routledge handbook of ethics and war. Routledge, London, pp 312–324

Dando M (2015) Neuroscience and the future of chemical-biological weapons. Springer

Dixon P (2009) "Hearts and minds?" British counter-insurgency from Malaya to Iraq. J Strateg Stud 32(3):353–381

Drescher KD, Foy DW, Kelly C, Leshner A, Schutz K, Litz B (2011) An exploration of the viability and usefulness of the construct of moral injury in war veterans. Traumatology 17(1):8–13

Eisler P (2007) The truck the Pentagon wants and the firm that makes it. http://usatoday30.usatoday.com/news/military/2007-08-01-force-protection-mraps_N.htm. Accessed 28 Aug 2016

Evans NG (2011) Emerging military technologies: a case study in neurowarfare. In: Tripodi P, Wolfendale J (eds) New wars and new soliders: military ethics in the contemporary world. Ashgate, Farnham, pp 105–116

Evans NG (2013) Contrasting dual-use issues in biology and nuclear science. In: Selgelid MJ, Rappert B (eds) On the dual uses of science and ethics. ANU E Press, Canberra, pp 255–273

Evans NG (2015) Dual-use decision making: relational and positional issues. Monash Bioeth Rev 32(3–4):268–283
Evans NG (2016) Ebola: from public health crisis to national security threat. In: Lentzos F (ed) Biological threats in the 21st century. Imperial College Press, London, pp 277–292
Evans NG, Moreno JD (2016) When neuroscience leads to neuroweapons. Bull At Sci. http://thebulletin.org/when-neuroscience-leads-neuroweapons9962. Accessed 19 Dec 2016
Fisher CE (2010) Brain stimulation and national security: considering the narratives of neuromodulation. AJOB Neurosci 1(2):22–24
Ford SB (2013) Jus Ad Vim and the just use of lethal force-short-of-war. In: Allhoff F, Henschke A (eds) Children of capital: eugenics in the world of private biotechnology. London, pp 63–75
Giordano J, Wurzman R (2011) Neurotechnologies as weapons in national intelligence and defense—an overview 2(1):T55–T71. http://www.synesisjournal.com/vol2_no2_t1/2011_2_1_Giordano_Wurzman_abstract.html
Gross ML (2010) Medicalized weapons & modern WAR. Hastings Cent Rep 40(1):34–43
Gryphon Scientific (2016) Risk and benefit analysis of gain of function research. Grphyon Scientific, Washington
Haas CN (2002) The role of risk analysis in understanding bioterrorism. Risk Anal 22(4): 671–677
Herington J (2012) The concept of security. In: Ethics and security aspects of infectious disease control. pp 7–26
Justo L, Erazun F (2007) Neuroethics and human rights. Am J Bioeth 7(5):16–18
Kamienski L (2016) Shooting up. Oxford University Press, New York
Kitcher P (2003) Science, truth, and democracy. Oxford University Press
Levy N (2007) Neuroethics. Cambridge University Press, Cambridge
Miller S, Blackler J, Alexandra A (2006) Police ethics. Allen & Unwin, Sydney
Miranda RA, Casebeer WD, Hein AM, Judy JW, Krotkov EP, Laabs TL et al (2015) DARPA-funded efforts in the development of novel brain–computer interface technologies. J Neurosci Methods 244:52–67
Moreno J (2011) Guest editorial: national security in the era of neuroscience. Synesis 2:T3–T4
Moreno JD (2012) Mind wars. Bellvue Literary Press, New York
Moreno JD (2013) Mind wars. Brain science and the military. Monash Bioeth Rev 31(2):83–99
Nadelhoffer T, Bibas S, Grafton S, Kiehl KA, Mansfield A, Sinnott-Armstrong W et al (2012) Neuroprediction, violence, and the law: setting the stage. Neuroethics 5(1):67–99
National Research Council, National Academy of Engineering (2014) Emerging and readily available technologies and national security. National Academies Press, Washington
Office of Science and Technology Policy (2015) Obama administration proposes over $300 million in funding for the BRAIN initiative. https://www.whitehouse.gov/sites/default/files/microsites/ostp/brain_initiative_fy16_fact_sheet_ostp.pdf. Accessed 28 Nov 2016
Organisation for the Prohibition of Chemical Weapons (1993) Convention on the prohibition of the development, production, stockpiling and use of chemical weapons and on their destruction. OPCW
Parens E, Johnston J (2007) Does it make sense to speak of neuroethics? Three problems with keying ethics to hot new science and technology. EMBO Rep 8(1S):S61–S64
Rawls J (2001) The law of peoples. Harvard University Press, Cambridge
Repantis D, Schlattmann P, Laisney O, Heuser I (2010) Modafinil and methylphenidate for neuroenhancement in healthy individuals: a systematic review. Pharmacol Res 62(3):187–206
Resnik DB (2007) Neuroethics, national security and secrecy. Am J Bioeth 7(5):14–15
Risen J (2015) Outside psychologists shielded U.S. torture program, report finds. http://www.nytimes.com/2015/07/11/us/psychologists-shielded-us-torture-program-report-finds.html?_r=0. Accessed 28 Aug 2016
Rose S (2011) A Faustian bargain? EMBO Rep 12(11):1086–1086
Roskies A (2002) Neuroethics for the new millenium. Neuron 35(1):21–23
Sayer NA, Noorbaloochi S, Frazier P, Carlson K, Gravely A, Murdoch M (2010) Reintegration problems and treatment interests among Iraq and Afghanistan combat veterans receiving VA medical care. Psychiatr Serv 61(6):589–597

Sehm B, Ragert P (2013) Why non-invasive brain stimulation should not be used in military and security services. Front Hum Neurosci 7:553

Selgelid MJ (2016) Gain-of-function research: ethical analysis. Sci Eng Ethics 22(4):923–964

Tennison MN, Moreno JD (2012) Neuroscience, ethics, and national security: the state of the art. PLoS Biol 10(3):e1001289

Tzu S (2015) The art of war. Ediciones LEA

United Nations (2012) Additional agreements reached by previous review conferences relating to each article of the convention. http://www.unog.ch/80256EDD006B8954/(httpAssets)/EBB7A76E3DC19651C1257B6D003A0028/$file/BWC%20&%20Additional%20Agreements%20Post%207RC.pdf. Accessed 28 Nov 2016

United Nations (n.d.) Convention on the prohibition of the development, production and stockpiling of bacteriological (biological) and toxin weapons and on their destruction

Walzer M (2015) Just and unjust wars. Basic Books, New York

Watterson C, Kamradt-Scott A (2016) Fighting flu securitization and the military role in combating influenza. Armed Forces Soc 42(1):145–168

Wheelis M (2012) The use and misuse of LSD by the U.S. Army and CIA. In: Tucker JB (ed) Innovation, dual use, and security. MIT Press, Cambridge, pp 289–302

Winkel W Jr (2009) Florence Nightingale. Epidemiology 20(2):311

Part III

The Future of Research Programs, Training, and International Neuroethics

Section Introduction: The Future of Research Programs, Training, and International Neuroethics

11

Eric Racine and John Aspler

Abstract

In this introduction to the section on the future of research programs, training, and international neuroethics, we explain how neuroethics pedagogy, research, and collaboration hinge on different views of the aims and academic boundaries of neuroethics. The six chapters in this section offer an array of contributions that critically reflect on different training programs and institutions around the world dedicated to neuroethics, with an eye toward the future.

For academic disciplines, the establishment of departments has long served as a means to formalize the existence of an academic area of inquiry. The curricula pursued within departments contribute to the perpetuation of traditions of scholarship and propel the progress and adaptation of disciplines in the face of new research objectives and methods. Depending on the nature of the field or discipline, the maintenance of rigid disciplinary departments can also lead to stagnation when scholars refuse to reject outdated or discredited approaches in favor of dogma and history. In contrast, important interdisciplinary fields of research, like gender studies, environmental sciences, and bioethics, have not developed following the traditional disciplinary approach. For example, in the case of bioethics, training

E. Racine (✉)
IRCM, Montréal, QC, Canada

Université de Montréal, Montréal, QC, Canada

McGill University, Montréal, QC, Canada
e-mail: eric.racine@ircm.qc.ca

J. Aspler
IRCM, Montréal, QC, Canada

McGill University, Montréal, QC, Canada
e-mail: John.Aspler@ircm.qc.ca

E. Racine, J. Aspler (eds.), *Debates About Neuroethics*, Advances in Neuroethics,
DOI 10.1007/978-3-319-54651-3_11

programs have long remained—and still are—largely interdisciplinary. Bioethics research still borrows extensively from the empirical methodologies of other disciplines, although there are perhaps theoretical approaches now more clearly associated with bioethics (e.g., narrative ethics, casuistry).

The recognition of the status of a discipline has advantages, but also has drawbacks. For example, the dynamism of interdisciplinary research can offer a younger generation of researchers and scholars a way out of potentially asphyxiating disciplinary environments and dogmatic traditions, allowing them to engage in more open-ended and open-minded collaborations. Sometimes, the problems at hand (e.g., gender inequalities, environmental threats) are simply not fully addressed by existing disciplines. In the best cases, interdisciplinary areas of inquiry like neuroethics can crystallize synergies across disciplines, can mobilize resources from otherwise incompatible perspectives, and can produce richer understandings of and approaches to unique sets of problems. Consequently, the creation of research centers and interdisciplinary training programs may be crucial to solidifying the notion that neuroethics is a truly interdisciplinary field. Likewise, the development of meaningful and impactful international research activities, which show distinctive value and impact, may be equally critical for neuroethics.

This third section of the book features the experiences of leaders involved in the establishment of dedicated training and research programs in neuroethics as well as other initiatives. They critically reflect on the progress realized thus far and explore where we should head next, notably pointing to the need to increase international diversity and representation in neuroethics. In Chap. 12, *Models of Engagement in Neuroethics Programs: Past, Present, and Future*, Laura Specker Sullivan, affiliated with the Neuroethics Thrust at the Center for Sensorimotor Neural Engineering of the University of Seattle, and Judy Illes, a neuroethics pioneer who led the Stanford Program in neuroethics (2001–2007) and then developed the Neuroethics Core at the University of British Columbia (2007–), reflect on ethics engagement in current neuroethics training programs. They review several models, their strengths and weaknesses, and discuss the programs in which they are themselves involved. In Chap. 13, *The Future of Neuroethics Research and Training*, philosopher of mind and ethicist Tom Buller reflects on the direction neuroethics ought to take and how training and education can help achieve these goals. He makes a number of critical observations about what it means for neuroethics that neuroscience is scaling up (i.e., in the form of the BRAIN Initiative in the USA and the Human Brain Project in Europe). In Chap. 14, *Growing Up with Neuroethics: Challenges, Opportunities and Lessons from Being a Graduate Student at a Disciplinary Crossroads*, Cynthia Forlini, an early-career neuroethics scholar trained in Canada and now based at the University of Sydney, reflects on her personal experience with the challenges and opportunities involved in studying in an emerging area of inquiry. She offers all readers, whether they are students, researchers, or observers of the development of neuroethics, a number of important questions to ponder in the development of new research and training programs. Next, Ralf Jox and Sabine Müller, two neuroethics leaders in Europe, report on European developments in neuroethics in Chap. 15, entitled *Neuroethics Research in Europe*. They discuss a number of important questions associated with the institutionalization of neuroethics in European academic contexts (e.g., less formal recognition of bioethics and neuroethics within academic institutions), which differ from North American contexts in significant and important ways. In Chap. 16, *Neuroethical Engagement on Interdisciplinary and International Scales*, philosophers John Shook and James Giordano review a series of international advances and argue for a more inclusive and international neuroethics. Giordano and Shook invite readers to think about several topical areas of neuroethics, with an eye for the international dimensions of a neuroethics that they already consider to be a discipline. Finally, in Chap. 17, *The Biopolitics of Neuroethics*, philosopher-bioethicist Fabrice Jotterand, an expert on psychiatric ethics from the University of Wisconsin, and the talented doctoral candidate Marcello Ienca, from the University of Basel, explore how neuroethics discourse can be shaped by political agendas—as well as how, in turn, neuroethics (especially the neuroscience of ethics) can shape political agendas and public discourse. They strongly recommend the recognition of diverse national research agendas in neuroethics and give an additional argument in favor of pluralism in neuroethics. The six chapters in this third section reflect a broad range of perspectives focused on the future of neuroethics, in the form of training programs and international inclusion.

12 Models of Engagement in Neuroethics Programs: Past, Present, and Future

Laura Specker Sullivan and Judy Illes

Abstract

This chapter surveys models for ethics engagement in neuroscience research and training and offers suggestions for future development. We review different ways in which ethics programs have sought to partner with neuroscience programs and critically align neuroscience with its ethical, legal, social, and policy implications. We compare the methods of neuroethics programs with other ethics programs associated with the biomedical sciences to elucidate current levels of interdisciplinary collaboration and the potential for expansion in the future. Based on our findings, we explore ways by which neuroethics can proactively seek new approaches to link with neuroscience and medicine.

12.1 Introduction

There are countless ways to account for differences between analytic science and humanistic values. David Hume perhaps encapsulated the divide best: an "ought" cannot be derived from an "is." Scientific statements about the world are descriptive, not prescriptive—they can provide objective facts about the natural world, but not proposals for future action. In response, guidance and training have

L.S. Sullivan (✉)
National Core for Neuroethics, University of British Columbia, Vancouver, BC, Canada

Center for Sensorimotor Neural Engineering, University of Washington, Box 37, 1414 NE 42nd Street Suite 204, Seattle, WA 98105-6271, USA

Department of Philosophy, University of Washington, Seattle, WA, USA
e-mail: specker@uw.edu

J. Illes
National Core for Neuroethics, University of British Columbia, Vancouver, BC, Canada
e-mail: jilles@mail.ubc.ca

E. Racine, J. Aspler (eds.), *Debates About Neuroethics*, Advances in Neuroethics,
DOI 10.1007/978-3-319-54651-3_12

become increasingly available to ensure that science engages social, political, and ethical values alongside advances in science themselves. Medical experimentation at Nuremberg, the hydrogen bomb, the Tuskegee Syphilis Studies, and Dolly the sheep are but a few examples of events in the history of biomedicine that have occasioned formal professional discussion and additional public debate. They have given rise to institutional review boards (also known as research ethics boards and research ethics committees), government regulation, and specific ethics training programs that, in this paper, we call Ethics Engagement Programs (EEPs). EEPs are different from Ethics *Education* Programs, such as those supported by the United Nations Education, Scientific, and Cultural Organization (UNESCO). Specifically, EEPs focus on active public and scientific engagement with particular ethical issues, while Ethics Education Programs focus on the development of pedagogical methods for teaching ethical principles and policies (Have 2006). In the 1990s, the programs we describe as EEPs focused on genetics; efforts in the 2000s concentrated on nanotechnology and synthetic biology; today, as the United States Presidential Commission for the Study of Bioethical Issues reported in *Gray Matters* (2014 and 2015), it is imperative for ethics to engage with neuroscience.

The *Gray Matters* reports explicitly acknowledge the importance of ethics engagement in neuroscience, neurology, and neural engineering research and training. Volume 1 offers three recommendations: (1) integrate ethics early and explicitly throughout research, (2) evaluate existing and innovative approaches to ethics integration, and (3) integrate ethics and science through education at all levels. Volume 2 focuses on the establishment and funding of multidisciplinary efforts to support neuroscience and ethics research and education. We take up the *Gray Matters*' recommendations as the benchmarks for evaluating the strengths and gaps in approaches to ethics integration broadly in the biosciences and in neuroscience specifically today.

12.2 Ethics Engagement Models Outside Neuroethics

We identify three types of models for ethics engagement with medicine, science, and technology. Organized roughly in the chronological order in which they first appeared, the first model connects researchers and the public, the second connects researchers and ethicists, and the third encourages ethically reflective research.

12.2.1 Connecting Researchers with Public Values

EEPs initially aimed to create a bridge between scientific expertise and deliberation about public values in two forms: (1) Ethical, Legal, and Social Implications (ELSI; ELSA [Ethical, Legal, and Social Aspects] in Europe) programs such as those implemented in the 1990s as part of the Human Genome Project, and (2) citizen engagement programs that aim to include the public in decisions about the development of new and emerging technologies. Both types of models are top-down: they are initiated by national governments and support ethical inquiry through grants for ELSI research and public engagement.

12.2.1.1 Ethical, Legal, and Social Implications (ELSI)

Established in the context of the Human Genome Project in the United States, ELSI programs originally operated downstream of, or in parallel with, scientific investigation, focusing on the social consequences of related innovation. These social consequences, described in terms of the implications of the scientific investigation and innovation, are then disseminated broadly to the public, government bodies, and even back to researchers themselves. The goals of these efforts were and remain to ensure that scientific advancement proceeds responsibly by balancing the potential for new knowledge with social welfare and possible harms (Juengst 1991). The early focus of these programs were described as broad and diffuse (McEwen et al. 2014), leading to criticism for failure to provide policy recommendations (McCain 2002). Based on this criticism, ELSI programs were refreshed in 2011 to: "foster basic and applied research on the ethical, legal, and social implications of genetic and genomic research for individuals, families, and communities" (https://www.genome.gov/10001618/the-elsi-research-program/).

A major benefit of ELSI programs has been the substantial research funding furnished to social scientists, humanists, and other experts to study issues related to genomics in health care, research, society, and law and public policy. The downside is that ELSI programs still: (1) concentrate research funding on a single field, in this case genomics (and genethics, which is to genomics what neuroethics is to neuroscience), and (2) operate separately from the scientific research itself. As to the former concern, some bioethicists have suggested that, "genethics scholarship is skewing research agendas and luring scholars away from the study of other topics that are as important, or more significant, than the careful consideration of genethics" (Turner 2003). These other topics include social issues such as malnutrition and poverty. Contemporary ELSI programs, however, continue to investigate niche fields of scientific innovation such as nanotechnology (Fisher 2005; Bennett and Sarewitz 2006) that some consider to be to the detriment of other worthy topics, such as meaningful solutions to homelessness (Turner 2003).

In regard to the latter concern, ELSI research often focuses on ethical issues that arise at the level of development, application, or use of scientific innovation, rather than the scientific decision-making process itself (Fisher 2005). ELSI programs might be conceived as responding to the scientist's query: "What should I know in order to conduct my (otherwise valuable) work in a socially responsible way?" (Juengst 1991: 68). There is an implicit assumption that the decision to pursue certain scientific projects rather than others is value-free, or is not an area where ethics ought to intervene. This raises questions about whether decisions concerning scientific priorities are less ethically significant than decisions about implementation of scientific developments.

12.2.1.2 Public Engagement

Public engagement models operate upstream of science by aiming to integrate citizens and public values into the decision-making process about the science itself. The Netherlands, Denmark, the United Kingdom (UK), Canada, and the United States (among others) have made progressive attempts, although not always successfully, to include the public in scientific decision-making (Leroux et al. 1998; Dickinson 2002; Abelson and Eyles 2004; Jensen 2005; Van Est 2011).

The Netherlands has the longest-running program for public engagement with science, beginning in 1981 with a series of government-organized public debates (Broad Societal Discussions, or BMDs) on energy policy that continued until 1984. This was an opportunity for citizens to state their opinions about nuclear energy. Unfortunately, the results of this discussion were not taken into account in policy formation; while the public rejected nuclear energy, it was nevertheless adopted (Van Est 2011).

The Dutch government tried again in the 1990s with a series of BMDs on cloning, xenotransplantation, and genetically modified food (Van Est 2011). While incorporating creative public engagement activities such as focus groups and local debates, this effort also did not affect policy: the Dutch government had already published a policy plan before the debates began. More recent attempts at public engagement in the Netherlands have worked to incorporate public engagement activities upstream of scientific decisions, such as through an independent committee on nanotechnology, but concrete engagement efforts have not yet taken place.

In contrast, the Danish Board of Technology, now the DBT Foundation, offers a more successful model than that of the Netherlands. Created in 1986, the DBT Foundation works to engage public citizens in debate and capacity building on a range of scientific issues. These include global warming, security technology, population growth (http://www.tekno.dk/theme/labor-market/?lang=en), and electronic patient records (Jensen 2005).

One way in which the DBT initially conducted ethics engagement was through consensus conferences in which experts provided introductory explanations of an issue to a citizen panel that raised questions about the issue. Other methods included dialogue meetings, future workshops on local issues, and citizen summits to test policy options. These questions were then brought to a separate set of experts who would respond at a live conference with the citizen panel. Finally, the citizen panel would create a consensus document for the Danish parliament to use as a policy recommendation. At the time, the Ministry of Research supervised the DBT and ensured that its citizen recommendations were considered by parliament (Jensen 2005). Most recently, the DBT Foundation has taken up issues in neuroethics, in particular, dual use issues in which a given technology can be applied towards both public and military uses and related to computational neuroscience and robotics in the context of the EU Human Brain Project.

Compared with the Netherlands and Denmark, the UK has been relatively late in instituting public engagement activities. Early efforts focused on public understanding of science (PUS) rather than public engagement with science (PES). The latter model did not take shape until 2003, with a series of public debates on the ethics of genetically modified food. However, it appears that the government largely ignored these debates in their deliberations about policy (Nature 2004). More recently, the UK has advocated an upstream model of public engagement, with the government describing a commitment "to enable [public] debate to take place upstream in the scientific and technological development process, and not downstream where technologies are waiting to be exploited, but may be held back by public skepticism brought about through poor engagement and dialogue on issues of concern" (Department of Trade and Industry quoted in Macnaghten et al. 2005). An interesting implication of this statement is that the UK views public engagement as not just a method for citizens to advise the government, but for citizens themselves to contribute to public education.

In contrast with the stops and starts of the Netherlands and the UK, Canada has had one of the most successful public participation programs (Dickinson 2002; Abelson and Eyles 2004), including micro-level public participation in regional health care programs, public consultation on xenotransplantation (Einsiedel and Ross 2002) and genomics governance (Burgess 2004; Page et al. 2016), and deliberative public engagement on biobanking (Burgess et al. 2008; O'Doherty and Hawkins 2010; Mascalzoni et al. 2016). For example, in the case of xenotransplantation, the transplantation of cells, tissues, or organs from one species to another, a deliberative democracy tool known as a citizen jury was used to create a citizen forum in which public citizens and experts came together to discuss scientific and ethical issues, ultimately concluding that Canada should not proceed with clinical trials in this area (Einsiedel and Ross 2002). While it is unclear what influence this conclusion had on policy in this case, a similar deliberative model for engagement about biobanking resulted in ongoing discussions between the BC biolibrary and the researchers responsible for initiating public engagement (Burgess et al. 2008).

Finally, the USA has been unusually slow compared to other countries with regard to implementing public engagement programs. There has been an emphasis on regulation and ethical expertise in deliberations about the direction of scientific progress in response to the perceived democratic challenges posed by scientific and technological innovation, although this may be changing. These challenges include possible infringement on human rights and values, disruption of established governmental categories, and an increasing gap between public and expert knowledge (Jasanoff 2011).

However, the USA may be seeing progress through public and private efforts. In 2003, the American Association for the Advancement of Science created the Center for Public Engagement with Science and Technology (Leshner 2003). Many of the Center's current public engagement efforts are training opportunities for scientists to learn how to engage with the public, rather than proactive public engagement programs per se.

The Building with Biology Program may provide more substantial engagement with the American public, being described as "a community of informal science educators, researchers, and scientists dedicated to developing innovative resources, practices, and processes to build the capacity of the field to use PES activities to extend STEM (science, technology, engineering, and math) learning about science, technology, and societal implications through public and scientist dialogue about synthetic biology" (http://buildingwithbiology.org). Similarly, the Jefferson Center, a private non-profit in Minnesota, has spearheaded public engagement efforts such as citizen juries since 1971 (http://jefferson-center.org/our-history/).

The benefits of public engagement programs are the direct connections they foster between scientists and the public. In theory, they integrate ethics by incorporating public values into scientific decision-making. In practice, integration has been less than complete, with public values acknowledged but not always utilized in policy decisions.

12.2.2 Connecting Researchers and Ethicists

The second type of model for ethics engagement in science works to connect researchers and ethics experts and also takes two forms: (1) research consultation service programs that facilitate scientists' ethical deliberation through consultation with ethics experts, and (2) either direct or indirect collaboration between science and ethics researchers on projects of mutual interest.

12.2.2.1 Research Consultation Services

Research consultation services (RCSs) operate midstream in scientific decision-making and scientific endeavors, in part to correct for the historically adversarial relationship between institutional review boards (IRBs) or research ethics boards (REBs) and researchers and the downstream position of many ELSI programs (Cho et al. 2008). Research ethics consultation services "provid[e] clinical and translational researchers with access to ethics specialists with the knowledge and expertise needed for addressing complex ethical issues in biomedical research" (Sharp et al. 2015: 615). They emphasize inter-professional discussion on concrete project-related issues. The RCS model was originally developed at Stanford University as a hybrid approach that combines individual and group consultations. It consciously draws from clinical ethics consultation models: to initiate a consult, a researcher must make a request to the team, which includes experts in philosophy, law, and biology. The initial team member contacted then decides whether to conduct an individual or team consultation. The content of consultations varies from narrow questions about particular research practices to broader questions about whether specific studies should even be initiated (Cho et al. 2008). While these types of conversations may have been taking place informally between researchers and bioethicists for decades, RCSs provide an opportunity to formalize them and measure the successes and failures of the interactions.

One downside of RCSs is the blurring of lines that distinguish them from IRBs or REBs. Another downside is the tenuous protection in place for confidentiality when team members have affiliations in multiple academic departments with conflicting obligations (Sharp et al. 2015: 615). A more complex limitation is that researchers must be able to recognize ethically salient issues in order to trigger a consult, a skill that presumes some degree of ethical fluency from the outset. Even without this ethics fluency, however, benefits arise from the ethics learning and mutual engagement across teams and disciplines that occurs once a consult is initiated.

12.2.2.2 Collaborative Adjacency

In contrast to the request-based model of RCSs, programs that aim at collaborative adjacency (Rabinow and Bennett 2012) bring scientists and humanists together to work on different dimensions of the same project. The paradigmatic (albeit failed) case for collaborative adjacency is SynBERC, a multidisciplinary synthetic biology research center funded by the National Science Foundation and created by four major research universities: University of California Berkeley, Massachusetts Institute of Technology, Harvard University, and University of California San Francisco.[1] From its inception, this center included a human practices thrust focused on critically examining "the ways in which synthetic biology is contributing or failing to contribute to the promised near future through its eventual input into medicine, security, energy, and the environment" (Rabinow 2009: 304). Rabinow and Bennett, the leaders of the ethics thrust, describe themselves as correcting the errors of the ELSI project. As with Cho's RCS program, they call their thrust "a post-ELSI program"; it is an attempt to organize, orient, and evaluate human practices. Ideally, this means that ethics work is conducted alongside science and engineering programs.

Rabinow and Bennett suggest that, "our challenge was to invent a new way of working that was not simply downstream and outside of the techno-science but upstream and adjacent to this new domain of biological engineering" (Rabinow and Bennett 2012: 5). At the beginning of this project in 2009, potential problems were noted: there were power differentials between bioscientists and human scientists, an unwillingness to collaborate, and implicit adhesion to the ELSI model. By 2012, progress in the attempt to implement collaborative adjacency was considered "at best stuttering and sporadic" (Rabinow and Bennett 2012: 7) and cooperation among unequal partners proved difficult. They concluded that "ethically…significant discordancies remain. Consequently, other forms of collaboration with other instances will have to be undertaken" (Rabinow and Bennett 2012: 7).

One successful example of collaborative adjacency can be drawn from an NIH-funded reproductive health lab (Campo-Engelstein and Rodriguez 2011). Two post-doctoral scholars from philosophy and history participated in joint research projects with scientists and graduate students in the lab, facilitated by their discussion of differences between the scientific process and historical and ethical research processes. They engaged in mutual education with scientists, a practice that allowed for the creation of a common language that made collaborative work possible: "It is this embedded nature of our fellowship that we suggest would make for a strong fellowship training in medical humanities and bioethics, for it not only provides a greater

---

[1] This type of project is not unique to the USA, and is also present in the UK in the context of synthetic biology and nanotechnology (Balmer et al. 2015).

understanding of science for us, but also engenders multidisciplinary research projects that few of us had contemplated" (Campo-Engelstein and Rodriguez 2011: 22).

While it is difficult to know the reasons for the success of one initiative over another, the status of the early career scholars, their openness to mutual education, and their willingness to find a new collaborative language are some of the features that might distinguish the reproductive health project from SynBERC. Overall, the demonstrated benefits of successful collaborative adjacency were the opportunities created for training young scholars and the support for collaborative research. However, the presumptions that ethics and science can proceed adjacently, and that institutional support for research in these two areas through separate funding mechanisms will be equal, may be difficult to realize.

12.2.3 Encouraging Reflective Research

The final type of ethics engagement model aims to encourage reflective research through structured midstream modulation or reflection programs. While methodologically similar to collaborative adjacency, its goal is not collaboration, but rather, one-way modification of the scientific research process by scientists themselves through ethicist- or philosopher-facilitated critical reflection.

12.2.3.1 Midstream Reflection (the Toolbox Project)

Midstream reflection programs create survey instruments and discussion guides to improve interdisciplinary scientific communication and collaboration (Fisher et al. 2015: 46). The paradigm of this form of ethics engagement, the Toolbox Project (based at the University of Idaho), "provides a philosophical yet practical enhancement to cross-disciplinary, collaborative science" (http://toolbox-project.org). Currently, the Toolbox Project is funded through the National Science Foundation (NSF).

The idea for the Toolbox Project first took shape in response to recognition of barriers to cross-disciplinary research, not necessarily between the sciences and humanities, but within diverse scientific and technological disciplines. In 2007, an interdisciplinary group of scientists, philosophers, and social scientists—both students and faculty—at the University of Idaho and Syracuse University identified six types of challenges that the Toolbox was designed to address: (1) levels of integration between disciplines, (2) linguistic and conceptual divides, (3) methods for validating evidence, (4) societal contexts of research, (5) perceived nature of the world as objective or evaluative, and (6) reductionist versus holistic schemas. The thesis of the Toolbox Project was that through philosophical inquiry, scientists can identify and explore differences in implicit or explicit assumptions across disciplines (Eigenbrode et al. 2007).

The form of philosophical inquiry used by the Toolbox Project "combines a philosophical perspective on values in science with quantitative data analysis techniques drawn from the sciences" (Robinson et al. 2016). The primary form of inquiry is a 2- to 3-h semi-structured dialogue among collaborators (O'Rourke and

Crowley 2013). Before and after each facilitated dialogue, participants are asked to complete a survey that focuses on the purposes, methodologies, values, and modes of communication in their scientific practices.

Benefits of using these instruments are that "the dialogues these instruments structure often include revelation of surprising difference, negotiation of conflicting perspectives, and an increase in mutual understanding" (Schnapp et al. 2012). The instruments can also highlight differences between researchers about the roles of values in science. In one study, social and behavioral scientists were less willing than life scientists to agree that value-neutral science is possible (Robinson et al. 2016). However, while the Toolbox Project seems to be effective at encouraging communication about disciplinary differences, it does not create opportunities for collaborative or transdisciplinary research or training itself. It is a means to encourage reflection about disciplinary barriers; it is not necessarily a means to remove these barriers.

12.2.3.2 Midstream Modulation (STIR)

Other midstream programs use embedded humanists to disrupt scientific routines by asking questions about how and why certain practices are undertaken (Fisher et al. 2015: 46). The paradigm of this model of ethics engagement, the socio-technical integration research (STIR) project at Arizona State University, conducts "a coordinated set of 20 laboratory engagement studies to assess and compare the varying pressures on—and capacities for—laboratories to integrate broader societal considerations into their work" (https://cns.asu.edu/research/stir). As with the Toolbox Project, STIR is also funded by the NSF.

In contrast with the Toolbox Project, STIR creates opportunities for humanists and scientists to engage directly in the laboratory setting. Initial efforts were based on the observation that upstream attempts at ethics engagement encourage public or social regulation of scientific decision-making, while downstream engagement efforts, like ELSI, focus on identifying the social implications of scientific decision-making. Neither, however, encourages scientists to participate in value-based reflection midstream of research. The thesis of the STIR project is that, for upstream and downstream engagement efforts to be effective, a midstream integration of technical elements and social values is needed (Fisher et al. 2006). The goal of the STIR project is to encourage scientist engagement in this midstream integration.

Based on a survey of past integration methods, the leaders of the STIR project suggest that "formal… or semiformal representation of participants' own activities and processes can encourage reflexive awareness" (Fisher et al. 2006: 492). To facilitate such awareness, graduate students are embedded in science or technology labs for 3 months, during which time they utilize a protocol that asks scientists (both PIs and trainees) questions about how and why they engage in certain practices. In doing so, they: "(1) identify and compare external expectations and demands for laboratories to engage in responsible innovation, (2) assess and compare the current responsiveness of laboratory practices to these pressures, and (3) investigate and compare how interdisciplinary collaborations may elucidate, enhance or stimulate responsiveness" (https://cns.asu.edu/research/stir).

Benefits of midstream modulation projects like STIR are that they train doctoral students in the social sciences to engage directly with scientists, they encourage scientists to reflect on what may be rote practices, and they facilitate collaborative inquiry between value theorists and scientists. Downsides reside in the limited timeframe of the inquiry—doctoral students are embedded in a lab for no more than one semester, which may not be enough time to create the types of integrated relationships needed for collaborative or transdisciplinary inquiry.

12.3 Ethics Engagement in Neuroscience

Ethics engagement models in neuroethics are growing. Here, we review a selection of current programs among more than 40 worldwide to date (Moses and Illes 2017), suggesting comparisons with the EEPs analyzed above. Finally, we will explore two cases of neuroethics engagement in detail: the National Core for Neuroethics at the University of British Columbia (UBC) and the Neuroethics Thrust at the Center for Sensorimotor Neural Engineering, the University of Washington (UW), with which both authors are affiliated.

12.3.1 Neuroethics Engagement Programs

At present, the International Neuroethics Society (INS) lists 25 organizations that work on issues in neuroethics on its website (http://www.neuroethicssociety.org/publications-organizations). Of these, five stand out as neuroethics EEPs, as opposed to educational or more classically bioethical programs. These are the National Core for Neuroethics at the University of British Columbia (UBC), the Neuroethics Program at Emory University, the Center for Neuroscience and Society at the University of Pennsylvania, the Centre for Neuroethics at Oxford University, and the Program in Ethics and Brain Sciences at Johns Hopkins University. Other relevant programs include the Neuroethics Workgroup of the USA BRAIN Initiative program and the Ethics Advisory Board of the European Union Human Brain Project.

The Emory Neuroethics Program describes itself as a community of scholars dedicated to investigating ELSI issues in neuroscience and neurotechnology. Its website implies that it may also conduct activities similar to the public engagement and RCS models described above (http://www.ethics.emory.edu/pillars/health_sciences/neuroethics.html). This suggests that much of their work takes place downstream of scientific inquiry. Likewise, the University of Pennsylvania program describes its work in terms of the "ethical, legal, and social implications of neuroscience" (http://neuroethics.upenn.edu). Its activities are primarily educational, including a neuroscience boot camp for professionals in law, ethics, and education, and a series of public talks, workshops, and student seminars. As with Emory and the University of Pennsylvania, the Oxford program is downstream of neuroscience research. It focuses on conceptual research more than education, emphasizing the usefulness of practical ethics for addressing advances in neuroscientific technology and clinical research and practice in neuroscience (http://www.neuroethics.ox.ac.uk/neuroethics_at_oxford/aims). It also offers public engagement in terms of a series of podcasts on issues in neuroethics and bioethics, termed "bio-ethics bites." The Johns Hopkins program stresses the importance of understanding the ethical and social issues in brain science, adding that philosophical and ethical analysis must possess an adequate understanding of the science behind advances in brain research. Its activities seem to be primarily educational and research-based, although it describes a consensus conference that "gathered key stakeholders to discuss the scientific and ethical challenges associated with deep brain stimulation for disorders of mood, behavior and thought" (http://www.bioethicsinstitute.org/research/science-ethics/program-in-ethics-and-brain-sciences).

Finally, the two neuroethics working groups sponsored by USA and EU government programs (the BRAIN initiative and the HBP) are both still at an advisory stage and have not yet established concrete ethics engagement programs, although the HBP has interfaced with the DBT Foundation as described above. The HBP ethics advisory group describes its goal as to "foster a lively debate on social and ethical issues within the HBP, the wider scientific community, and also with the general public" (https://www.humanbrainproject.eu/faq/ethics). This allows room for a wide variety of programs, from ELSI research to public engagement to RCSs.

From this brief overview of neuroethics programs, two points immediately stand out to us: (1) these neuroethics programs currently operate downstream of scientific research, largely following an ELSI model in which ethical issues emerge from scientific research, and (2) neuroethics programs focus on education and research activities more than on some of the proactive bioethics EEPs described above, such as collaborative adjacency and midstream modulation. There seems to be room for increased exploration of alternative methods of ethics engagement in neuroethics, including deliberative democratic models of public engagement, RCSs, collaborative adjacency programs, and midstream reflection and modulation activities. These would create even greater opportunities for neuroethics program engagement between ethicists and scientists and for the encouragement of ethically reflective research.

One reason for the focus of neuroethics programs on ELSI research and education activities may be that neuroethics as a discipline is still at a relatively early stage with only 15 years of history (compared, for example, to over a 100 years for a field like radiology), and methods for engaging ethics with neuroscience are still being developed. However, while neuroethics certainly includes unique content about topics such as cognitive liberty and cognitive enhancement (Roskies 2016), there are many structural similarities between EEPs in bioethics and neuroethics. EEPs in both areas are faced with a perceived gap between axiological or value-based inquiry and objective or fact-based inquiry, different institutional and governmental mechanisms for funding research, and divergent educational priorities between science and the humanities or social sciences.

In the following section, we describe two neuroethics programs with which both authors are affiliated. This description explores the possibilities for—and barriers to—ethics engagement activities in neuroscience, neurology, and neurotechnology.

12.3.2 National Core for Neuroethics, UBC

The National Core for Neuroethics is dedicated to tackling the ethical, legal, policy, and social implications of frontier technological developments in the neurosciences and focuses on protecting and ensuring the public good by aligning innovations in the brain sciences with human values. As a center that values ethics engagement, empirical ethics research is closely tied to its daily activities. As with other ELSI programs, the Core engages in programs and education for the public and for scientists on the ethical implications of contemporary neuroscience. The majority of publications are in medical, clinical, and science journals, reaching the researchers and scientists who are the stakeholders and beneficiaries of study results, not only other scholars and interlocutors from within the neuroethics field, per se. Many of the trainees come directly from science and medicine programs and hope to specialize in the ethical aspects of this work. Further, through its work in the domain of both cross-cultural and environmental neuroethics, the Core serves to collaboratively bridge research goals based on the general priorities of Western societies and the priorities and values of populations such as First Nations in Canada (Butler et al. 2009; Stevenson et al. 2013; Illes et al. 2014; Cabrera et al. 2015, 2016; Stein and Illes 2015; Crooks et al. 2015).

Benefits of the ethics activities undertaken at the Core include a fit between ELSI funding opportunities and research activities, direct communication of ethics research in science journals, and training for scientists interested in ethics analysis. Drawbacks are that, in fitting ethics messages to audiences trained in the biomedical sciences, new knowledge generated may be better appreciated by the scientific community than by colleagues in conceptual ethics.

12.3.3 Neuroethics Thrust, Center for Sensorimotor Neural Engineering, UW

The Neuroethics Thrust is superficially similar to Rabinow and Bennett's human practices project, one of the collaborative adjacency EEPs described above. It is based at an NSF-funded engineering research center (ERC), represents core scholarship for the center, and aims to realize some level of "collaborative adjacency." Unlike SynBERC, the work of the Neuroethics Thrust is divided into three groups of activities: (1) education, (2) research, and (3) scientific engagement.

Members of the thrust create pedagogical resources, teach ethics classes, lead ethics roundtables for researchers, offer ethics journal clubs to graduate students, and provide fellowships for students in the sciences and humanities who wish to develop skills in neuroethics. The neuroethics thrust conducts original empirical and theoretical research on ethical issues connected to neural engineering (Klein et al. 2015, 2016; Klein 2015; Klein and Ojemann 2016; Goering forthcoming; Specker Sullivan et al. under review). Scientific engagement activities include embedding graduate students and postdoctoral fellows into neural engineering labs, providing research ethics consultations to scientists at the CSNE, and working collaboratively with students and scientists on research projects. The embedded ethicists especially provide opportunities for the realization of joint projects involving ethicists and researchers.

In addition, the thrust enjoys collaboration with the National Core of Neuroethics that enables the two centers to combine strengths in conceptual neuroethics with empirical neuroethics. This includes joint recruitment of postdoctoral fellows in neuroethics based at UW, but with regular time on-site at UBC, which facilitates collaborative projects with leaders of the two centers (Specker Sullivan and Illes 2016; Specker Sullivan et al. under review) and coordinate joint lab meetings.

The EEP activities undertaken at the CSNE have both benefits and drawbacks. Benefits include the development of an ethically reflective scientific community through frequent collaboration and interaction between ethics and science faculty and graduate students. Drawbacks include different funding sources and divergent expectations about research dissemination. There may also be a temptation to transfer responsibility for addressing ethical aspects of research to ethicists on staff, rather than scientific researchers taking on this responsibility themselves.

12.3.4 Distinguishing Neuroethics Engagement and Plotting Neuroethics Development

Neuroethics engagement programs build on past efforts to engage science and human values. The Core at UBC makes empirical work useful and accessible to scientists and clinicians, engaging directly and regularly with scientists and clinicians through research, public programs, and education. The Core also proactively identifies policy applications of its neuroethics work and has applied them to date in the areas, for example, of regenerative medicine, neurodevelopmental disorders, and e-health in aging and dementia (Illes et al. 2011, 2016; Kirschen et al. 2014; Longstaff et al. 2015; Robillard et al. 2015; Robillard 2016). Likewise, the neuroethics thrust at UW's CSNE combines elements of all three types of ethics engagement models described above. The thrust (1) studies the conceptual ethical issues and implications raised by the scientific study of brain computer interfaces, (2) provides ethics consultation and engages in collaborative work with scientific researchers at the CSNE, and (3) creates opportunities for researchers to critically reflect on their own work.

Taken together, these two programs offer complementary approaches to ethics engagement in neuroscience, neurology, and neural engineering. They conduct both empirical and conceptual work in neuroethics, engage with researchers and scientists on a small-scale interpersonal level and on a large-scale public engagement level, and provide guided dialogue and reflection on science and values. Both programs have advantages and drawbacks; a common dilemma is that scholars working to create bridges between science and ethics may find themselves in an interdisciplinary space in which funding opportunities, publication priorities, and career prospects are more uncertain than for their colleagues in traditional disciplinary tracks.

12.4 Looking Ahead to the Future of Neuroethics Engagement

While great strides have been made in ethics engagement research and training programs, there is more to be done in the field of neuroethics. The recommendations of the Gray Matters Report for ethics engagement in Volume 1 and Volume 2 serve as the basis from which to chart the future direction of neuroethics engagement. The call for early and frequent integration suggests the need for upstream methods of integrating science with human values. This may involve, for example, more public engagement in science decision-making, the early availability of research consultation services, or the development of collaboratively adjacent research teams. Ongoing conversations about innovative approaches to ethics integration in neuroscience are essential.

Furthermore, integrating ethics not just with science practice, but also with science education and training is vital to the creation of a sustainable culture of ethics in science. As we have discussed, while ethics engagement programs tend to focus on research, an emphasis on both research and pedagogy requires stakeholders to work to incorporate ethics on all possible fronts. Finally, acknowledgement in Volume 2 of the difficulty of conducting engagement work within traditional funding mechanisms has been upheld by the separation of ethics and science in the ELSI program, the failure of public engagement uptake, the barriers to researcher use of RCSs, the power differentials in collaborative adjacency, and the advisory nature of midstream modulation programs. As a result, there is an ever greater need for transdisciplinary generation of, support for, and maintenance of ethics engagement programs. The United States National Institutes of Health Brain Initiative's recent request for guidance on neuroethics (http://grants.nih.gov/grants/guide/notice-files/NOT-MH-16-014.html) and the response offered by the INS and others may indicate that these funding mechanisms are on the horizon. Among other issues, the INS response stresses the need for participatory models of policy decision-making, integration of diverse stakeholder perspectives when designing and setting priorities for new technologies, and researcher responsibility for reflecting on ethical issues and communicating these issues to the public.

Conclusion

Review of EEPs outside of neuroethics can shape the direction of engagement efforts within neuroethics. While there are aspects of neuroethics that are distinct from ethics engagement with genetics, nanotechnology, and synthetic biology, such as conceptual questions about the self and cognitive liberty, institutional realities about the methods and goals of neuroscientific research, and the democratic implications of research and technology that influence the mind, overlapping challenges remain. In particular, as with programs in bioethics more generally, neuroethics faces decisions about downstream, upstream, and midstream engagement, challenges in creating effective training models and making policy recommendations, and barriers to funding multidisciplinary and transdisciplinary projects.

Based on the successes and failures of previous models and reflections on lessons of the past, we suggest that neuroethics engagement efforts in the future ought to:

1. Seek opportunities for early integration of ethics and other value perspectives with scientific inquiry. This includes engaging with upstream decision-making in basic science, as well as training early career and student scientists in value-based reflection and social engagement.
2. Focus on ethics engagement activities that engage scientists as participants, create lasting relationships between the public and scientific research, and have clear policy recommendations that can be utilized by groups responsible for policy formation.
3. Advocate for the creation of funding mechanisms to facilitate transdisciplinary work. This is necessary for ensuring that scientists, humanists, and social scientists work collaboratively and can mutually enrich each other's knowledge and critical capacities.

Appreciating the successes and failures of past ethics engagement efforts allows present neuroethics programs to evaluate their current practices and identify areas for future innovation and development.

Acknowledgments JI is Canada Research Chair in Neuroethics and President of the International Neuroethics Society. Her work and the National Core for Neuroethics have been supported by a range of government and private funders including the Canadian Institutes of Health Research, the BC Knowledge Development Fund, the Canadian Foundation for Innovation, and Vancouver Coastal Health Research Institute. LSS is a Postdoctoral Fellow at the Center for Sensorimotor Neural Engineering, University of Washington, and at the National Core for Neuroethics. This work is supported by the National Science Foundation, Award Number EEC-1028725. The content is solely the responsibility of the authors and does not necessarily represent the official views of the National Science Foundation.

References

Abelson J, Eyles J (2004) Public participation and citizen governance in the Canadian health system. Changing Health Care in Canada: The Romanow Papers, vol 2, pp 279–311

Balmer AS, Calvert J, Marris C, Molyneux-Hodgson S, Frow E, Kearnes M et al (2015) Taking roles in interdisciplinary collaborations: reflections on working in post-ELSI spaces in the UK synthetic biology community. Sci Technol Stud 28(3):3–25

Bennett I, Sarewitz D (2006) Too little, too late? Research policies on the societal implications of nanotechnology in the United States. Sci Cult 15(4):309–325

Burgess M (2004) Public consultation in ethics: an experiment in representative ethics. J Bioeth Inq 1(1):4–13

Burgess M, O'Doherty K, Secko D (2008) Biobanking in British Columbia: discussions of the future of personalized medicine through deliberative public engagement. Pers Med 5(3):285–296

Butler R, Dwosh E, Beattie BL, Guimond C, Lombera S, Brief E et al (2009) Genetic counseling for early-onset familial Alzheimer disease in a large aboriginal kindred from a remote community in British Columbia: unique challenges and possible solutions. J Genet Couns 5(4):90–99

Cabrera L, Beattie LB, Dwosh E, Illes J (2015) Converging approaches to understanding early onset familial Alzheimer disease: A First Nation study. SAGE Open Medicine 3
Cabrera L, Tesluk J, Matthews R, Chakraborti M, Illes J (2016) Brain matters: from environmental ethics to environmental neuroethics. Environ Health 15:20–25
Campo-Engelstein L, Rodriguez SB (2011) Two chicks in a lab with eggs. Hast Cent Rep 41(3):21–23
Cho MK, Tobin SL, Greely HT, McCormick J, Boyce A, Magnus D (2008) Strangers at the benchside: research ethics consultation. Am J Bioeth 8(3):4–13
Crooks A, Li N, Snyder J, Dharamsi S, Benjaminy S, Jacob KJ et al (2015) "You don't want to lose that trust that you've built with this patient…": (dis)trust, medical tourism, and the Canadian family physician-patient relationship. BMC Fam Pract 16:25
Dickinson HD (2002) How can the public be meaningfully involved in developing and maintaining an overall vision for the health system consistent with its values and principles? Commission on the Future of Health Care in Canada
Eigenbrode SD, O'Rourke M, Wulfhorst JD, Althoff DM, Goldberg CS, Merrill K et al (2007) Employing philosophical dialogue in collaborative science. BioScience 57(1):55–64
Einsiedel EF, Ross H (2002) Animal spare parts? A Canadian public consultation on xenotransplantation. Sci Eng Ethics 8:579–591
Fisher E (2005) Lessons learned from the ethical, legal and social implications program (ELSI): planning societal implications research for the National Nanotechnology Program. Technol Soc 27(3):321–328
Fisher E, Mahajan RL, Mitcham C (2006) Midstream modulation of technology: governance from within. Bull Sci Technol Soc 26(6):485–496
Fisher E, O'Rourke M, Evans R, Kennedy EB, Gorman ME, Seager TP (2015) Mapping the integrative field: taking stock of socio-technical collaborations. J Respons Innov 2(1):39–61
Goering S (forthcoming) Thinking differently: neurodiversity and neural engineering. In: Rommelfanger K, Johnson LS (eds) Routledge handbook of neuroethics. Routledge, New York
Have HT (2006) The activities of UNESCO in the areas of ethics. Kennedy Inst Ethics J 16(4):333–351
Illes J, Reimer J, Kwon BK (2011) Stem cell clinical trials for spinal cord injury: readiness, reluctance, redefinition. Stem Cell Rev Rep 7(4):997–1005
Illes J, Davidson J, Matthews R (2014) Environmental neuroethics: changing the environment—changing the brain. J Law Biosci 1:221–223
Illes J, Owen AM, Byram AC; the MCS Working Group (2016) Operationalizing neuroimaging for disorders of consciousness in the face of uncertainty and contingency: a view for the Canadian landscape. Can J Neurol Sci 1–3
Jasanoff S (2011) Constitutional moments in governing science and technology. Sci Eng Ethics 17(4):621–638
Jensen CB (2005) Citizen projects and consensus-building at the Danish Board of Technology. Acta Sociol 48(3):221–235
Juengst ET (1991) The human genome project and bioethics. Kennedy Inst Ethics J 1(1):71–74
Kirschen MP, Topjian AA, Abend NS, Illes J (2014) Neuroprognostication after pediatric cardiac arrest. Pediatr Neurol 51(5):663–668
Klein E (2015) Informed consent in implantable BCI research: identifying risks and exploring meaning. Sci Eng Ethics 22(5):1299–1317
Klein E, Ojemann J (2016) Informed consent in implantable BCI research: identification of research risks and recommendations for development of best practices. J Neural Eng 13(4):043001
Klein E, Brown T, Sample M, Truitt AR, Goering S (2015) Engineering the brain: ethical issues and the introduction of neural devices. Hast Cent Rep 45(6):26–35
Klein E, Goering S, Gagne J, Shea CV, Franklin R, Zorowitz S et al (2016) Brain-computer interface-based control of closed-loop brain stimulation: attitudes and ethical considerations. Brain Comput Interfaces 3(3):140–148
Leroux T, Hirtle M, Fortin LN (1998) An overview of public consultation mechanisms developed to address the ethical and social issues raised by biotechnology. J Consum Policy 21:445–481

Leshner AI (2003) Public engagement with science. Science 299:977
Longstaff J, Krahmova V, Portalis E, Illes J (2015) Sharing with more caring: coordinating and improving the ethical governance of data and biomaterials obtained from children. PLoS One 10(7):e0130527
Macnaghten P, Kearnes M, Wynne B (2005) Nanotechnology, governance, and public deliberation: what role for the social sciences? Sci Commun 27(2):1–24
Mascalzoni D, Dove ES, Rubinstein Y, Dawkins HJ, Kole A, McCormack P et al (2016) International Charter of principles for sharing bio-specimens and data. Eur J Hum Genet 24(7):1096
McCain L (2002) Informing technology policy decisions: the US human genome Project's ethical, legal, and social implications programs as a critical case. Technol Soc 24(1):111–132
McEwen JE, Boyer JT, Sun KY, Rothenberg KH, Lockhart NC, Guyer MS (2014) The ethical, legal, and social implications program of the National Human Genome Research Institute: reflections on an ongoing experiment*. Annu Rev Genomics Hum Genet 15(1):481–505
Moses T, Illes J (2017) Ethics, ethicists, and professional organizations in the neurological sciences. Am J Bioeth Neurosci 8(1):3–11
Nature (2004) Editorial: going public. Nature 431:833
O'Doherty K, Hawkins A (2010) Structuring public engagement for effective input in policy development on human tissue biobanking. Public Health Genomics 13(4):197–206
O'Rourke M, Crowley SJ (2013) Philosophical intervention and cross-disciplinary science: the story of the toolbox project. Synthese 190(11):1937–1954
Page A, Baker D, Bobrow M, Boycott K, Burn J, Chanock S et al (2016) Genomics. A federated ecosystem for sharing genomic, clinical data. Global Alliance for Genomics and Health. Science 352(6291):1278–1280
Rabinow P (2009) Prosperity, amelioration, flourishing: from a logic of practical judgment to reconstruction. Law Lit 21(3):301–320
Rabinow P, Bennett G (2012) Designing human practices: an experiment with synthetic biology. The University of Chicago Press, Chicago and London
Robillard JM (2016) The online environment: a key variable in the ethical response to complementary and alternative medicine for Alzheimer disease. J Alzheimer Dis 51(1):11–13
Robillard JM, Illes J, Arcand M, Beattie BL, Hayden S, Lawrence P et al (2015) Scientific validity and ethics of online tests for Alzheimer disease. Alzheimers Dement (Amst) 1(3):281–288
Robinson B, Vasko SE, Gonnerman C, Christen M, O'Rourke M, Fosl PS (2016) Human values and the value of humanities in interdisciplinary research. Cogent Arts Humanit 3(1):1123080
Roskies A (2016) Neuroethics. The Stanford encyclopedia of philosophy (Spring 2016 Edition). http://plato.stanford.edu/archives/spr2016/entries/neuroethics/
Schnapp LM, Rotschy L, Hall TE, Crowley S, O'Rourke M (2012) How to talk to strangers: facilitating knowledge sharing within translational health teams with the Toolbox dialogue method. Transl Behav Med 2(4):469–479
Sharp RR, Taylor HA, Brinich MA, Boyle MM, Cho M, Coors M et al (2015) Research ethics consultation: ethical and professional practice challenges and recommendations. Acad Med 90(5):615–620
Specker Sullivan L, Illes J (2016) Beyond communication and control: towards ethically complete rationales for brain-computer interface research. Brain Comput Interfaces 3(3):156–163
Specker Sullivan L , Klein E, Brown T, Sample M, Pham M, Tubig P et al (under review) Keeping disability in mind: a case study in implantable brain-computer interface research
Stein DJ, Illes J (2015) Beyond scientism and skepticism: an integrative approach to global mental health. Front Psychiatry 6:166
Stevenson S, Beattie BL, Vedan R, Dwosh E, Bruce L, Illes J (2013) Neuroethics, confidentiality, and a cultural imperative in imperative in early onset Alzheimer disease: a case study with a First Nation population. Philos Ethics Humanit Med 8(15):1–6
Turner L (2003) The tyranny of 'genethics'. Nat Biotechnol 21(11):1282–1282
Van Est R (2011) The broad challenge of public engagement in science: commentary on: "constitutional moments in governing science and technology". Sci Eng Ethics 17(4):639–648

13 The Future of Neuroethics Research and Training

Tom Buller

Abstract

A number of major national and international projects have recently been launched to meet the "Grand Challenge" of understanding the brain. With substantial funding, these projects aim to reveal the neurophysical processes underlying our mental states and seek to develop technologies to restore brain function when it has been impaired. The projects also include explicit investigation of the ethical, social, and legal implications of neuroscience and the ethical conduct of neuroscience research. The advent of these projects, therefore, offers an extraordinary opportunity for neuroethics to play an integral role in what may be the defining scientific program of our time. To fulfill this role, neuroethics must contribute fully to framing the investigation of ethical issues and the identification of those that need to be addressed. This chapter focuses on four issues that warrant particular attention as we move forward: (1) the ethical conduct of research; (2) responsible research and innovation; (3) responsibility, crime, and punishment; and (4) the identity of neuroethics.

13.1 Introduction

In the last few years, a considerable number of major national and international projects have been launched with the goal of understanding how the brain works.[1] These projects seek to develop innovative neurotechnologies; to

[1] Australian Brain Alliance, BRAIN Initiative, Human Brain Project, Canada Brain Research Fund, China Brain Project, Cuban Human Brain Mapping Project (CHBMP), Israel Brain Technologies, Latin American Brain Mapping Network (LABMAN), Brain Mapping by Integrated Neurotechnologies for Disease Studies (Brain/MINDS), Korean Brain Initiative, Blue Brain Project.

T. Buller
Department of Philosophy, Illinois State University, Normal, IL, USA
e-mail: tgbulle@ilstu.edu

E. Racine, J. Aspler (eds.), *Debates About Neuroethics*, Advances in Neuroethics,
DOI 10.1007/978-3-319-54651-3_13

understand how the brain creates our thoughts, motivations, and feelings; and to develop ways to restore cognition and brain function when they have been impaired. The advent of these projects and their level of public and (in some cases) private funding support the claim that understanding the brain is the scientific inquiry of the current era—the next great scientific journey of exploration comparable to the Apollo program and the Human Genome Project. As we move forward, a host of ethical questions will be raised, and it is reasonable to think that neuroethics will play a role in answering them; since these projects are still in their early stages, however, precisely what role this will be is still very much to be determined.

One way neuroethics can help to clarify the matter is by taking the lead in setting the ethical agenda—by helping to map out the ethical domains and identify the major questions that each of the projects need to address. Accordingly, the main goal of this chapter is to identify topics that I believe should be a part of this agenda. In the first section of the chapter, I provide a brief description of the BRAIN Initiative and the Human Brain Project and the major ethical issues identified in these projects. In the second section, I consider what will become a core element as the projects move forward, namely, the ethical conduct of research. The third section focuses on "responsible, research, and innovation" and identifies important global, social, and moral issues that should be considered part of this notion. The fourth and fifth sections discuss the impact of neuroscience on the law and the identity of neuroethics itself, respectively.

13.2 The "Grand Challenge"

In April 2013, President Barack Obama announced the launch of the *Brain Research through Advancing Neurotechnologies (BRAIN) Initiative*, a multibillion dollar project to understand and map the human brain. In his remarks at the launch of the project, the President said:

> Imagine if no family had to feel helpless watching a loved one disappear behind the mask of Parkinson's or struggle in the grip of epilepsy. Imagine if we could reverse traumatic brain injury or PTSD for our veterans who are coming home. Imagine if someone with a prosthetic limb can now play the piano or throw a baseball as well as anybody else, because the wiring from the brain to that prosthetic is direct and triggered by what's already happening in the patient's mind. What if computers could respond to our thoughts or our language barriers could come tumbling down… That's the future we're imagining. That's what we're hoping for. That's why the BRAIN Initiative is so absolutely important (President Obama 2013).

The extraordinary nature of the project and its significance can be seen in the preamble to *BRAIN 2025: A Scientific Vision*:

> We stand on the verge of a great journey into the unknown—the interior terrain of thinking, feeling, perceiving, learning, deciding, and acting to achieve our goals—that is the special province of the human brain. These capacities are the essence of our minds and the aspects

> of being human that matter to us. Remarkably, these powerful yet exquisitely nuanced capacities emerge from electrical and chemical interactions among roughly 100 billion nerve cells and glial cells that compose our brains.... We believe this to be a moment in the science of the human brain where our knowledge base, our new technical abilities, and our dedicated and coordinated efforts can generate great leaps forward in just a few years or decades. Like other great leaps in the history of science—the development of atomic and nuclear physics, the unraveling of the genetic code—this one will change human society forever (BRAIN 2025: A Scientific Vision: 9).

The US Administration has awarded the BRAIN Initiative substantial federal funding to achieve its goals: approximately $100 million in FY14, $200 million in FY 2015, $300 million in FY16, and more than $434 million in FY17. In terms of cost projections, it is predicted that the BRAIN Initiative will receive $400 million a year for the next 5 years (2015–2020) and $500 million a year for the last 5 years (2020–2025). The total funding for the project is expected to be approximately $4.5 billion (BRAIN 2025: 122). In comparison, total funding for the previous major scientific mapping initiative, the Human Genome Project (HGP), has been approximately $2.7 billion (in 1991 dollars/$4.7 billion in 2016 dollars (McEwen et al. 2014)). The US Congress at the time mandated that "not less than" 5% of the HGP be set aside for research on the ethical, legal, and social implications (ELSI) of genomic science. The HGP's ELSI budget increased from $1.57 million to $18 million in FY13 and has awarded $317 million in funding total. (A similar 5% of BRAIN Initiative funds for an ELSI program for neuroscience would amount to $225 million). Although the BRAIN Initiative has not so far allocated a specific percentage or amount of funds toward the investigation of ethical issues, the project has recently announced the first funding opportunity for ethics-related research, "Research on the Ethical Implications of Advancements in Neurotechnology and Brain Science" (RFA-MH-17-260 2016).

In the same year that the BRAIN Initiative commenced, the European Commission launched the Human Brain Project (HBP) as part of its Research Framework Program, *Horizon 2020*. The report to the Commission states:

> Understanding the human brain is one of the greatest challenges facing 21st century science. If we can rise to the challenge, we can gain fundamental insights into what it means to be human, develop new treatments for brain diseases, and build revolutionary new information and communication technologies (ICT).... We find that the major obstacle that hinders our understanding of the brain is the fragmentation of brain research and the data it produces. Our most urgent need is thus a concerted international effort that can integrate this data in a unified picture of the brain as a single multi-level system. To reach this goal, we propose to build on and transform emerging ICT technologies (The Human Brain Project 2012, Abstract).

The estimated total cost of the Human Brain Project is 1.1 billion Euros ($1.3 billion) spread over a period of 10 years (HBS-PS Consortium 2012).

The Human Brain Project and the BRAIN Initiative both explicitly recognize the importance of the ethical, legal, and social implications of neuroscience research and the necessity to ensure that this research is conducted in an ethical fashion. For example, one of the 12 subprojects (subproject 12) of the Human Brain Project is "Ethics and Society":

> The HBP Ethics and Society Subproject explores the Project's social, ethical, and philosophical implications. It promotes engagement with decision-makers and the general public, fosters responsible research and innovation by raising social and ethical awareness among project participants, and ensures that the Project complies with relevant legal and ethical norms (www.humanbrainproject.eu).

The Human Brain Project has specifically allocated 2.1% of the project's total budget—24.5 million Euros—($27.4 million) toward subproject 12.

For its part, BRAIN 2025 identifies two broad "equally important" categories of ethical issue that fall within the BRAIN Initiative: (1) the ethical conduct of research and (2) the ethical and social implications of new technologies and scientific discoveries (BRAIN 2025: 118). According to a familiar description, neuroethics has two main areas of focus: the ethics of neuroscience and the neuroscience of ethics (Roskies 2002). The ethics of neuroscience broadly matches the second of the BRAIN Initiative's categories, and within this area we find topics that have received considerable discussion since the field of neuroethics emerged, for example, cognitive enhancement, neuroimaging, and the implications of neuroscience for the law. Interestingly, however, the BRAIN Initiative's other category—the ethical conduct of research—has received rather less attention.

The first volume of *Gray Matters* calls for ethics "integration at all levels" (p. 16), and the second volume makes specific recommendations regarding informed consent and research participation (p. 7) (President's Commission for the Study of Bioethical Issues, 2014, 2015); nevertheless, there is a need for more detailed discussion of these topics in order to develop the appropriate guidelines, principles, and safeguards as these projects move forward (Greely et al. 2016). Why the ethical conduct of research should have received less attention is not immediately obvious, but one reason is that in many cases these technologies are still in their early stages—far too early to fully realize some of the ethical issues that have already received considerable attention. Since these technologies (and other types of interventions) are some way off, pragmatic questions regarding the conduct of research have not yet come to the fore. The launch of the BRAIN Initiative, Human Brain Project, and other similar projects suggests that these practical questions need to be addressed sooner rather than later.

13.3 The Ethical Conduct of Research

As President Obama said in his remarks launching the BRAIN Initiative, it is the hope and goal of the project to help find treatments for a range of mental disorders (e.g., neurodegenerative disorders like Alzheimer's and Parkinson's diseases, mental illnesses like schizophrenia and depression). The project will also seek to develop novel technologies to help those individuals who suffer from conditions for which current interventions are very limited (e.g., amyotrophic lateral sclerosis (ALS), severe spinal cord injury). At its first meeting, the BRAIN Initiative's Neuroethics Workgroup identified "four priority ethical concerns: the ethics of research with invasive technologies; altering, enhancing and/or manipulating the self and agency;

privacy/discrimination concerns and the need for privacy protections; and the translation to contexts beyond the clinic/bench: ethics of commercialization, private-publication partnerships, and wider application of imaging technologies for commercial purposes" (Neuroethics Workgroup 2016). According to the Workgroup, the first topic—the ethics of research with invasive technologies—can be further divided into three elements: "the validity of informed consent and competency determinations; obligations regarding long-term upkeep of implanted devices, and follow-up with research participants; and participants' physical and perceived non-physical harms, and how to manage such issues."

A number of factors make future neuroscientific research into the above conditions challenging. First, there is the rather obvious fact that all of these conditions pertain to brain function. The exact nature of the relationship between mind, brain, and body is, of course, yet to be determined; nevertheless, whether one thinks that this relationship is one of identity, reduction, or causality, there is general agreement that the brain plays a unique role regarding our thoughts, actions, and identity. In this regard, interventions into the brain are more profound and more controversial than interventions into other parts of the body. Second, many of the conditions for which neuroscience research will be conducted are conditions in which the person's competency and autonomy have been compromised, sometimes greatly so. Third, and what follows from the previous point, is that a substantial proportion of research participants will be appropriately described as "vulnerable" and, therefore, deserving of special protections. The majority of research participants will suffer from conditions that are likely to adversely affect rationality and decision-making capacity, either directly in the aforementioned cases of Alzheimer's disease or schizophrenia or indirectly as a result of depression and anxiety. Furthermore, it is likely that future research will also involve children, given the prevalent desire for parents and society to address behavioral issues through pharmacology. Fourth, many of the current and future interventions can appropriately be described as "novel," that is to say, they are either significantly different than current practice or the current practice is still in its early days. This novelty has obvious implications for informed consent, since the newness or distinctiveness of the intervention makes it difficult to determine its risks and benefits. Fifth, and finally, since many of the conditions for which research will be conducted are those that affect higher cognitive function, the likelihood of being able to conduct earlier stage research on animal models is low.

13.3.1 Research on Neurodegenerative Disorders

To highlight the challenges ahead, it is helpful to consider research into treatments for Alzheimer's and Parkinson's diseases (Barker and de Beaufort 2013). At present, the main treatments for Alzheimer's disease (cholinesterase inhibitors and memantine) have some efficacy in early stages and are able to delay the progression of the disease for about a year. Currently, there are a number of trials for next-generation drug therapies being undertaken that target key components of the disease, for example, the formation of plaques and tangles, inflammation, and insulin resistance. Other potential treatments are the delivery of nerve growth factor (NGF) to affected neurons (Eyjolfsdottir et al. 2016) and the transplantation of neural stem cells into the damaged areas of the brain (Blurton-Jones et al. 2009). In July 2016, Stemedica announced the first stem cell clinical trial to treat Alzheimer's disease, a phase II single-blind, placebo-controlled, crossover trial in which 40 participants with mild to moderate dementia receive a single intravenous dose of human mesenchymal stem cells (hMSC) (www.stemedica.com).

All five factors mentioned in the previous section apply to the search for potential treatments for neurodegenerative disorders, particularly the second and fourth factors: the diminished competence of the participants and the novelty of the treatment, respectively. In one sense, of course, all (legitimate) research is new since there must be some degree of clinical equipoise, some information to be discovered that will show whether the new treatment is better or worse than existing treatments; but in the case of "novel" interventions—those for which there is no comparable existing treatment—the degree of clinical equipoise will be that much greater. In regard to competence, if the decision-making capacity of participants in a clinical trial (or research more broadly) is diminished to a degree that prevents them from consenting themselves, consent must be obtained through a proxy, which presents special challenges (Grill and Karwalish 2010; Kim et al. 2010; Appelbaum 2010). Although research involving individuals with impaired decision-making capacity is not unique to research on neurodegenerative disorders, what makes research for these disorders challenging is that the decision-making capacity of research participants is not simply diminished but *diminishing*, that is to say, it is deteriorating *over time*. Unless we impose the restriction that a competent person cannot consent to participate in research at a later time once decision-making capacity has been lost, research into neurodegenerative disorders will involve participants who were able to give their voluntary informed consent at the start of the trial but may not be able to do so at the end. Alternatively, and more optimistically, if we were to discover potential treatments that show a marked ability to restore decision-making capacity, then our research guidelines would have to deal with cases in which participants were unable to consent at the start of the trial but were able to do so at its conclusion.

As we become better able to diagnose Alzheimer's disease in its earliest stages, and perhaps even presymptomatically, we will be in a better position to consult with the individual who has been diagnosed regarding their opinions about present and future research participation. One option that might be considered to help further empower such individuals and facilitate research, and that might address the matter of diminishing competence, is the use of advance directives for research (Buller 2014). This type of instrument would enable a person to consent to research that will occur at a later time once decision-making capacity has become significantly impaired. For example, a person recently diagnosed with Alzheimer's disease who is still competent may wish to participate in the trial because they very much want to help find a cure for the disease. If advance directives for research do become a more widespread instrument in these contexts, a further question to ask is whether they should be limited to research that involves no greater than minimal risk.

Another challenge raised by the transitional nature of competence relates to identity and authenticity. One of the issues with neurotechnologies such as deep brain stimulation (DBS) is the effect that such technologies will have on the psychology and personality of the person undergoing the treatment or intervention (Lipsman and Glannon 2013). For the most part, this issue tends to be raised at the more abstract, rather than the practical level (e.g., in regard to the authenticity of an individual's psychological states posttreatment). But there are clear practical applications if we consider the matter from the point of view of the person participating in the research. For example, a person who is considering whether to participate in a trial of a potential new treatment for Alzheimer's disease may be worried about the effect of the drug on his memory and might ask, "Will I recover the memories that I have been unable to retrieve or will the treatment just enable me to form new memories?" or "What memories will I have of the time when I was suffering from dementia?"

One final issue I will raise regarding Alzheimer's disease is vulnerability, the third factor listed above (Labuzetta et al. 2011). This term is somewhat opaque, but in broad terms, it can be understood to identify those individuals who warrant special protections on account of their age, mental capacity, or situation. Accordingly, and respectively, the following groups are typically considered vulnerable: children, individuals with impaired decision-making capacity, and prisoners (Bell et al. 2014). Individuals suffering from Alzheimer's disease are vulnerable in different ways and to different degrees throughout the course of the disease, and this presents specific challenges to caregivers and clinicians, from diagnosis through treatment (Gauthier et al. 2013).

Many of these issues also apply to patients suffering from Parkinson's disease, but there is one further issue that pertains particularly to Parkinson's patients: the therapeutic use of DBS. As described in the report by the Nuffield Council on Bioethics: "DBS involves implanting electrode arrays into a deep subcortical brain nucleus using image-guided stereotactic neurosurgical techniques… DBS electrode leads are connected to battery-driven stimulus generators (IPGs) which are implanted subcutaneously, such that the system is located entirely within the patient's body" (Nuffield Council on Bioethics 2013: 23). DBS is used to treat a variety of conditions including depression, addiction, and other neuropsychiatric disorders. DBS was first developed in France in 1987 and is now the most widely used form of brain stimulation (Nufffield 2013: 23). DBS has been shown to be an effective treatment for Parkinson's disease in many cases, but as Clausen describes, there are a number of important ethical issues with this type of treatment (Clausen 2010).

To some degree, any individual with a chronic medical condition may feel herself dependent upon a medical treatment or intervention, and this dependence may lead the individual to feel that her autonomy is constrained; however, it is plausible to contend that there is something qualitatively different about the feeling of dependence on a device that has been implanted into the body, whose function is sometimes (literally) vital to maintaining the individual's life (e.g., a cardiac pacemaker). In the case of neural implants, although the implanted device may not be "vital" in the above sense, its presence may be felt as more peculiar since its purpose is to regulate the person's brain—the organ that is the basis of the person's self and identity. It is, perhaps, one thing to recognize that one's heart rate is controlled by a device and quite another to recognize that one's emotional and physical movements might be. As DBS becomes more commonplace, we may come to regard such devices as simply differently constructed functional units, as we do with artificial knees, and become more comfortable with the idea that the brain's functions can be variously realized. There is reason to think that as our ability to manipulate and modify the brain and body increases, so will we have to consider the effects that such modifications have on self-perception and our notions of health, autonomy, and disability.

13.4 Responsible Research and Innovation: Global, Social, and Moral

13.4.1 Global Values

As part of the discussion surrounding the BRAIN Initiative and other similar projects, the notion of "responsible research and innovation" (RRI) has gained increasing attention. As Garden et al. describe (2016: 643–644), RRI incorporates a number of elements, such as the need to include diverse stakeholders throughout the research process; the identification of ethical, legal, and social implications of innovative technologies; and the guidance of research through appropriate regulatory frameworks. RRI can legitimately apply to research and innovation broadly; however, within the context of the major brain projects, the emphasis has been on technological development.

One of the important factors to consider in assessing the ethical nature of research is the degree to which the benefits and harms are distributed fairly. This can be achieved by ensuring that research participants are also research beneficiaries through the inclusion of diverse stakeholders, as suggested above. But this does not directly address another type of ethical consideration, namely, the extent to which the benefits of the research—and the newly discovered interventions, if the research is successful—are available broadly. According to the World Health Organization (WHO 2016a, b), 47.5 million people worldwide have dementia and roughly half (58%) are living in low- or middle-income countries. In regard to depression, 350 million people are affected, over 800,000 commit suicide each year, and suicide is the second leading cause of death in 15- to 29-year-olds (www.who.int). In China alone, it is estimated that 260 million people (one-fifth of the population) suffer from chronic neuropsychiatric or degenerative diseases (Poo et al. 2016: 592). Since technology will play such an integral part in major brain projects, it is understandable that discussion of RRI has focused on neurotechnologies and innovation, but in our search to discover new treatments for dementia and depression, we should not lose sight of the fact that many who suffer from these diseases have limited access to healthcare and even less to state-of-the-art technology. This means that ethical questions remain, even if, for example, we developed an effective treatment for dementia using implanted neural cells or for depression using DBS, for it is likely that only a small percentage of those suffering from these conditions would ever be able to receive these treatments.

A different way in which we need to pay attention to the diversity of values concerns technology itself and a perspective that underlies the major brain projects. As quoted above (p. 1), the preamble to BRAIN 2025 speaks of "a great journey into the unknown" and how the project will (it is hoped) enable us to discover "the essence of our minds and the aspects of being human that matter to us." Many find such language uncontroversial for they readily accept what is implicit in the above, namely, that our minds matter most to us, that the brain is the realizer of our mental lives (if not identical to it), and that to understand the mind, we need to understand the brain. But for a variety of religious, cultural, and philosophical reasons, many people believe that what is essential to humanity cannot be understood through brain research, even if such research could explain the neurophysical basis of the mind. In the context of the brain projects, this provides another reason for discussion of their ethical, legal, and social implications, so that they can also be informed by a diverse set of communities, particularly those from cultures that may have a rather different perspective on the primacy of empirical science (Fadiman 2012; Poo et al. 2016: 593).

13.4.2 Environmental Factors

When we look at the world from the perspective of neuroscience and, to an extent, neuroethics, there is a tendency to focus exclusively on the internal operations of the brain, rather than the external factors that might affect these operations. In discussing research in light of the BRAIN Initiative, Human Brain Project, and the other projects, it is customary to frame discussion in terms of novel neurotechnologies: how these technologies will help us better explain brain function and, hopefully, enable us to repair or restore lost function. However, other important factors that can influence brain function tend to receive less attention than the internal and technological elements, such as socioeconomic status (Hackman and Farah 2009), nutrition (Prado and Dewey 2014), and other environmental factors (Toste et al. 2015; Kim et al. 2013). These present external, indirect, and perhaps more mundane influences on brain function, but in our goal to find treatments for a variety of mental disorders, it is important that they be given full attention.

In one sense, of course, the brain functions as a "bottleneck" for all external factors exerting influence through the brain (Greene and Cohen 2004: 1781). Accordingly, an argument could be made that we should focus our attention on the internal elements rather than the distal causes. In response, one might argue that although understanding the neurological basis of depression, for example, may help us discover new treatments, we may be able to aid those who suffer from depression and other forms of mental illness by also paying attention to external factors. If successful, the major projects may help us to understand the neurophysical basis of a number of mental disorders and could lead to the discovery of psychopharmacological and other types of internal intervention; but we should not lose sight of the external factors that contribute to the onset of these conditions in particular individuals and, equally importantly, that a positive change in these external factors may have a significant beneficial effect.

13.4.3 Animal Research

A common and historically influential view is that humans have higher moral status than all other living creatures. The basis for this might be that we believe only humans are created in the image of God, or because we believe that humans are unique in having a well-developed prefrontal cortex and are capable of self-consciousness and ratiocination. In either case, these unique attributes make all the moral difference—for it is held that they provide sufficient justification for treating the interests of all other creatures as subordinate to our own.

One of the clearest examples of this moral framework is in our use of animals in research. Many people believe that it is morally appropriate to test drugs, techniques, and devices on animals before they are tested on humans. The position is justified on the grounds that the welfare of a human is more valuable than the welfare of an animal, other things being equal. Animal research will be an important part of many, if not most, of the major brain projects.

As the Human Brain Project states:

> The HBP would allow clinical researchers to gain a new understanding of brain disease, design new diagnostic tools and develop new treatments. We therefore argue that the potential contribution to human wellbeing and the potential long-term reduction in animal experimentation amply justify the project's limited and responsible use of animals (HBS-PS Consortium 2012, p. 91).

For their part, the Chinese and Japanese brain projects incorporate significant research on nonhuman primates (Poo et al. 2016: 593–594; Okano et al. 2016: 583–586). As described by Poo et al. in regard to the China Brain Project, there is an increasing interest among Chinese researchers in the use of nonhuman primates in neuroscience research, and several new research facilities have been built. Furthermore, there is an easily accessible source of nonhuman primates in China, particularly macaques. The Japanese Brain/MINDS project identifies research on nonhuman primates as a necessary step to gain an accurate understanding of the human brain. Specifically, the project is engaged in "developing the common marmoset as a model animal for neuro-science, the project aims to build a multiscale marmoset brain map, develop new technologies for experimentalists, create transgenic lines for brain disease modeling, and integrate translational findings from the clinical biomarker landscape" (Okano et al. 2016: 582).

There is extensive discussion regarding the moral and mental status of animals (Farah 2008; Dennett 1995; Carruthers 1992; DeGrazia 1996; Dawkins 2012; Shriver 2006) and the use of animals in neuroscientific research (Bateson 2011; Blakemore et al. 2012; Bennett and Ringach 2016). It is beyond the scope of this chapter to make any contribution to this discussion other than to make the following points. At first glance, it would seem that if one believes that it is morally acceptable to use animals to discover new treatments for cancer, for example, then it is no less acceptable to do so in regard to Alzheimer's disease—for there would seem to be no clear moral difference regarding the degree of invasiveness, the net benefit to be gained, or the individual, social, and economic costs to be saved. Furthermore, the argument can be made that the use of nonhuman primates is specifically (and perhaps uniquely) justified in neuroscience research precisely because of their proximity to humans. But this raises the question of whether we should attempt to limit the use of animals in research by distinguishing interventions according to whether they are for "cosmetic" or "medical" purposes (Chatterjee 2006).

Many of those who support the use of animal in scientific research are less permissive in regard to their use in the testing of cosmetics and other consumer products, and presumably this is because we believe that cosmetic testing has less value and does not so readily outweigh the welfare of the animals. It could be claimed that, just as someone might try to enhance self-image or gain self-confidence by getting a tattoo, piercing, or changing hair color, someone might try to achieve the same goal through psychopharmacology. Accordingly, one would require stronger moral justification for the use of animals in cases of "cosmetic" neuroscience research than would be required in the medical cases. More broadly, one might appeal to the (albeit flawed) distinction between treatment and enhancement and argue that stronger justification is required to use animals to investigate ways to enhance cognition above the normal range than to bring people up to this range. It might reasonably be objected that the same type of intervention can be used in both cases, and hence it is not possible to differentiate cosmetic from medical research and treatment from enhancement. For example, Ritalin (methylphenidate) is used both as treatment for ADHD and as cognitive enhancer. In response, it can be claimed that what matters morally is the goal and nature of the research, rather than how the medication or device is used.

A final point to consider is whether the major brain projects will make us more, rather than less, aware of the similarities between human and nonhuman primates. As has been noted before, there is a certain irony and paradox in the fact that the very reason for why nonhuman primates are used in research, namely, their similarity to humans, is also presented as reason for them not to be used in this way. As Colin Allen has argued, disagreements between supporters and critics of the use of nonhuman primates in research are unlikely to be resolved soon, for both similarities and dissimilarities are taken to matter morally (Allen 2006). To a certain degree, this disagreement may be lessened, perhaps, as more is learned about brain function, since it seems that questions such as whether animals feel pain, are self-conscious, or have a theory of mind will, at least in principle, be informed by advances in neuroscience. One may go so far as to say that it is likely that the differences between the primates are differences in physical degree rather than mental kind (Shriver 2006) and that we do not need to know "what's it like" to be a cow, macaque, or marmoset in order to know whether the animal is in pain or depressed.

13.4.4 Education and Training

A final element of RRI pertains to education and training. The major brain projects that have been launched are highly complex, multidisciplinary ventures with far-reaching implications. As stated in Gray Matters, ethics needs to be integrated "early and explicitly" and "at all levels" (p. 43–45). The integration of ethics in this fashion is an essential part of the commitment to RRI, but it is important that when we talk about "integrating ethics," we understand this phrase broadly as encompassing not only the inclusion of nonscientists on review boards and the development of the relevant research protocols but also to include discussion of the sort of global, social, and environmental concerns described briefly above. We should take seriously the charge to integrate ethics "at all levels" and provide education and training for both those who are already established researchers and those who are interested in contributing to the projects in the future (National Institutes of Health 2014). Furthermore, the integration of ethics education and training is only one half of the picture, for there is an equally important need for scientific training for those with limited neuroscientific knowledge. In order for neuroethics to make a significant contribution to the major projects, education and training about current and future neuroscience research and methodologies, brain function, and neurobiology need to be available for those working in neuroethics. An example of a highly regarded program that attempted to address the situation is "Neuroscience Boot Camp," a program established by the Center for Neuroscience and Society at the University of Pennsylvania. In addition, the Human Genome Project funded workshops focusing on the ethical, legal, and social implications of genome research, which also helped provide nonscientists with an introduction to genetics. As Green has argued, there is a need for a similar program to be part of the BRAIN Initiative (Green 2014).

13.5 Responsibility, Crime, and Punishment

The implication of neuroscience for the law and for morality has generated great interest in neuroethics. One reason why this topic has received so much attention pertains to the following line of thought: as we learn more about brain function and the way in which neurophysical states realize psychological states, we may come to see a tension between deterministic neuroscience and normative systems like the law and morality. Whereas the law and morality assume that people are, for the most part, intentional rational agents capable of voluntary action, neuroscience suggests that our actions and behaviors are caused by neurophysical, not psychological, states and events. Accordingly, we cannot really hold people to be responsible for their actions, and we should, therefore, revise our moral and criminal systems.

Within the neuroethics literature, there are two influential responses to this "global challenge" to agency. On the one hand, there are those like Stephen Morse who argue that neuroscience presents no threat because the law is essentially compatibilist and, therefore, unaffected by any deterministic claims (Morse 2004). In assessing whether a person is responsible for her actions, the core question for the law is whether the person was rational and their actions were uncoerced at the time the action was performed. The law is simply uninterested in whether the person "really did it"—if by "really did it" we mean that the person's actions were wholly unconstrained by deterministic causes. A seemingly contrasting position is defended by Greene and Cohen who argue that "for the law, neuroscience changes everything and nothing" (Greene and Cohen 2004). According to their view, although it may be the case that the law is compatibilist, our moral intuitions are libertarian, that is to say, what we want is genuine free will: to say that a person acted voluntarily is to say that there were no deterministic influences. As empirical neuroscience reveals the deterministic nature of human mental life, there will be increasing tension between the law and morality. Eventually, Greene and Cohen contend, we will no longer be concerned with asking whether a person "really" intended to act that way, for we will know intentions do not cause actions. Despite this conclusion, however, they argue that we will continue to praise or blame people for their actions, for although this practice can no longer be justified on deontological grounds, it can be justified on the utilitarian grounds that punishment has social benefits.

Greene and Cohen's view amounts to saying that, even though we know that people are not really responsible for their actions, it is still beneficial and useful in consequentialist terms to hold them so; as Bernard Williams has argued, however, this only works if the practice is not widely known (Smart and Williams 1973: 123). In order for punishment to be effective, it must be closely linked to what is fair or unfair, for if it is arbitrary, it is less likely to appropriately direct behavior: a person, "Simon," will be deterred from wrongdoing only if he believes that he will be punished for wrongdoing, but not for acting rightly (i.e., if he believes that punishment is tied to what is justified or fair). If this condition is not met, then it is difficult to see what reason Simon would have for being deterred from wrongdoing, since he could equally be punished for acting rightly.

The problem for a consequentialist justification of punishment is that there is no direct relation between what is efficacious and what is fair or unfair, since it might be efficacious on consequentialist terms to "punish" those who have not committed an act of wrongdoing in order to act as a deterrent to those who might. If we assume that Simon is aware of this fact, then he has no clear reason to refrain from wrongdoing, since he could be punished either way. In other words, either we have to construct a framework in which people will continue to believe that they will be punished only if they commit acts of wrongdoing, or we have to abandon the attempt to justify punishment on consequentialist grounds.

If the above analysis is correct, then this suggests, contrary to what Greene and Cohen claim, that unless we are prepared to keep people in the dark, neuroscience will change everything. If Greene and Cohen are correct to claim that our moral intuitions are libertarian, then it is reasonable to suppose that these intuitions permeate our society and our social systems. Accordingly, neuroscience will place pressure not only on the law but also on education, government, and any aspect of human society in which notions of agency and responsibility are applicable. Thus, we will be faced with the very real and important question of how to proceed in the future: do we continue to praise and blame people for their actions, even though we all know that free will is an illusion, or do we radically revise the way that we go about leading our lives?

A third option is a version of the compatibilist stance advocated by Morse, one that is more empirically informed and accepts that responsibility claims do say something about the relationship between a person's psychological and neurophysical states. There is a difference between saying the law is only concerned with the question as to whether the accused was rational at the time of committing the crime and saying that neuroscience plays no role in the determination of rationality. If we learn that the behavior of psychopaths is caused by a brain abnormality that impairs their ability to be empathetic, this information could be seen as relevant to the question of whether psychopaths are capable of rational behavior (Levy 2007: 250).

In the future, the problem we may well face is not how to justify punishment given that agency or free will is an illusion but how to frame and define voluntary behavior in a way that is empirically informed. In one sense, deterministic neuroscience can no more reveal free will to be an illusion than it can reveal it to be genuine, since the very question at hand is whether or not determinism and free will are compatible. What we are left with, therefore, is the very practical task of framing and defining our notion of rationality in light of what we learn about brain function.

13.6 The Identity of Neuroethics

In this final section, I wish to change tack and consider the future of neuroethics on a rather more abstract, speculative level. One might think that neuroethics is but the latest entry in the long list of subdisciplines found under the general category of "applied ethics"—a term used to identify any field or subdiscipline that focuses on the ethical issues of a particular profession or social-cultural institution or our obligations to a particular entity or set of entities. Accordingly, biomedical ethics, nanoethics, and genethics focus on the ethical issues found in biomedicine, nanotechnology, and genetics, respectively; organizational ethics pertains to the ethical behavior of institutions, and environmental ethics and animal ethics focus, in turn, on our obligations to the environment and nonhuman animals. All of these fields—albeit to different degrees—are examples of practical ethics since they are empirical and seek to provide ethical guidance in "real life," i.e., actual, particular cases; in this regard, applied ethics is inherently normative and practical.

Owen Flanagan has described what he refers to as the two "grand images of who we are, the humanistic and the scientific" (Flanagan 2002). According to the first of these images, we are, as humans, intentional, rational, conscious agents capable of voluntary action: we are aware of our surroundings and our psychological states; we have thoughts and emotions; we can decide to act one way rather than another, and our actions follow directly from these decisions. This humanistic image underlies our belief that people are, for the most part, agents who can be held responsible for their actions and is therefore an essential component of our normative systems of law and morality.

In contrast, the second of these images describes humans as physical systems exhaustively determined by physical laws of cause and effect. Accordingly, all the causal work occurs at the physical rather than the psychological level. This is not to rule out the possibility that we are conscious, sentient beings, but it is to claim that the mental is either epiphenomenal or reducible to the physical; in other words, the image is mechanistic and the mental has little causal or explanatory role. If we adopt the scientific image, then it would seem that notions like free will, agency, and responsibility are in need of substantial revision. For if the real causal work occurs at the sub-personal, neural level rather than at the conscious personal level, then it seems that we are not really the instigators of our thoughts and actions.

What distinguishes neuroethics from the other types of applied ethics is that it so clearly incorporates both images: on the one hand, neuroethics considers issues that are contingent upon the legitimacy of the humanistic image; on the other hand, the field adopts the scientific image and explores issues that challenge the validity of agency and the very notion that we are conscious, intentional agents. The ethics of neuroscience fits the broad notion of applied ethics as the examination of ethical issues regarding a profession, practice, or discipline (Roskies 2002). In this case the discipline is neuroscience, and neuroethics examines the ethical implications of advances in this discipline, for example, whether advances in neuroimaging pose a threat to privacy or whether harmful memories should be erased through pharmacological means. In contrast, the neuroscience of ethics is consistent with the scientific image and is what sets neuroethics apart from the other types of applied ethics, for all of these other fields take it for granted that we are intentional, voluntary agents. Practical, normative fields like biomedical ethics, business ethics, and environmental ethics have limited scope for they focus on the morality of human actions within particular professions or contexts; there is no suggestion that scientific research in these fields is going to provide us with reason to challenge the assumption of agency.

As the work of the BRAIN Initiative and the other projects discussed move forward, and as we discover more about brain function and the way this function realizes our mental states and directs our behaviors, it is reasonable to surmise that the scientific image will gain greater prominence and may, to some, appear to be the more accurate image. Accordingly, there may appear to be good reason to radically revise our concepts of agency and responsibility and not just within the framework of the law and morality. The discovery that our actions are the result of brain processes rather than conscious thoughts should have an impact on our educational practices and consumer practices and upon all aspects of human behavior. As we move forward, however, we should recognize that across the world, the humanistic view may well have considerably more support than the scientific view. For much of the world, therefore, the two images will remain as they are, namely, two very different and incommensurate perspectives on human nature and who we are.

Conclusion

In 30 years' time, our understanding of the brain could be revolutionized. By 2047 perhaps we will have almost complete knowledge of brain function, of the neurological basis of thought, feeling, and consciousness, and by this time repairing, restoring, and enhancing brain function will have become a routine. The main goal of this chapter was to identify some of the future directions for neuroethics. In so doing, I have deliberately tried to focus on some topics that I believe warrant particular attention as we move forward. The BRAIN Initiative, Human Brain Project, and the other major brain projects are remarkable ventures—and the scientific journey that they are engaged in is appropriately seen to parallel the Apollo program. These projects also raise a host of ethical issues that range from macrolevel questions about public funding to microlevel questions about patient care. In order for societies to have the best chance of providing informed, reasoned, and compassionate answers to these questions, it is vital that these projects include a wide variety of disciplines, participants, and opinions. Neuroethics is well placed to play a substantial and significant role in helping to provide answers to these questions; in fact, one might say that neuroethics bears considerable responsibility to help ensure that the projects' research is conducted in an ethical fashion and that the ethical, legal, and social implications of the research are properly addressed.

References

Allen C (2006) Ethics and the science of animal minds. Theor Med Bioeth 27:375–394

Appelbaum PS (2010) Consent in impaired populations. Curr Neurol Neurosci Rep 10:367–373

Barker RA, de Beaufort I (2013) Scientific and ethical issues related to stem cell research and interventions into neurodegenerative disorders of the brain. Prog Neurobiol 110:63–73

Bateson P (2011) Review of research using non-human primates. The Wellcome Trust, London

Bell E, Racine E, Chiasson P, Dufourcq-Brana M, Dunn LB, Fins JJ et al (2014) Beyond consent in research: revisiting vulnerability in deep brain stimulation for psychiatric disorders. Camb Q Healthc Ethics 23:361–368

Bennett AJ, Ringach DL (2016) Animal research in neuroscience: a duty to engage. Neuron 92:653–657

Blakemore C, Clark JM, Nevalainen T, Oberdorfer M, Sussman A (2012) Implementing the 3Rs in neuroscience research: a reasoned approach. Neuron 75:948–950

Blurton-Jones M, Kitazawa M, Martinez-Coria H, Castello NA, Müller F-J, Loring JF et al (2009) Neural stem cells improve cognition via BDNF in a transgenic model of Alzheimer disease. PNAS 106(32):13594–13599

Buller T (2014) Advance consent for dementia research. J Med Ethics 41:701–707

Carruthers P (1992) The animals issue: moral theory in practice. Cambridge University Press, Cambridge

Chatterjee A (2006) The promise and predicament of cosmetic neurology. J Med Ethics 32(2):110–113

Clausen J (2010) Ethical brain stimulation—neuroethics of deep brain stimulation in research and clinical practice. Eur J Neurosci 32:1152–1162

Dawkins MS (2012) Why animals matter: animal consciousness, animal welfare, and human well-being. Oxford University Press, Oxford

DeGrazia D (1996) Taking animals seriously: mental life and moral status. Cambridge University Press, Cambridge

Dennett DC (1995) Animal consciousness: what matters and why. Soc Res 62(3):691–710

Eyjolfsdottir H, Eriksdotter M, Linderoth B, Lind G, Juliusson B, Kusk P et al (2016) Targeted delivery of nerve growth factor to the cholinergic basal forebrain of Alzheimer's disease patients: application of a second-generation encapsulated cell biodelivery device. Alzheimers Res Ther 8:1–11

Fadiman A (2012) The spirit catches you and you fall down: a Hmong child, her American doctors, and the collision of two cultures. Farrar, Straus and Giroux, New York

Farah MJ (2008) Neuroethics and the problem of other minds: implications of neuroscience for the moral status of brain-damaged patients and non-human animals. Neuroethics 1:9–18

Flanagan O (2002) The problem of the soul: two visions of the mind and how to reconcile them. Basic Books, New York

Garden H, Bowman DM, Haesler S, Winickoff DE (2016) Neurotechnology and society: strengthening responsible innovation in brain science. Neuron 92:642–646

Gauthier S, Leuzy A, Racine E, Rosa-Neto P (2013) Diagnosis and management of Alzheimer's disease: past, present and future ethical issues. Prog Neurobiol 110:102–113

Greely HT, Ramos KM, Grady G (2016) Neuroethics in the age of brain projects. Neuron 92:637–641

Greene J, Cohen J (2004) For the law neuroscience changes everything and nothing. Philos Trans R Soc Lond B 359:1775–1885

Green RM (2014) The need for a neuroscience ELSI program (page inside back cover). Hastings Center Rep 44(4)

Grill JS, Karwalish J (2010) Addressing the challenges to successful recruitment and retention in Alzheimer's disease clinical trials. Alzheimers Res Ther 2(6):1–11

Hackman DA, Farah MJ (2009) Socioeconomic status and the developing brain. Trends Cogn Sci 13(2):65–73

HBS-PS Consortium (2012) The Human Brain Project: a report to the European Commission. www.humanbrainproject.eu. Accessed 30 Oct 2016

Kim SY, Uhlmann RA, Appelbaum PS, Knopman DS, Kim HM, Damschroder L et al (2010) Deliberative assessment of surrogate consent in dementia research. Alzheimers Dement 6:342–350

Kim P, Evans GW, Angstadt M, Ho SS, Sripada CS, Swain JE et al (2013) Effects of childhood poverty and chronic stress on emotion regulatory brain function in adulthood. Proc Natl Acad Sci U S A 110(46):18442–18447

Labuzetta JN, Burnstein N, Pickard J (2011) Ethical issues in consenting vulnerable patients for neuroscience research. J Psychopharmacol 25(2):205–210

Levy N (2007) Neuroethics: challenges for the 21st century. Cambridge University Press, Cambridge

Lipsman N, Glannon W (2013) Brain, mind, and machine: what are the implications of deep brain stimulation for perceptions of personal identity, agency, and free will? Bioethics 27(9):465–470

McEwen JE, Boyer JT, Sun KY, Rothenberg KH, Lockhart NC, Guyer MS (2014) The ethical, legal, social implications program of the National Human Genome Research Institute: reflections on an ongoing experiment. Annu Rev Genomics Hum Genet 15:481–505

Morse SJ (2004) New neuroscience, old problems. In: Garland B (ed) Neuroscience and the law: brain, mind, and the scales of justice. Dana Press, New York, pp 157–198

National Institutes of Health (2014) BRAIN 2025: a scientific vision. Braininitiative.nih.gov

Neuroethics Workgroup Meeting (9 Feb 2016) braininitiative.nih.gov. Accessed 30 Oct 2016

Nuffield Council on Bioethics (2013) Novel neurotechnologies: intervening in the brain

Obama BH (2013) Remarks made by the President on the BRAIN Initiative and American Innovation. www.whitehouse.gov. Accessed 30 Oct 2016

Okano H, Sasaki E, Yamamori T, Iriki A, Shimogori T, Yamaguchi Y et al (2016) Brain/MINDS: a Japanese national brain project for marmoset neuroscience. Neuron 92:582–590

Poo M, Du JL, Ip NY, Xiong ZQ, Xu B, Tan T (2016) China brain project: basic neuroscience, brain diseases, and brain-inspired computing. Neuron 92:591–596

Prado EL, Dewey KG (2014) Nutrition and brain development in early life. Nutr Rev 72(4):267–284
President's Commission for the Study of Bioethical Issues (2014) Gray matters: integrative approaches for the study of neuroscience, ethics, and society, vol 1
President's Commission for the Study of Bioethical Issues (2015) Grey matters: integrative approaches for the study of neuroscience, ethics, and society, vol 2
President's Commission for the Study of Bioethics Issues (2015) Gray matters: topics at the intersection of neuroscience, ethics, and society
RFA-MH-17-260 (2016) BRAIN Initiative: Research on the Ethical Implications of Advancements in Neurotechnology and Brain Science. https://grants.nih.gov/grants/guide/rfa-files/RFA-MH-17-260.html
Roskies A (2002) Neuroethics for the new millennium. Neuron 35:21–23
Shriver A (2006) Minding animals. Philos Psychol 19(4):433–442
Smart JJC, Williams B (1973) Utilitarianism for and against. Cambridge University Press, Cambridge
Stemedica (2016) www.stemedica.com. Accessed 30 Oct 2016
Tost H, Champagne FA, Meyer-Lindenberg A (2015) Environmental influence in the brain, human welfare and mental health. Nat Neurosci 18:1421–1431
World Health Organization (2016a) http://www.who.int/mediacentre/factsheets/fs362/en/ (Dementia). Accessed 20 Nov 2016
World Health Organization (2016b) http://www.who.int/mediacentre/factsheets/fs369/en/ (Depression). Accessed 20 Nov 2016

Growing Up with Neuroethics: Challenges, Opportunities and Lessons from Being a Graduate Student at a Disciplinary Crossroads

14

Cynthia Forlini

Abstract

As neuroethics grew as a field of study, training and research progams produced a cohort of pioneer graduate students. These students helped define the foundations and develop neuroethics scholarship as we know it. In this chapter, I describe three persisting experiences that marked my graduate studies in neuroethics: (1) training and researching at a disciplinary crossroads, (2) establishing a knowledge base in a nascent and interdisciplinary field and (3) practising responsible communication of topical research. Each experience presented unique challenges, opportunities and lessons that marked my journey as a graduate student during the emergence of neuroethics and influenced my trajectory as an early career researcher. I conclude by considering how growing as a researcher alongside a developing field contrasts with the experience of current students within specialized neuroethics training and research programs.

14.1 Introduction

Located at the 'intersection of bioethics and neuroscience' (Racine and Illes 2008), neuroethics is a high traffic area of scholarship. The specific aims are manifold and have practical underpinnings that impact the care of patients, in addition to theoretical and philosophical underpinnings that weave an interdisciplinary understanding of neuroscience (Racine 2008). Different definitional approaches are driven by the ever-expanding range of neuroethical issues such as knowledge generated by and about neuroscience (Roskies 2002), technologies that are being developed and

C. Forlini, Ph.D.
Centre for Values, Ethics and the Law in Medicine, The University of Sydney, Sydney, NSW, Australia
e-mail: cynthia.forlini@sydney.edu.au

E. Racine, J. Aspler (eds.), *Debates About Neuroethics*, Advances in Neuroethics,
DOI 10.1007/978-3-319-54651-3_14

applied to the brain (Wolpe 2004) and the impact of neuroscience on healthcare (Racine and Illes 2008). Pluralism in the definitions of neuroethics implies that it is more than bioethicists and neuroscientists that are populating this disciplinary intersection. They are joined by healthcare professionals of all types, philosophers, policy makers and science communication specialists, to name a few other stakeholders. Neuroethics has become more than an intersection of just two paths, but a crossroads of many avenues of research inquiry.

Like many of my predecessors and peers, I came to neuroethics from a science background. The biochemistry program I was enrolled in mandated a bioethics unit during the last year of study. Having reignited my interest in the responsible communication of science to the public, that bioethics unit single-handedly determined the course of my graduate training. My initial interest was in genetic testing. In 2006, through a series of fortuitous circumstances, I would be introduced to a newly formed neuroethics research unit and proposed a project that examined stakeholder perspectives and public understanding of the non-medical use of stimulants for cognitive enhancement.

The ethics of cognitive enhancement are complicated, to say the least. At the heart of the debate is a tension between, on the one hand, the rights and liberties of individuals to make choices about how to best achieve their goals and, on the other hand, the social outcomes (e.g., peer pressure and dishonesty) of enhanced performance in the context of competitive academic and professional environments (Racine and Forlini 2009). Other issues such as distributive justice and authenticity add to the ethical and social complexity of cognitive enhancement (Forlini and Racine 2012). During my Master of Arts (Forlini 2008) and PhD (neuroscience) (Forlini 2013), I worked to untangle these issues as my contribution to a more productive debate. Couching my research within neuroethics, a disciplinary crossroads, offered the flexibility to approach these issues with an arsenal of knowledge and evidence drawn from ethics, philosophy, social sciences, journalism and neuroscience.

The perspective that I present in this chapter is a reflection upon the challenges that I faced as a graduate student at a time when neuroethics was still the 'new kid' on the bioethics block. Difficult as they were, these challenges revealed themselves as opportunities to become a more skilful researcher and remain with me as lessons that continue to shape the way I conduct research in neuroethics as an early career academic. I do not pretend to speak for all students in my cohort or early career researchers in neuroethics. Experiences would vary according to academic background, country, university, supervisor and available resources (e.g., funding and institutional support), among many other factors. Rather, the goals of this chapter are to acknowledge the journey taken by trainees and to testify to the impressive expansion of neuroethics scholarship over the last decade.

In order to illustrate the challenges, opportunities and lessons that became significant training milestones for me, I embark on a tiny bit of nostalgia tinged with the kind of clarity and wisdom that only hindsight can provide. I focus on three persisting experiences that marked my graduate studies in neuroethics: (1) training and researching at a disciplinary crossroads, (2) establishing a knowledge base in a nascent and interdisciplinary field and (3) practising responsible communication of

Table 14.1 The challenges, opportunities and lessons for the future derived from characteristic experiences of a graduate student in the early days of neuroethics

Experience	Challenge	Opportunity	Lesson for the future
Training and researching at a disciplinary crossroads	Determining appropriate enrolment in a disciplinary program	Creating space in different university departments for neuroethics scholarship	Forming strong research groups that engage multiple university departments
	Justifying methods and relevance of neuroethics to intersecting fields	Adapting presentation of research to other disciplinary contexts	Clearly communicating neuroethics research to extra-disciplinary audiences
Establishing a knowledge base in a nascent and interdisciplinary field	Few pedagogical guidelines and training opportunities that help to create a foundation for building expertise	Exploring and refining knowledge base with mentors, colleagues and peers	Fostering collaboration and dialogue that contributes to shared understanding of neuroethics scholarship
Practising responsible communication of topical research	Satisfying the appetite for answers to socially and clinically pressing questions	Honing communication skills to effectively interact with public audiences and media requests	Directly engaging with other sources of expertise that can shed light on blind spots arising in neuroethics scholarship

topical research. Table 14.1 synthesizes these experiences and their respective challenges, opportunities and lessons that will be systematically addressed in the rest of this chapter. I conclude with some thoughts on the trend toward specialization in neuroethics and the effect that might have on more recent cohorts of graduate students and the future researchers they will become.

14.2 Same, Same but Different: Training at a Disciplinary Crossroads

Neuroethics' position at a disciplinary crossroads creates the perfect venue for dialogue, debate and deliberation between scholars as they grapple with theoretical and empirical questions. For a student, this rich environment offers many opportunities to learn, explore and innovate. However, keeping a permanent station at the crossroads can be lonely. Consider a driver approaching a crossroads. The driver has two options: (1) to change directions and consider another route to their destination or (2) to orient themselves with regard to other places and landmarks as indicated by the road signs. The driver does not and cannot remain at the crossroads. The crossroads is a place for transition and orientation, but not for stopping or building a foundation. The same might be said of scholars who come toward neuroethics. Under the auspices of neuroethics, they can engage with others, but they ultimately return to their home discipline. Where does that leave students who want to train and specialize in neuroethics, a field that is by definition at a disciplinary crossroads?

Training at the disciplinary crossroads that neuroethics represents poses two challenges that are not typically present in fields with longer histories. The first challenge relates to picking a training program within the university, and the second is making research relevant to that chosen discipline. I was fortunate enough to train within a group whose research was and still is focused on neuroethics. Daily interactions were with colleagues who were working toward the same goals and generally through the same challenges as I. At the university program level, things were different. The 'neuro' prefix would prove to be a label that I often had to explain and defend.

Coming up through the Master's level, I was enrolled in a bioethics program. This enrolment was relatively uncontroversial. Bioethics was an appropriate venue in which to learn about the principles and methodological approaches that I would use in my research. However, throughout this period, the position of neuroethics with respect to bioethics (i.e., whether it was within or outside the purview of bioethics) was an area of contention. Parens and Johnston (2007) famously asked whether neuroethics was 'reinventing the bioethics wheel' by unnecessarily overspecializing and perpetuating the same limitations as the rest of bioethical inquiry. After all, neuroethics did not offer any frameworks or methods distinct from what I was learning in the bioethics curriculum. What was the added value of identifying as a neuroethics researcher within a bioethics program? Research focused on improving outcomes for research participants, patient care, and public understanding of neuroscience, some of the tenets of neuroethics (Racine 2010), seemed to be as productive an approach to ethics research as any. However, I wondered whether seeking empirical outcomes overshadowed the normative questions at the core of ethical challenges in neuroscience and whether, ultimately, my empirical experience could satisfy the expectations of jobs in bioethics departments after my studies. Might my work be more relevant if it were couched in a neuroscience program?

At the PhD level, I enrolled in a neuroscience program at another university, but anchored my work within the same neuroethics research group. In the neuroscience program, the coursework was less flexible, and I found myself in lectures about neurotransmission and cognitive neuroscience. My peers were molecular biologists, healthcare professionals, neuroimagers and psychologists—in other words, neuroscientists with a biological approach to neuroscience. These were the practitioners that neuroethics research aims to engage with and help, yet I found myself feeling disconnected. My work in cognitive enhancement was appreciated for its pertinence to the university setting, but my approach was far from understood. Few of my neuroscience colleagues had encountered the issues at the heart of neuroethics inquiry. Even fewer colleagues were familiar with the ethical concepts and qualitative methods that I used in my research. Studies on the attitudes of neuroscientists have since shown that my impression was correct. For example, ethical issues in neuroscience are not consistently addressed in neuroscience training programs (Morein-Zamir and Sahakian 2010). Even at the more senior level, neuroscientists may not be sensitized to the salient ethics issues related to neuroscience or the impact of neuroethics research on their day-to-day work (Brosnan and Cribb 2014). Whereas bioethics and neuroethics appeared too similar, I now faced an epistemological chasm dividing neuroscience from ethics even though they ought to have been intersecting.

The disciplinary pivot that I have described highlights the opportunities to refine neuroethics scholarship in bioethics, neuroscience and other intersecting fields. In addressing the numerous critiques of the emergence of neuroethics, Racine observed that 'several controversies surrounding the birth and evolution of neuroethics are based on some straightforward misunderstandings and caricatures' (Racine 2010: 95). Imperative to dispelling these misunderstandings and caricatures is demonstrating to the intersecting disciplines the value of coming to the crossroads where neuroethics operates. This task begins by recognizing how disciplinary divides affect perceptions of the pursuit(s) of 'ethics'. Brosnan and Cribb (2014) propose that the gap between bioethics and neuroscience might be due in part to the diverging ethical pursuits of both groups of researchers, with the former focusing on the broader permissibility of neuroscience and the latter on its day-to-day implementation. For now, disciplinary divides and divergent ethical pursuits may influence neuroethics trainee enrolment in university programs and perhaps even how their research is received, but they must not prevent these trainees from engaging with neuroethics scholarship at the disciplinary crossroads.

The task of inviting intersecting disciplines to engage with neuroethics continues with the development of neuroethical tools to analyse issues that arise in neuroscience. A recent example is the 'Operational Neuroethical Risk Analyses and Management Paradigm (ON-RAMP)' proposed by Giordano (2016) that considers emerging neuroscience and technology multidimensionally. The ON-RAMP is a four-step process that (1) queries the type and objectives of the science and technology, (2) frames the intended applications in their context of use, (3) defines the domains (mostly sociocultural) that are likely to be impacted by the neuroscience and technology and (4) models and plots trajectories of effect. Adapting mandatory ethics content such as responsible conduct of research courses to the neuroscience context would be another approach that would sensitize trainees in a way that they could more readily recognize and engage with the potential ethics issues they will face in their research (Bell 2015). Both these strategies could create space for neuroethics so that it might be a mainstay in university programs and disciplines where it was once a novelty or an outlier.

The direct, although long-term, benefit of broadening the disciplinary reach of neuroethics at the program level is the eventual diversification of research groups. Leefmann et al. (2016) recently showed that the majority of authors publishing in neuroethics have scientific backgrounds. Furthermore, they reported that the 'bulk of neuroethics research since 1995 has been published in journals that usually do not conceive of moral, legal and social questions of biomedical research as their proper topics' (Leefmann et al. 2016). These results are paradoxical given the data cited above showing limited awareness of neuroethics among neuroscientists. Such strong representation of the scientific community is encouraging, but also indicative of underrepresentation and lack of crossover with other intersecting disciplines. By offering opportunities to engage with neuroethics early on in training programs, universities can create common ground between programs, departments and disciplines that would naturally meet at the neuroethics crossroads.

14.3 Building a Knowledge Base at the Crossroads: A Team Effort

Creating common ground for the different stakeholders in neuroethics scholarship begins with establishing a knowledge base of core competencies developed during neuroethics training. Many consider the birth of neuroethics as we know it to have coincided with the Neuroethics: Mapping the Field conference in 2002 (Safire 2002). Though the issues addressed by neuroethics scholarship, the concept of an area of ethics dedicated to issues of neuroscience and even the term 'neuroethics' all predate that meeting, it is fair to consider 2002 the moment where 'neuroethics' entered the ethics lexicon in a lasting manner (Racine 2010). As a result, the term 'neuroethics' was put into play before many of the seminal texts and lectures we now consider to be foundational appeared. Producing the reference material that makes up the knowledge base would necessarily be a stepwise endeavour. Whereas the previous section addressed the experience of determining the breadth of neuroethics training, this section will consider the experience of creating depth and laying the foundation for a common knowledge base for trainees. What is the appropriate knowledge base for a nascent and interdisciplinary field? Pedagogically speaking, what are the priorities for neuroethics training? Perhaps, more importantly, how is this knowledge base best accessed?

When I joined a newly formed neuroethics research group in 2006, neuroethics had become a rich context for researchers. There was a wealth of concepts, ideas and projects to develop, from theoretical and conceptual development of the burgeoning field to empirical investigations of pressing issues in neuroscience research, clinical care and public understanding. Coming to neuroethics from biochemistry with no background in neuroscience or ethics and philosophy, the challenge was finding the tools that would help guide my pursuit of knowledge to build a foundation for my research. I was welcomed to neuroethics with much enthusiasm for the possibilities that lay ahead and with a stack of publications to read—both articles and important volumes of the time (Glannon 2007; Illes 2006). Neuroethics would not be an option in the coursework offered by my Master of Bioethics program until a few years later (2008). Other than publications, there were few pedagogical options that curated the salient topics and approaches useful for the study of neuroethics issues. Without the handbooks, academic journals, units of study and conferences that exist today, neuroethics appeared to be more of a subculture seeded within bioethics and neuroscience than an emerging field in its own right. My knowledge base was initially cobbled together from many types of literature, but still separated by disciplinary fault lines.

Two valuable opportunities emerged from self-curating my knowledge base in neuroethics during my graduate studies. First, I learned how to delve into the literature of other disciplines (e.g., behavioural psychology, sociology and linguistics) unfamiliar as they may have been, to find the theories and evidence that would help me to better understand my area of research. From this exploration of different disciplines, a simple literature review turned into a full-fledged project that I would base my Master's thesis on. The result was a novel finding that the bioethics literature, public health literature and print media had diverging discourses on the ethics of the non-medical use of stimulants for cognitive enhancement (Forlini and Racine 2009; Racine and Forlini 2010). Each of these literatures constructed a unique paradigm to address exactly the same phenomenon. The project was a testament to not only the interdisciplinary nature of neuroethical issues but also the role neuroethics inquiry can play in creating common ground for related disciplines. By integrating the knowledge generated in disciplines relevant to my research (bioethics, epidemiology and science communication), I identified blind spots in the ethics debate that were in need of attention.

Second, being part of a research unit meant that I was not alone in my quest to build a knowledge base in neuroethics at the time of its infancy. I had a set of peers in close proximity that were on the same journey to learn about what neuroethics was and contribute to what it should be. A variety of research topics pulsed through the group. I learned about them as my colleagues shared their self-curated approach to other specific topics in neuroethics. We gained a deeper understanding of the diverse issues that fell within the purview of the neuroethics that we saw growing up around us and the outcomes that solid inquiry can generate. Through our exchanges, my peers were my curators of neuroethics knowledge, and I, theirs. Through journal clubs, seminars and spontaneous chats about methodological and theoretical road blocks, I learned about deep brain stimulation, neuroimaging, disorders of consciousness, brain death and neurodevelopmental disorders, just to name a few topics. This learning was ongoing as we followed the progress of each other's projects. These organized and sometimes spontaneous meetings were our classroom, laboratory and, ultimately, road to amassing knowledge and developing expertise at a time when there were few guideposts to determine the course of our training.

At the very core of individual and joint efforts to build a knowledge base during my neuroethics training was an understanding that much of the work we were doing was giving voice to a group or piece of knowledge that needed to be amplified and integrated into debates about ethical issues related to the brain. Collaboration was key for this type of endeavour. The first meeting of the Neuroethics Society[1] (INS) was held in Washington, D.C., in 2008. No one there at the time wore the label of 'neuroethicist', as they might now be referred to. It was a meeting of minds from many different disciplines that were interested in exploring what a new field could do to help identify and solve the distinct ethical challenges that arise as we seek to understand more about the brain and mind. More than anything, it was a dialogue with the objective of starting to create common ground. Even as a trainee at that meeting, I felt I was part of the movement to characterize the shared endeavour of neuroethics scholarship. The recurrence and success of subsequent INS meetings is a testament to how neuroethics scholarship lends naturally to interdisciplinary collaboration. Numerous resources have since emerged from dialogues and collaborations such as this conference, which have helped to establish a knowledge base for neuroethics: bibliographies (Buniak et al. 2014; Darragh et al. 2015; Martin et al. 2016), handbooks (Illes and Sahakian 2011; Clausen and Levy 2015), journals such as *Neuroethics* and *AJOB: Neuroscience* and meetings (e.g., International

[1] Now known as the *International Neuroethics Society*: http://www.neuroethicssociety.org/

Neuroethics Society and *Brain Matters*). Graduate students are no longer self-curating to the extent that my colleagues and I once had to because neuroethics has now developed for itself more intricate scaffolding.

14.4 Baptisms by Fire: Practising Responsible Communication of Neuroethics Research

A chapter about training in neuroethics would not be complete without acknowledging an intersection of the interdisciplinary crossroads with the public sphere. Neuroscience is pervasive in public health policy across the lifespan (Broer and Pickersgill 2015). Indeed, stakeholders outside academia assign personal and professional meaning to discourses about the brain that changes the way they might approach neurological and psychiatric diseases or roles as healthcare professionals (Pickersgill et al. 2015). It is inevitable that stakeholders, including the media, want to report on cutting-edge research and flesh out conflicts and issues that occur as new technology or new ways of thinking are introduced into clinical practice and society. There is, however, a risk of creating ambiguity and misinformation if facts are not responsibly and accurately reported. Notable examples include (1) confusion regarding the prognosis of Terri Schiavo in reporting about the ethics of withdrawing artificial hydration and nutrition in patients with disorders of consciousness (Racine et al. 2010), (2) colloquial use of the term 'brain death' incongruent with the determination of death by neurologic criterion (Daoust and Racine 2014), (3) assumption that neuroimaging can provide tangible evidence of human consciousness (Racine et al. 2005) and (4) the myth that donepezil, and medication used to treat Alzheimer's disease, is an effective cognitive enhancer for healthy individuals (Wade et al. 2014). Discussion of neuroethics issues needs to be framed by responsible communication and interpretation of neuroscience findings as well as careful consideration of the accompanying neuroethical deliberation. The process of translating research into accurate and balanced information for general audiences is the responsibility of scientists, ethicists and members of the media alike (Forlini et al. 2015).

Delivering accurate and balanced accounts of one's research is a skill that can only be honed through practice. Given the incremental nature of research, it is challenging to satisfy the appetite for answers to the socially and clinically pressing questions that nourish neuroethics inquiry. Neuroethics addresses cutting-edge issues deriving from technological innovation, legal ambiguities, or the evolution of philosophical concepts and is expected to contribute solutions to constantly shifting problems. Audiences outside ethics may not appreciate the difficulty of turning an 'is' into an 'ought'. With so many examples of miscommunication or misrepresentation of research findings within neuroethics, trainees might not feel comfortable engaging with audiences outside neuroethics. It might be equally daunting for them to (1) communicate authority and expertise when engaging with audiences outside neuroethics or (2) say 'we don't know yet' instead of providing hasty and desirable answers. Both of these points reflect necessary skills for researchers committed to engaging meaningfully with other disciplines and stakeholders.

The media coverage of cognitive enhancement over the last decade has been impressively unrelenting. Waves of enthusiasm are cyclic, swelling with the publication of new research (Wade et al. 2014). I've been called upon numerous times by the media, even as a trainee, to comment on research findings or human interest stories about cognitive enhancement by virtue of my research. I have also served on panels for science communication events directed at engaging students and members of the public. Gaining first-hand experience engaging with audiences outside neuroethics and mostly outside academia brought about two opportunities. First, it required me to periodically reflect on the most important messages for different audiences that emerged from my research. Formulating a clear and concise stance helps to open the channels of communication more effectively. Second, engaging with groups outside neuroethics provides opportunities to implement communication recommendations that have been developed as a part of neuroethics scholarship (Illes et al. 2010; Racine et al. 2005). I continue to seek these opportunities for engagement and appreciate them as occasions to examine the import of my research and learn from wisdom beyond neuroethics and academia.

Being an expert is not only about providing information and correcting misconceptions, but it is also about seeking out and attending to other sources of knowledge. The parallel lesson to the importance of responsible communication of research is the willingness to listen to other voices and sources of knowledge both within and outside the disciplinary crossroads. Brosnan and Cribb (2014) showed that 'the ethical issues that were seen as relevant by our neuroscientists were not those receiving attention by neuroethics' (p. 114). Neuroethics should not be missing the mark with key stakeholder groups. The same is true of audiences and stakeholders outside academia. They are not one homogenous 'public' to be painted with the same brush. Nor do they conform to a 'deficit model' of public understanding of science (Wynne 1993) that presumes 'the public usually has considerable lack of knowledge on and understanding of science' (Schicktanz et al. 2012). The distinction between researcher (ethicist) and stakeholder (lay citizen) is perhaps more usefully reframed as a difference between a group that 'knows something' and another that 'knows something else', respectively. Enthusiasm for the promises of neuroscience and neurotechnology (i.e., neurohype) is perhaps not something to be immediately corrected but rather a phenomenon to be considered as an indication of which topics and issues are likely to be prioritized by certain groups (van de Werff et al. 2016). Neurohype may avail itself to be a road map that allows researchers working in neuroethics to reflect on the ethical implications of neuroscience and how it supports or refutes current norms and values.

14.5 Toward a Neuroethics Specialty: All Grown Up?

Many of the challenges, opportunities and lessons described in this chapter are intrinsic to graduate training regardless of program and discipline. These aspects of graduate training were compounded given the context of emergence and growth for neuroethics. Students in my cohort may have experienced discomfort from being in disciplinary contexts that were not well suited to developing expertise in neuroethics. This feeling will likely become increasingly rare given the sharp increase of specialized neuroethics training opportunities that have sprouted in the last decade on an international scale. Buniak et al. (2014) published an extensive list of universities, think tanks and governmental agencies in six regions spanning the globe that house and nurture neuroethics program. These programs are a testament to the valuable normative and empirical contributions neuroethics makes to intersecting disciplines. They are also recognition of the need for the function and role of a neuroethicist in diverse settings (Fischbach and Mindes 2011) and therefore the need to 'cultivate a cadre of multi-disciplinary and task agile subject matter experts that can operate on local to global levels' (Giordano 2016). With dedicated neuroethics programs offering curated training, it seems that neuroethics is retreating from the crossroads to instead become a cornerstone of neuroscience and medical specialities. This progress is the direct result of the tireless work and advocacy of scholars that I admire and take as models for my journey as researcher in neuroethics.

I conclude with a few thoughts from the perspective of an early career researcher about some effects that the specialization of neuroethics training may have. The lessons I have described in this chapter have served me well and continue to inspire me to work in neuroethics and engage with stakeholder groups. As I progressed after my studies to seek fellowships, funding and university positions, I discovered that the fundamental disciplinary divides persist. Neuroethics is theoretically tethered to philosophy, ethics and the social sciences while being practically linked to neuroscience and medicine. Neuroethics research is also difficult to box into the field of research codes one must assign a project in funding applications to designate the relevant subjects (and reviewers). More often than not, these codes create a piecemeal portrayal of our research activities as they reflect the distinct disciplinary streams rather than the integrative and analytic endeavour of neuroethics. Publishing neuroethics research outside of specialized neuroethics journals poses similar challenges to funding. While universities are building capacity for neuroethics, other infrastructures essential to research are not evolving with the same enthusiasm or speed. Specialization is still very much a work in progress beyond the training context. As a result, academic positions in neuroethics proper are much more limited than the opportunities to engage with neuroethics from an intersecting discipline, especially those of neuroscience and clinical practice.

The numerous university training programs and research groups dedicated to neuroethics are fortifying the structure of neuroethics as a discipline in its own right. It is wonderful to be able to offer the next cohorts of trainees a curated pedagogical approach to neuroethics that provides the background knowledge and tools to situate themselves and their interests within neuroethics. However, I do wonder, as I now design and deliver a graduate seminar in neuroethics for the first time in my career, whether to some extent I am doing a disservice to students in presenting an edited snapshot of neuroethics. Might the specialization of neuroethics training be creating an insular environment away from the complexity of the disciplinary crossroads upon which neuroethics was built? Instead of colonizing new territory, perhaps the objective of neuroethics training would be better oriented toward ensuring that trainees are at ease operating at a disciplinary crossroads. In this fashion, the 'neuroethicist' is not simply another actor at the crossroads, but rather, the traffic conductor with both the breadth and depth of knowledge to ensure free flowing ideas and progress in an area that will only continue to pose more complex questions and challenges.

References

Bell E (2015) A room with a view of integrity and professionalism: personal reflections on teaching responsible conduct of research in the neurosciences. Sci Eng Ethics 21(2):461–469.

Broer T, Pickersgill M (2015) Targeting brains, producing responsibilities: the use of neuroscience within British social policy. Soc Sci Med 132:54–61.

Brosnan C, Cribb A (2014) Between the bench, the bedside and the office: the need to build bridges between working neuroscientists and ethicists. Clin Ethics 9(4):113–119.

Buniak L, Darragh M, Giordano J (2014) A four-part working bibliography of neuroethics: part 1: overview and reviews—defining and describing the field and its practices. Philos Ethics Humanit Med 9:9.

Clausen J, Levy N (eds) (2015) Handbook of neuroethics. Springer, Dordrecht

Daoust A, Racine E (2014) Depictions of 'brain death' in the media: medical and ethical implications. J Med Ethics 40(4):253–259.

Darragh M, Buniak L, Giordano J (2015) A four-part working bibliography of neuroethics: part 2—neuroscientific studies of morality and ethics. Philos Ethics Humanit Med 10:2

Fischbach R, Mindes J (2011) Why neuroethicists are needed. In: Illes J, Sahakian BJ (eds) The Oxford handbook of neuroethics. Oxford University Press, New York, pp 343–376

Forlini C (2008) Examining discourses on the ethics and public understanding of cognitive enhancement with methylphenidate M.A. Université de Montréal, Montréal

Forlini C (2013) Examining stakeholder perspectives and public understanding of the ethical and social issues of cognitive enhancement using methylphenidate. McGill University, Montréal

Forlini C, Racine E (2009) Disagreements with implications: diverging discourses on the ethics of non-medical use of methylphenidate for performance enhancement. BMC Med Ethics 10:9

Forlini C, Racine E (2012) Added value(s) to the cognitive enhancement debate: are we sidestepping values in academic discourse and professional policies? AJOB Prim Res 3(1):33–47

Forlini C, Partridge B, Lucke J, Racine E (2015) Popular media and bioethics scholarship: sharing responsibility for portrayals of cognitive enhancement with prescription medications. In: Clausen J, Levy N (eds) Handbook of neuroethics. Springer, Dortecht, pp 1473–1486

Giordano J (2016) Toward an operational neuroethical risk analysis and mitigation paradigm for emerging neuroscience and technology (neuroS/T). Exp Neurol 287(Pt 4):492–495.

Glannon W (2007) Bioethics and the brain. Oxford University Press, Oxford

Illes J (2006) Neuroethics: defining the issues in theory, practice, and policy. Oxford University Press, Oxford

Illes J, Sahakian BJ (eds) (2011) Oxford handbook of neuroethics. Oxford University Press, New York

Illes J, Moser MA, McCormick JB, Racine E, Blakeslee S, Caplan A et al (2010) Neurotalk: improving the communication of neuroscience research. Nat Rev Neurosci 11(1):61–69

Leefmann J, Levallois C, Hildt E (2016) Neuroethics 1995-2012. A bibliometric analysis of the guiding themes of an emerging research field. Front Hum Neurosci 10:336.

Martin A, Becker K, Darragh M, Giordano J (2016) A four-part working bibliography of neuroethics: part 3—"second tradition neuroethics"—ethical issues in neuroscience. Philos Ethics Humanit Med 11(1):7.

Morein-Zamir S, Sahakian BJ (2010) Neuroethics and public engagement training needed for neuroscientists. Trends Cogn Sci 14(2):49–51.
Parens E, Johnston J (2007) Does it make sense to speak of neuroethics? Three problems with keying ethics to hot new science and technology. EMBO Rep 8 Spec No:S61–S64
Pickersgill M, Martin P, Cunningham-Burley S (2015) The changing brain: neuroscience and the enduring import of everyday experience. Public Underst Sci 24(7):878–892.
Racine E (2008) Comment on "Does it make sense to speak of neuroethics?". EMBO Rep 9(1):2–3
Racine E (2010) Pragmatic neuroethics: improving treatment and understanding of the mind-brain. Basic bioethics. MIT Press, Cambridge
Racine E, Forlini C (2009) Expectations regarding cognitive enhancement create substantial challenges. J Med Ethics 35(8):469–470
Racine E, Forlini C (2010) Cognitive enhancement, lifestyle choice or misuse of prescription drugs? Ethical blindspots in current debates. Neuroethics 3(1):1–14
Racine E, Illes J (2008) Neuroethics. In: Singer PA, Viens AM (eds) The Cambridge textbook of bioethics. Cambridge University Press, Cambridge, pp 495–504
Racine E, Bar-Ilan O, Illes J (2005) fMRI in the public eye. Nat Rev Neurosci 6(2):159–164
Racine E, Karczewska M, Seidler M, Amaram R, Illes J (2010) How the public responded to the Schiavo controversy: evidence from letters to editors. J Med Ethics 36(9):571–573.
Roskies A (2002) Neuroethics for the new millennium. Neuron 35(1):21–23
Safire W (2002) Visions for a new field of neuroethics. In: Marus SJ (ed) Neuroethics: mapping the field. Dana Press, Washington, DC
Schicktanz S, Schweda M, Wynne B (2012) The ethics of 'public understanding of ethics'—why and how bioethics expertise should include public and patients' voices. Med Health Care Philos 15(2):129–139.
van de Werff T, Slatman J, Swierstra T (2016) Can we "remedy" neurohype, and should we? Using neurohype for ethical deliberation. AJOB Neurosci 7(2):97–99
Wade L, Forlini C, Racine E (2014) Generating genius: how an Alzheimer's drug became considered a 'cognitive enhancer' for healthy individuals. BMC Med Ethics 15:37.
Wolpe PR (2004) Neuroethics. In: Post SG (ed) The encyclopedia of bioethics. MacMillan Reference, New York, pp 1894–1898
Wynne B (1993) Public uptake of science: a case for institutional reflexivity. Public Underst Sci 2(4):321–337

Neuroethics Research in Europe

15

Ralf J. Jox and Sabine Müller

Abstract

The emerging field of neuroethics attracts researchers from diverse disciplines to study the ethical implications of neuroscience, its technological applications, and its impact on manifold practices in society. During the last two decades, this neuroethical endeavor has developed remarkably, spurred by funding agencies and policy makers who not only promoted transnational neuroethics research but also the building of networks and scientific platforms, such as neuroethics conferences, societies, and journals. However, except in North America, traditional academic institutions like universities and private research organizations have been slow to commit themselves to neuroethics. This is the case in Europe. Looking ahead, the future of neuroethics will undoubtedly be fueled by future advances in the neurosciences, but its acceptance and usefulness will depend on the pursuit of rigor and quality, an intensified interdisciplinary integration, and a truly translational approach.

15.1 Introduction

Historically, the beginnings of neuroethics were predominantly shaped by scholars in Canada and the United States. It was here that the term was first coined by Anneliese Pontius in New York in the 1970s and that the legendary conference in

R.J. Jox, M.D., Ph.D. (✉)
Institute of Ethics, History and Theory of Medicine, Ludwig-Maximilian University Munich, Lessingstr. 2, 80336 Munich, Germany
e-mail: ralf.jox@med.lmu.de

S. Müller, M.Sc., Ph.D.
Department for Psychiatry and Psychotherapy, CCM, Charité—Universitätsmedizin Berlin, Charitéplatz 1, 10117 Berlin, Germany
e-mail: mueller.sabine@charite.de

E. Racine, J. Aspler (eds.), *Debates About Neuroethics*, Advances in Neuroethics,
DOI 10.1007/978-3-319-54651-3_15

San Francisco in the year 2002 propagated neuroethics as a new research field (Pontius 1973; Safire 2002; Marcus 2004). Not even 15 years have passed since this conference, but neuroethics has indeed developed considerably—not only in North America but also in Europe and elsewhere (Racine 2010). We define neuroethics as a research *field*, to distinguish it from homogeneous *disciplines* with their specific methods and exclusive object areas. Neuroethics instead draws from several disciplines, uses several methods, and does not even have a unanimously and agreed upon definition yet. In our eyes, it is a subset of applied ethics with a large interface with bioethics that also has interfaces with neurophilosophy, military ethics, and political ethics (Jox et al. 2010). An emerging and relatively small research field like neuroethics will benefit considerably from efforts to improve communication, knowledge exchange, and networking among its researchers on a global scale. Therefore, we think that it is of value to characterize the state of neuroethics research in Europe, not only to inform the "New World" what is going on in the "Old World" in this respect but also to promote a better integration of European neuroethicists among themselves.

We will structure our chapter in three parts: First, we will present the existing infrastructures for European neuroethics research in terms of institutions, societies, and journals, without neglecting to point out the deficiencies and challenges in this area. Second, we will highlight some major examples of European neuroethics research projects with their topics, aims, methods, funding sources and outcomes. Third, we will look at the current state of training opportunities for young neuroethics researchers, both the formal academic training that is offered by master's or PhD programs and the informal training that can be achieved by scientific conferences, symposia, summer schools, and similar meetings.

15.2 Neuroethics Infrastructures in Europe

15.2.1 Neuroethics Institutions

Universities in Europe are traditionally organized in faculties and departments. Within these larger structures there are institutions or chairs whose denominations commonly derive from their teaching portfolios, which are, however, often described in terms of traditional academic disciplines. Thus, it is not surprising that there are almost no institutions, chairs, or professorships that explicitly contain the term "neuroethics" or "ethics in neuroscience" in their names (given that there are no specific undergraduate degrees in neuroethics). The few exceptions that exist, such as the Oxford Centre for Neuroethics or the department of "Ethics in the Neurosciences" at the Jülich Research Center in Germany, are interdisciplinary platforms or small subgroups of institutes. On the face of it, there is a significant difference when compared to North America, where the number of academic centers devoted solely to neuroethics is higher.

Most neuroethics researchers with high publication output are based in the United States, followed by Canada, Great Britain, Germany, and Australia

(Leefmann et al. 2016). There exist special neuroethics research centers or larger neuroethics research units. Some of the well-known examples are the Center for Neuroscience and Society, University of Pennsylvania (USA); the Mind, Brain Imaging and Neuroethics Research Unit, University of Ottawa; the Canadian National Core for Neuroethics, University of British Columbia; and the Neuroethics Research Unit, IRCM, Montréal (all Canada) (Leefmann et al. 2016).

Yet, this does not mean that there is no neuroethics research activity elsewhere in the world, including in Europe (Fukushi et al. 2007; Greely et al. 2016; Erbguth and Jox 2017). As university professors and senior scholars are free to choose their topics of research, some of them indeed focus on neuroethics notwithstanding their broader affiliations and teaching portfolios in bioethics, psychiatry, cognitive psychology, neuroscience, criminal law, or practical philosophy. According to a recent bibliometric study on the authors of neuroethics journal publications on a global scale, the majority of neuroethicists work inside the institutional structures of either the biomedical sciences or bioethics (Leefmann et al. 2016). Only a minority is based in the traditional humanities and the social sciences (Leefmann et al. 2016). They predominantly have an academic background in neuroscience, psychology, or medicine, whereas researchers from philosophy, bioethics, and the social sciences are marginal in this field (Leefmann et al. 2016).

Europe also hosts a number of non-university research institutions, such as the Max Planck institutes, many of which are devoted to neuroscience and conduct excellent research. Yet, most of them have been reluctant to launch research on the ethical, legal, and social implications of neuroscience. Notable exceptions are the French National Institute of Health and Medical Research (INSERM) and the already mentioned Jülich Research Center in Germany.

The fact that European neuroethicists are usually embedded in non-neuroethics surroundings has both advantages and disadvantages. It may enable them to establish cross-links to other discourses in ethics, neuroscience, medicine, or law and use insights from these areas to approach neuroethics topics. Yet, it can also hamper the depth and speed of neuroethics research in Europe, put neuroethics scholars in a rather isolated situation, and prevent predictable career advancement due to this risky specialization within a more generalist research context.

15.2.2 Neuroethics Societies

A small meeting held in Asilomar, California, in 2006, led to the idea to launch a *Neuroethics Society*. The Asilomar attendees felt the need for such an organization to promote the kind of sustained interaction, learning, and critical discussion that will strengthen the field of neuroethics and draw new people into this field. In 2010, they aligned with the *International Neuroethics Network (INN)*, an interdisciplinary group of scholars, scientists, clinicians, and other professionals who share an interest in the social, legal, ethical, and policy implications of advances in neuroscience. Its mission is to promote the development and responsible application of neuroscience through interdisciplinary and international research, education, outreach and public engagement for the benefit of people of all nations, ethnicities, and cultures. In 2011, the name of the *Neuroethics Society* was changed to the *International Neuroethics Society* (INS) in order to underline its outreach beyond the United States and Canada.

While this society is still largely influenced by North American scholars, its theoretical scope is global. There are, however, also European neuroethics organizations, both on a national and international level. In 2007, a *European Neuroscience and Society Network (ENSN)* was founded at the *London School of Economics*, funded by the *European Science Foundation* (http://www.kcl.ac.uk/sspp/departments/sshm/research/ENSN/European-Neuroscience-and-Society-Network.aspx). Its aim was to foster an interdisciplinary European network of social scientists, bioethicists, and neuroscientists working on the ethical, legal, and social aspects of neuroscience. The ENSN was directed by a steering committee consisting of representatives from Austria, Denmark, Estonia, Finland, Germany, the Netherlands, Norway, Portugal, Switzerland, and the United Kingdom, chaired by Professor Nikolas Rose from King's College, London. The funding period for this network activity ended in 2012, but since then, the network carries on its mission as the *Neuroscience and Society Network (NSN).*

The ENSN supported interdisciplinary discussions by organizing six interdisciplinary conferences, four workshops, and four "NeuroSchools" for early career scholars. The workshops and conferences, which were held in both Europe and North America, led to the publication of annual volumes in international journals. Four key theme areas were investigated: (1) neuroethics and beyond (setting an agenda for Europe), (2) public health and the politics of the neurosciences, (3) neuroeconomies (markets, choice, and neurotechnologies), and (4) sources of the neurochemical self (consciousness, personhood, and difference).

The ENSN aimed specifically to develop an empirical social science research base, since many of the researchers involved stressed the importance of empirical scholarship rather than abstract speculation. The ENSN suggested that European researchers might be leaders in the development of an alternative research agenda (1) by building an empirical knowledge base about the use and context of modern neuroscientific technology, (2) by contributing to a realistic understanding of present and expected neuroscientific advances, and (3) by incorporating the methodological and analytic strengths of social science traditions into neuroethics.

An important result of ENSN activities was that it provided many researchers with a sense of academic community, which they often lack at their home institutions, and helped them to overcome their isolation. Many long-lasting collaborative partnerships have resulted from these contacts, which have in some cases resulted in collaborative publications and research projects.

Another European-wide initiative that is more focused on the legal aspects of the neurosciences is the *European Association for Neuroscience and Law* (http://www.neurolaw-eanl.org). This society constituted itself in 2010 in Pavia, Italy, with original participants from Belgium, Germany, Italy, Spain, and the United Kingdom, as well as Australia and the United States. It holds annual meetings and sometimes also summer schools for young scholars. The two existing European societies for bioethics, the European Society for Philosophy of Medicine and Healthcare (ESPMH) and the European Association of Centres of Medical Ethics (EACME), also offer platforms for the presentation of neuroethics research at their annual conferences held in various locations throughout Europe, but they do not have specific neuroethics subsections or working groups.

In addition to these European societies, there are smaller organizations at the national or regional level. For example, there is the Italian society of neuroethics and philosophy in neuroscience (http://societadineuroetica.it), founded in 2013 with its headquarters in Milan. It is highly active and organizes yearly conferences in Padova, to which it also invites non-Italian neuroethicists from Europe and beyond (held in English and Italian). The society also awards prizes to renowned neuroethicists worldwide. Moreover, it organizes summer schools and workshops.

For the German-speaking countries (Germany, Austria, and Switzerland), the *Academy of Ethics in Medicine (AEM)* in Göttingen, Germany, founded a working group on neuroethics in 2015. This group, led by the authors of this article, meets twice a year and is about to conduct a questionnaire survey among neuroethics researchers in the German-speaking area, aiming to describe the state of neuroethics research, topics, methods, funding sources, and qualifications.

15.2.3 Neuroethics Publication Platforms

Scientific articles on neuroethics topics can be published in a wide variety of peer-reviewed journals in the areas of neuroscience, medicine, bioethics, philosophy, psychology, law, and the social sciences. According to the already mentioned bibliometric study, neuroethics articles tend to appear much more frequently in bio medical journals than in journals of the humanities or social sciences (Leefmann et al. 2016), except for topics related to philosophy of mind, legal studies, social neuroscience, and enhancement.

There are very few journals solely devoted to neuroethics, and those that exist are international but dominated by researchers from English-speaking countries: *Neuroethics*, published by Springer and founded in 2008 (impact factor 2015 of 1.305), and the *American Journal of Bioethics Neuroscience* (AJOB Neuroscience, no impact factor), published by the Taylor & Francis Group and introduced in 2010. Another journal whose focus is a bit larger, but which also carries the term "neuroethics" in its title is the *Journal of Cognition and Neuroethics*, a completely open access journal founded in 2013 that appears only online (no impact factor yet). All three of these journals, supplemented by the journal Biosocieties, are partner journals of the International Neuroethics Society. The Cambridge Quarterly of Healthcare Ethics recently announced that it will publish an annual issue called "Clinical Neuroethics."

In addition to various monographs on neuroethics and several large academic handbooks with comprehensive scopes (Illes and Sahakian 2011; Clausen and Levy 2015), the publisher Springer has recently launched a book series "Advances in Neuroethics," edited by a transatlantic team of four neuroethicists from the United States, Canada, and Germany and supported by a diverse editorial board with members from various continents (http://www.springer.com/series/14360).

Interestingly, a search for internet blogs on neuroethics yields quite a large number of hits (Table 15.1). While short texts on neuroethics topics can also be found in neuroscience or bioethics blogs, there seems to be an active blogging scene specifically dedicated to neuroethics, which, however, is primarily led by US-based researchers.

Table 15.1 Active internet blogs with exclusive or predominantly neuroethics content (selection)

Name and web address (URL)	Hosted by	Frequency
The Neuroethics Blog http://www.theneuroethicsblog.com/	Center for Ethics, Neuroethics Program at Emory University, USA	Weekly to daily
Neuroethics & Law Blog http://kolber.typepad.com/	Prof. Adam Kolber, Brooklyn Law School, USA	Every few days
The Law and Neuroscience Blog http://lawneuro.org/blog/	MacArthur Foundation Research Network, Vanderbilt University, USA	Every few weeks
Blog of the Oxford Uehiro Centre for Practical Ethics http://blog.practicalethics.ox.ac.uk/neuroethics/	Philosophy Faculty, University of Oxford, UK	Every few weeks
Neuroethics at the Core: The Blog of the National Core for Neuroethics https://neuroethicscanada.wordpress.com/	National Core for Neuroethics, Vancouver, Canada	Roughly every few months
Bioethics.net, blog posts neuroethics http://www.bioethics.net/topics/neuroethics/	American Journal of Bioethics, USA	Every few months
NeuroEthicsWomen Leaders https://neuroethicswomenleaders.com/	NEW Leaders, Karen Rommelfanger, USA	Every few weeks
News and views on what's happening in the world of brain science, neuroethics, and education from the Dana Foundation https://danablog.org/category/neuroethics/	Dana Foundation, USA	Every few months
Gruppo di neurobioetica. Le neuroscienze que amano l'uomo. Il blog http://neurobioetica.blogspot.de/ (Italian)	UNESCO Chair in Bioethics and Human Rights, Rome, Italy	Every few weeks
Blog Menschen-Bilder: Mensch, Gesellschaft und Wissenschaft http://scilogs.spektrum.de/menschen-bilder (German)	Spektrum, Stephan Schleim, Germany/ the Netherlands	Every few weeks to months

15.3 Neuroethics Research

Having looked at some of the infrastructures that exist for neuroethics in Europe, we will now present a few examples of recent neuroethics research from this part of the world. Inevitably, this selection will be biased by personal knowledge, accessibility of information, and understanding of different European languages. There are also several previous summaries of neuroethics research from other European countries, such as Spain, Italy and France (Bonete 2013; Caffo 2013; Baertschi 2009).

In general, the fact that research in ethics is influenced by the underlying philosophical traditions of local cultures should be considered. As opposed to Asian philosophies and religions like Buddhism, Hinduism, Taoism, or Confucianism, Western philosophy largely originated in Europe and spread to other parts of the world via colonization, migration, and globalization. The dividing lines concerning philosophical and ethical approaches do not fall between Europe and America, but within Europe itself. The British Islands have an ethical tradition that is particularly distinct from the continental European one, being characterized mainly by utilitarianism. Similar tendencies can be found in Northern Europe (e.g., in Scandinavian countries, in the Netherlands).

In contrast, the western, eastern, and southern parts of the European continent have been influenced more by deontological ethics theories, Christian ethics, and hermeneutic approaches. Due to the English language, political history, and the commonwealth, the situation in North America and Australia has been impacted much more by British philosophical traditions than by continental European ones. The consequences can often be seen in bioethics debates and biopolitical regulations. Some good examples are the different biopolitical laws concerning stem cell research, assistive reproduction techniques, or assisted suicide across Europe (Jox et al. 2013; Harper et al. 2014; Hovatta et al. 2010). This also holds true for neuroethics. For example, a comparative questionnaire survey among Canadian and German neurologists has found much higher approval rates for withholding and withdrawing life-sustaining treatment in disorders of consciousness among Canadian neurologists than among their German counterparts (Kuehlmeyer et al. 2014).

Against this backdrop of different philosophical and ethical traditions, however, the recent developments of globalization, intensified scientific exchange, and researcher mobility have more and more blurred these lines. Nowadays, research topics, positions, and arguments in neuroethics are not so different between European and North American researchers. This will become clear as we discuss the most recent lines of research in European neuroethics.

15.3.1 Neuroethics Research in Great Britain

At the *University of Oxford*, the *Oxford Centre for Neuroethics* is engaged in neuroethics research on the following topics: (1) cognitive enhancement; (2) borderline consciousness and severe neurological impairment; (3) free will, moral responsibility, and addiction; and (4) the neuroscience of morality and decision-making. Furthermore, the *Oxford Centre for Neuroethics* conducts applied neuroethics research in other areas in light of input from practicing clinicians and scientists. According to their own statement, they plan to analyze technological advances as they occur and respond to ethical issues arising from basic and clinical neuroscience (http://www.practicalethics.ox.ac.uk).

15.3.2 Neuroethics Research in Germany

In Germany, the Federal Ministry for Education and Research (Bundesministerium für Bildung und Forschung, BMBF) is the primary funding organization for research projects on ethical, legal, and social aspects (ELSA) of the modern life sciences. A main goal is that the results of funded projects will serve as a basis for societal discourse, research and development, medical practice, and decisions made by politicians and legislators.

The BMBF has been supporting ELSA research in a program-based manner since 1997. Current support amounts to around 4.5 million euros annually. Different lines of funding exist: "research projects," "support for young researchers," and "public discourse." The BMBF generally supports interdisciplinary research in the format of joint research groups. Since 2006, research has often been developed in cooperation with international partners, among them notably Canada. Summer schools are funded to give young researchers the opportunity for interdisciplinary and international work and to promote network building. The funding for public discourse projects supports projects in schools, in the general public, or online in order to raise public awareness and inform people about neuroethics issues.

Since 2004, some of the funded ELSA projects have focused on the impact of the modern neurosciences on society. The following national but interdisciplinary ELSA research projects have been funded in the area of neuroethics: (1) chances and risks of neuroenhancement; (2) philosophical, ethical, and legal issues of dementia; (3) ethical and social issues of neuroplasticity; and (4) prenatal, predictive, and presymptomatic diagnosis of neurodegenerative disorders.

Furthermore, several neuroethics summer schools were funded, focusing on topics such as the genome and the brain, Alzheimer's dementia, neuroenhancement, the impact of the neurosciences on psychiatry, the mechanization of the brain, the persistent vegetative state, deep brain stimulation and personal identity, and neuroimaging and neuromarketing (http://www.gesundheitsforschung-bmbf.de/de/5590.php). Many of these summer schools have impacted the field via seminal journal articles, special issues of journals, or edited book volumes (Jox et al. 2012).

Finally, two discourse projects with young people were performed on neuroenhancement and the chances and risks of the modern neurosciences.

15.3.3 International Collaborations for Neuroethics Research

In 2007, a trinational neuroethics research initiative was started by Germany, Finland, and Canada. Five bi- or trinational projects, respectively, were funded by the national state funding organizations of the respective countries (BMBF for Germany, the Academy of Finland, and the Canadian Institutes of Health Research (CIHR)). These projects investigated ethical, legal, and societal issues on the following topics: (1) neuroimaging, (2) deep brain stimulation, (3) chronic disorders of consciousness, (4) intersubjectivity, and (5) neuroenhancement.

In 2015, a call for proposals for "European Research Projects on Ethical, Legal, and Social Aspects (ELSA) of Neuroscience" was published within the European program "ERA-NET Neuron." This is a collaboration of funding organizations across Europe, Israel, Turkey, and Canada, which set out to promote basic, clinical, and translational research in disease-related neuroscience in order to find better treatments for diseases of the brain and the nervous system. For the first time since its initiation in 2003, this initiative decided to fund ELSA research focusing on the neurosciences. The call for funding published in 2015 targeted researchers in many European countries, among them Germany, Belgium (Wallonia), Finland, Spain, and Portugal, as well as Canada.

The aim of the call was to facilitate multinational, collaborative research projects that address important questions regarding ethical, philosophical, legal, and socio-cultural aspects related to neuroscientific research and recent advances in the field. Five projects are currently receiving funding (for more details about the projects, see http://www.neuron-eranet.eu/en/630.php):

1. Enhancing the Informed Consent Process: Supported decision-making and capacity assessment in clinical dementia research.
2. The Integration of cross-disciplinary research in Neuroscience and social science—a methodological case study on economic policies and the neuroscience of agency (INSOSCI).
3. Intelligent Neuro-Technologies Restoring Functions of Action and Communication: an Evaluation Study (INTERFACES, www.bci-ethics.de).
4. Psychiatric Neurosurgery—Ethical, Legal, and Societal Issues (PNS, https://mindandbrain.charite.de/).
5. Ethical and Legal Framework for Predictive Diagnosis of Alzheimer's Disease Quality of Life of Subjects at Risk and their Close Others (PreDADQoL).

The principle investigators of all projects are affiliated with German institutions. The authors of this paper are the principle investigators on two of these projects (Ralf Jox: INTERFACES, Sabine Müller: PNS).

15.3.4 Ethical Research Within the Human Brain Project (HBP)

The Human Brain Project (HBP) is a highly ambitious flagship project funded by the European Union (https://www.humanbrainproject.eu/2016-overview). It aims to put in place a cutting-edge, information-and-communications-technology-based scientific research infrastructure for brain research, cognitive neuroscience, and brain-inspired computing. The project promotes collaboration across the globe and is committed to driving European industry forward.

The Human Brain Project includes social and ethical research on its potential risks, because it recognizes that these risks have to be addressed in an open and transparent manner—preferably a long time before they materialize. To accomplish that, the HBP has installed the HBP's Ethics and Society Program (SP 12), which will use 3% of the project's budget. The HBP Ethics and Society Subproject explores the project's social, ethical, and philosophical implications. It promotes engagement with decision-makers and the general public, fosters responsible research and innovation by raising social and ethical awareness among project participants, and ensures that the project complies with relevant legal and ethical norms.

15.4 Neuroethics Research Training

15.4.1 Formal Academic Degree Programs

As neuroethics is not yet an established field of academic activity, it is not surprising that the number of specific training programs is still small. As with every new and emerging field of research, the pioneers come from different disciplines and have various background qualifications. In the already cited bibliometric study, Leefmann et al. found that 40% of the authors of neuroethics articles had a background in medicine, 24% were trained in psychology or the neurosciences, while rather few had a background in philosophy or ethics (Leefmann et al. 2016).

And yet, 17% of neuroethics authors had a double qualification or some additional training in philosophy or ethics. Depending on the kind of research and the methodological approach, neuroethicists may need more competencies in philosophy, social sciences, law, neuroscience, or medicine. Therefore, it seems that the field is currently open to graduates from all of these (and even more) diverse academic camps. The neuroethics newsletter "Brainstorm" that appears every few weeks includes a highly recommendable section where neuroethics scholars reflect on their career paths and give suggestions to young scholars (http://www.ircm.qc.ca/LARECHERCHE/axes/neuro/neuroethique/Pages/Groupe.aspx?PFLG=1033&lan=1033).

Since the overlap between neuroethics and bioethics is relatively large, bioethics is often the preferred academic route (e.g., a masters of arts). Good bioethics or medical ethics master's courses will also encompass neuroethics topics, some of which may have special tracks with the option to concentrate on neuroethics. With the rise of specific neuroethics institutes at universities, there are even some specific neuroethics training programs developing. Interestingly, the only neuroethics master's program proposed by the European Commission's website on learning opportunities and qualifications in Europe is in Belgrade, Serbia. It is a 1-year program that is taught in Serbian and costs € 2500 (http://ec.europa.eu/ploteus/en/content/neuroethics-2). In addition, the Berlin School of Mind and Brain offers a master's in "Mind and Brain" that combines philosophy and neuroscience (http://www.mind-and-brain.de/master), as does the Munich Graduate School of Neurosciences (http://www.gsn.uni-muenchen.de/index.html). Renowned neuroethicists, however, often recommend following traditional disciplinary training first, before adding another qualification or specializing more in neuroethics (see http://neuroethics.upenn.edu/career-corner/).

In contrast to the undergraduate and master's level of education, there are more and more opportunities to do a PhD in neuroethics. Basically, every academic center with stable activity in neuroethics offers PhDs in that area. This does not mean, however, that these PhDs are embedded in structured PhD programs with an exclusive focus on neuroethics. Examples of structured PhD programs in Europe are the doctoral programs of the already mentioned interdisciplinary neurophilosophy schools in Berlin and Munich, Germany. A similar program is hosted by the University of Pavia in Italy (http://www.iusspavia.it/dott.php?id=18#.WA5up_mLTak).

15.4.2 Conferences and Courses

Young scholars who want to qualify in neuroethics do not need to undergo formal training in that specific field, but they will often choose formal qualifications in traditional fields and further specialize by attending summer schools, workshops, conferences, or other meetings about neuroethics. Here, they can choose from a range of options.

With regard to scientific conferences, there is the annual conference of the INS, which has so far always taken place in the United States, often attached to the meeting of the Society for Neuroscience. The INS conference usually lasts 1–2 days and offers a range of different formats, from poster presentations and flash talks to plenaries and roundtables. Another conference series that began in 2009 in Halifax, Canada, is called "Brain Matters." It is not tied to an academic society but has been drawing a multiprofessional group of neuroethicists to almost yearly conferences in Montreal, Cleveland, Vancouver, and Scottsdale, Arizona.

Another regular conference that is a bit smaller in size, but is attended by senior experts in the field, always takes place in Europe: the conference of the Neuroethics Network organized by Thomasine Kushner and Yves Agid, sponsored by Cambridge University Press and the Paris Brain and Spine Institute (Hôpital Pitié-Salpetrière). The 2–3-day conference is always held in June in Paris (http://www.icmbioethics.com). We have already mentioned the annual conference of the Italian Society for Neuroethics and Philosophy of the Neurosciences in Padova and the Neurolaw conferences of the European Association for Neuroscience and Law. In Portugal, the Catholic Institute of Bioethics of Porto organized its second international conference on neuroethics in 2016 (http://www.bioetica.porto.ucp.pt/icone).

In addition to large conferences, there are smaller meetings and training schools for young academics, also in Europe. The BMBF, for example, funds 1-week summer and winter schools on bioethics for 10–20 young academics from mostly European countries. In this context, there have been meetings on various neuroethics topics, such as the vegetative state, cognitive enhancement, pediatric neuroenhancement, neurotechnologies, the genome and the brain, Alzheimer's dementia, personal identity, neuroimaging, and neuroeconomics (http://www.gesundheitsforschung-bmbf.de/de/5590.php). In Italy, the University of Padova often hosts neuroethics study days and summer schools (http://societadineuroetica.

it/sine-formazione-convegni.html). Another European venue that frequently hosts bioethics workshops (including neuroethics workshops) is the Brocher Foundation on the shores of Lake Geneva (Lac Léman) in Switzerland (http://www.brocher.ch). Individual meetings for young academics have recently been held in Marseille, France (http://neuroethics2016.sciencesconf.org/); Zurich, Switzerland (http://www.bmel.uzh.ch/en/med/events-news/events/neuroethics.html); and Berlin, Germany (http://www.nncn.de/en/news/events/pdfs/winterschuleflyer).

We are convinced that the vivid neuroethics scene in Europe will develop further and gradually also find its way into academic institutions, because the rapid development in many neuroscientific areas raises many new ethical issues. The emerging field of neuroethics will, according to our view, benefit most if scholars from all parts of the world, especially Europe and North America, work together and meet and exchange their ideas.

References

Baertschi B (2009) La neuroéthique: ce que les neurosciences font à nos conceptions morales. edition la découverte, Paris

Bonete E (2013) Neuroethics in Spain: neurological determinism or moral freedom? Neuroethics 6:225–232

Caffo L (2013) La neuroetica in Italia: una rassegna filosofica. Riv Int Filos Psicol 4:385–389

Clausen J, Levy N (2015) Neuroethics handbook. Springer, Amsterdam

Erbguth F, Jox RJ (eds) (2017) Angewandte Ethik der Neuromedizin. Springer, Berlin

Fukushi T, Sakura O, Koizumi H (2007) Ethical considerations of neuroscience research: the perspectives on neuroethics in Japan. Neurosci Res 57:10–16

Greely HT, Ramos KM, Grady C (2016) Neuroethics in the age of brain projects. Neuron 92:637–641

Harper J, Geraedts J, Borry P, Cornel MC, Dondorp WJ, Gianaroli L et al (2014) Current issues in medically assisted reproduction and genetics in Europe: research, clinical practice, ethics, legal issues and policy. Hum Reprod 29:1603–1609

Hovatta O, Stojkovic M, Nogueira M, Varela-Nieto I (2010) European scientific, ethical, and legal issues on human stem cell research and regenerative medicine. Stem Cells 28:1005–1007

Illes J, Sahakian BJE (2011) The Oxford handbook of neuroethics. Oxford University Press, Oxford

Jox RJ, Ulmer S, Reiter-Theil S (2010) From brain death to brain chip—do we need "neuroethics"? Schweiz Arch Neurol Psychiatr 161:290–295

Jox RJ, Kuehlmeyer K, Marckmann G, Racine E (2012) Vegetative state—a paradigmatic problem of modern societies. LIT, Münster

Jox RJ, Horn RJ, Huxtable R (2013) European perspectives on ethics and law in end-of-life care. Handb Clin Neurol 118:155–165

Kuehlmeyer K, Palmour N, Riopelle RJ, Bernat JL, Jox RJ, Racine E (2014) Physicians' attitudes toward medical and ethical challenges for patients in the vegetative state: comparing Canadian and German perspectives in a vignette survey. BMC Neurol 14:119

Leefmann J, Levallois C, Hildt E (2016) Neuroethics 1995-2012. A bibliometric analysis of the guiding themes of an emerging research field. Front Hum Neurosci 10:336

Marcus SJ (2004) Neuroethics: mapping the field. Dana Press, New York City

Pontius AA (1973) Neuro-ethics of "walking" in the newborn. Percpet Mot Skills 3:235–245

Racine E (2010) Pragmatic neuroethics. The MIT Press, Cambridge and London

Saffire W (2002) Visions for a new field of "neuroethics". Dana Press, San Francisco

Neuroethical Engagement on Interdisciplinary and International Scales

16

John R. Shook and James Giordano

Abstract

After a decade of growth and development, neuroethics as a defined discipline is establishing domains of inquiry and action, a defined canon, and set(s) of practices. Neuroethical address and discourse must engage the realities forged and fostered by brain science no matter where they emerge and deliberate upon neurotechnological applications on the international scale. The invention and application of neurotechnologies are raising questions of ethics, to be sure. Neuroscientific innovations are also altering and challenging how we regard ourselves as moral beings worthy of ethical standing. Neuroethical investigations, at the empirical levels of experimental research and clinical application or the philosophical levels of exploring moral capacities or ethical issues, concern ideas of what it means to be human and ideals of humanity-wide importance. As a discipline and in practice, neuroethics must heed the subjective realities of the people who take part in neuroscientific research and therapy. No single method could do justice to understanding ourselves as persons, nor could any single country monopolize the meaning of self-identity and self-worth. Therefore, neuroethics must become "disciplined" to be realized as genuinely intercultural, as well as thoroughly interconnected. We find that these goals and tasks are already being achieved through multidisciplinary and multinational networked teams that conduct collaborative inquiries in specific areas of both local and global concern. These teams deserve attention and appreciation as exemplars for future disciplinary progress in neuroethics.

J.R. Shook, Ph.D.
Philosophy Department and Graduate School of Education, University at Buffalo, Buffalo, NY 14260, USA
e-mail: jrshook@buffalo.edu

J. Giordano, Ph.D., M.Phil. (✉)
Departments of Neurology and Biochemistry, Neuroethics Studies Program, Pellegrino Center for Clinical Bioethics, Georgetown University Medical Center, Washington, DC, USA
e-mail: james.giordano@georgetown.edu

E. Racine, J. Aspler (eds.), *Debates About Neuroethics*, Advances in Neuroethics,
DOI 10.1007/978-3-319-54651-3_16

16.1 Neuroethics: Becoming a Discipline

The establishment of neuroethics as an emerging field initially suggested its categorization as yet another domain of applied bioethics. The fast growth of the field in recent years has involved extensive multidisciplinary enterprises. Neuroethics is developing foci that while based, at least in part, upon longstanding philosophical questions and issues important to bioethics at large takes something of a unique perspective, given the way that neuroethics is inextricable from the neural and cognitive sciences (Giordano 2010, 2014). Bioethics has struggled to transcend its American matrix, with much success. If bioethics will no longer be simply an "applied ethics," how could neuroethics be confined in that manner? Going further, we anticipate that neuroethics should become a disciplinary field in its own right.

We are observing, and describing in this chapter, the sorts of neuroethical research that relieve neuroethics from the status of an applied subfield. The only possible proof that neuroethics should prioritize interdisciplinary and international dimensions is to point out the demonstrably admirable research energizing much of neuroethics and guiding its development. Neuroethics was never going to be delimited by a few older disciplines, a couple of cultures, or any single country's ethics. There is no paradox inherent to the idea that the fulfillment of a multidisciplinary and multinational field of inquiry can eventually result in a new discipline. (Most other disciplines were birthed from that kind of heritage.) We have no deductive argument dictating that future. It is not inevitable. However, if our observations are sound, that disciplinary development and destiny for neuroethics seem probable.

The ethical centrality of the person serves as a primary illustration of this development. Bioethics had to presume much about the locus of moral value in the "person" and had to compare moral and legal views on personhood from established professions (e.g., medicine and law primarily), in order to manage conflicts among inflexible principles respecting persons. Neuroethics need not be so subordinate to such encumbrances. The biological, psychological, and anthropological nature of the individual, and the cognitive bases for the agency and responsibility that are expected of persons, are matters for ever-deepening study. How and why people become capable of moral judgment, and come to expect dignity and respect, are matters amenable to empirical inquiry in tandem with ethical insight.

Transcendent, trans-historical, and essentialist notions of personhood can retreat to their proper scale as important constructs that are embedded in socio-cultural contexts. That sort of context does not reduce those ideas to hollow words or superficial vanities. Understandings of persons, their moral features, and their ethical values are realizable in those instances in which they develop and establish their effectiveness in the world. In fact, they are so real that the social and life sciences can simultaneously investigate their many dimensions and implications. Neuroethical inquiry and discourse will not (and arguably should not) seek to duplicate the pursuits of other disciplines, but it will borrow investigational methods and background knowledge as needed for its more specific areas of concern. Along the way, neuroethical address must assume a responsible stance toward the ethical standards it will uphold. That responsibility cannot be left to any country's laws or one culture's traditions. As one of us (Giordano) has affirmed

> [R]esearch is being conducted—and highly scrutinized—in a number of developing and non-developed countries. Moreover the contexts of such research are changing. Thus, there is a need to address not only nations and cultures' needs, values and mores, but those ethical ideals and systems that are operative in the countries and cultures in which said research is being undertaken. So, while it is important to ensure that research is conducted ethically, an equally—if not more—important question is "by which ethical standards?" (Giordano 2013: 3).

To be sure, neuroethical discourse has never been bereft of ethical standards. As a field, neuroethics could easily become "disciplined" in a rigid manner, by incorporating mechanical application of self-standing ethical rules in relation to sterile clinical scenarios involving objective physiological interventions for measured therapeutic outcomes. Translatable neuroscience can take center stage, and neuroethical debates would shift from technique to technique as they are envisioned and put into practice. Hopefully, neuroscientific discoveries and neurotechnological inventions will "translate" in a practical sense into health-care diagnostics and treatments, and perhaps into consumer applications as well. But their ethical implications must "translate" in a second sense—an ethical sense. It is an ethical question to ask how those advances may be compatible (or not) with the values, self-conceptions, and life plans of intended recipients, and other members of society.

Any neurological intervention, no matter how medically impressive, should also be ethically acceptable, especially to those who will be most deeply affected. Medical ethics, and bioethics more generally, have held biomedical advances to high ethical standards, and neuroethical analyses can do no less. All the same, such expectations remain external to the patient's own moral sensibilities, and indifferent to an individual's unique perspective on what it means to be that particular person. Ethical principles apply to persons, and should aim to protect persons, in a generic fashion; it suffices to be classified as a person in order to be covered by those principles. How any individual may happen to personally exemplify personhood and live one's life as a special self are just details, largely irrelevant to the universal rights and rules for all people. Human rights and fundamental laws should apply to all persons equally, without regard for individual differences.

As persons, however, we each discover and shape our individuality in our own ways, through the social and cultural resources accessible to us. No one tries to be an individual in the abstract; each person's own sense of individuality, along with one's sense of self and self-worth, is precious and irreplaceable. Bioethics need not deny these matters, but neuroethics has an extra responsibility to them. Neuroscience and neurotechnologies, by centering on neurological functions, more easily affect and potentially impact the ways people experience being themselves. There is no need to embrace a dualism of body and mind, or to reduce the 'self' or 'individuality' to the brain, in order to grasp how the brain is not just another organ of the body. A brain working better is not just a better-working brain. An individual's subjective world is also potentially at stake.

Walter Glannon of the University of Calgary and Nir Lipsman of Toronto Western Hospital have explored the medical case of an accountant receiving successful deep brain stimulation, illustrating the primacy of the subjective perspective:

> Any perceived threat the device poses to the accountant's experience of control is more acute than how people with cardiac arrhythmias such as atrial fibrillation might perceive implanted pacemakers or defibrillators. Unlike these other devices, DBS can directly modify mood and behaviour by directly altering the neural bases of unconscious and conscious mental states. The fact that a functioning DBS system modulates his brain and mind at the unconscious level, and that he becomes consciously aware of the system and its effects on his mood only when it malfunctions, reinforces the idea that he may have only limited ability to control his mental states and how they issue in his actions (Lipsman and Glannon 2013: 468).

Neuroethical inquiry must ask—and try to answer—questions of how experiments and treatments affecting the brain translate into that subjective world, a person's self-conception, and each individual's own values.

We propose that the future of neuroethical engagement on interdisciplinary and international scales will track and enhance the capacity to ensure than neurological advances can "translate" for all people. Only highly selective illustrations of this view can be gathered here, and we regret that citations are limited to English-language writings (but see Buniak et al. 2014; Darragh et al. 2015; Martin et al. 2016; Becker et al. 2017). Yet, we find that even this brief survey reveals a degree of useful subcategorization for areas of neuroethical investigation. Those subordinate areas, we believe, can help to integrate what may otherwise appear to be distantly related inquiries.

16.2 Persons as Research Participants and Individual Subjects

In 2002, the year that "neuroethics" was thrust into the academic spotlight, a special issue of the journal *Brain and Cognition* was devoted to this new discipline. The opening editorial by Judy Illes and Thomas Raffin (2002) surveyed the papers in this issue, noting how they cover ethical challenges in both basic and clinical research. Topics addressed included protecting human research subjects, patient privacy, communicating diagnostic findings, adequately interpreting findings from new brain imaging techniques, revising views of cognitive functioning as findings are interpreted, forming prognoses and predictions for patients, raising hopes about cognitive enhancements, and counseling patients about making meaningful and informed choices.

These and related topics still form much of the core of neuroethical focus, and rightly so. Like bioethics, neuroethics must support a robust and proactive stance toward safeguarding medical information, and minimizing risks of experimental research. Diagnostics about one's neurological and cognitive functioning are among the most sensitive and private types of medical information, perhaps only comparable to one's genetic information (Heinrichs 2012). Deep brain stimulation has rightly received intense ethical scrutiny. The emergence of transcranial stimulation deserves the same level of scrutiny (Rossi et al. 2009; Horvath et al. 2013). Research participants must receive due protection as subjects by the medical and legal fields. That proactivity must extend to the "subjective" realm of the first-person standpoint in lived experience, if there is to be a fully disciplinary neuroethics. The moral problems and questions of ethics that flow from the stages of cutting-edge research and clinical application amply illustrate central neuroethical concerns for the meaning and morality of neurotechnologies *as viewed by subjects themselves*.

After moving from Stanford to develop the National Core for Neuroethics at the University of British Columbia, Illes has continued to guide investigations into needed engagements between implementations of novel neurotechnologies and analyses of their ethical implications. These investigations can remedy the tendency to overlook the individuality of participants, and their subjectivity, while focusing on technical soundness. For example, Illes and coauthors have surveyed debates on neuroimaging, finding that those debates do not take much interest in individual subjects and their personal histories:

> As reviewers frequently note, fMRI studies in this area often include patients with different clinical symptoms, disease subtypes, ages of onset, illness duration, severity of symptoms, medication status (including dosage and side effects), comorbid conditions, and substance abuse histories. Since it has been shown that these factors are related to brain functioning, uncontrolled variation of these factors may well compromise the internal validity of studies. Sample heterogeneity, thus, constitutes a serious threat to the knowledge value of studies in this area (Anderson et al. 2012).

The neuroethics team led by Eric Racine at the Institut de recherches cliniques de Montréal, associated with the Université de Montréal and McGill University, has raised similar concerns about neuroimaging practices in Canada (Deslauriers et al. 2010). Racine's team has also examined the procedures for obtaining and insuring informed consent for neuroimaging research in Canada. They reported:

> There appears to be significant variability and inconsistency in important areas in neuroimaging research such as risk reporting and the management of incidental findings. For example, we found evidence of substantial between-site and within-site variability in the strategies approved to handle incidental findings. There were also significant variations in the disclosure of risks associated with MRI and fMRI consent forms, especially with respect to psychological risks and risks associated with dizziness, discomfort, stress, and fatigue. We also encountered a few [consent] forms where the risks of MRI were downplayed (e.g., "no-risk" statements) and incidental findings were described as a benefit (Palmour et al. 2011: 5).

Downplaying risks, especially where no regard is shown for a participant's medical history or current state of health, must raise serious concerns in any country.

The matter of incidental findings is also concerning; the variability found within Canadian research is not uncommon. Divergent views among researchers are also found within the medical community in Japan, according to an investigation by Misao Fujita, of the Uehiro Research Division for iPS Cell Ethics at Kyoto University, and her colleagues at Kyoto, and the University of Tokyo (Fujita et al. 2014). Illes and colleagues at the University of British Columbia and the Stanford

Center for Biomedical Ethics agree that neuroimaging researchers themselves express concerns about adequate clarity and availability of ethical guidance about incidental findings (Borgelt et al. 2013; see also Kehagia et al. 2012; and Brosnan et al. 2013).

A survey of numerous countries, conducted by a team composed of specialists at various UK institutions such as the Royal Free Hospital NHS Trust, SFC Brain Imaging Research Centre at the University of Edinburgh, and Imperial College Healthcare NHS Trust, further confirmed this absence of standardization across all types of medical imaging. A key recommendation of their report was that "discourse of national and international practice is required in the light of existing legal and ethical frameworks to develop robust and practical guidelines for both research centres and ethics committees considering proposals for imaging research" (Booth and Boyd-Ellison 2015: 462). Members of that team, notably Thomas Booth and J. M. Wardlaw, have coauthored further work pursuing the ethical ramifications of discovering and reporting incidental neurological findings (Wardlaw et al. 2015; Ulmer et al. 2013).

16.3 Subjects as Relational

Due ethical concern for each subject's individuality and personal values is entirely consistent with ample recognition of membership in a group category. Cordelia Fine, at the University of Melbourne in Australia, focuses on the dangers, and opportunities, of studies of neurological similarities and differences that are putatively relevant to gender (Fine 2013). Fine has also teamed with coauthors in the UK, USA, and Switzerland to formulate guidelines for conducting and interpreting neuroimaging relating to sex/gender studies (Rippon et al. 2014). Cultural and religious heritage are further contexts in which due sensitivity to an individual's background and priorities is essential (e.g., Semrud-Clikeman and Bledsoe 2014). Wael Al-Delaimy, at the Division of Global Health of University of California, San Diego, sets out a valuable perspective for neuroethics:

> Western secular philosophy can direct ethical questions to standard ethical concepts of autonomy, beneficence, nonmaleficence, and justice, but other traditional non-Western populations may require evaluations of ethical questions from the specific contexts of their faith and culture. The religion of Islam with more than 1.3 billion followers worldwide is the second largest after Christianity and a unique global subpopulation. New intrusive technologies or instrumentation such as fMRI are likely to have ethical and religious implications for this large percentage of the world population because such technologies may also offer insight into the process of consciousness, spirituality and thought (Al-Delaimy 2012: 510).

An appreciation for the special circumstances and psychological consequences inherent to neuroimaging small children should also have an important place in research neuroethics (Shechner et al. 2013). Children are simultaneously neurological subjects and developing subjective mentalities. Ilina Singh, professor of neuroscience and society at the University of Oxford, has led interview sessions with these patients. She reports that

> [It] is not the case that encounters with neuroscience discourse or technologies necessarily lead children to construct neurological subjectivities. Rather, children are seen to be active and creative participants in discursive power negotiations among social, biological and technological forces. Throughout these negotiations many children sustain a resilient sense of self and agency, perhaps because the embodied experience of moral struggle over self-control outweighs reductive explanations for behaviour that they may hear from adults (Singh 2013: 825).

Adults can be vulnerable subjects, no less than children. However, classifying "vulnerability" can obscure ethical problems. Eric Racine's team has called into question the usual attribution of vulnerability by biomedical ethics according to diagnosed condition. Such subjective vulnerability is largely due to a host of individual factors relevant to one's capacity to meaningfully participate in treatment decisions (Bracken-Roche et al. 2016). Joining in additional inquiry to the ethical issues of informed consent is a number of senior neuroethicists: Joseph Fins of Cornell University, Paul Ford of the Cleveland Clinic, and Walter Glannon at the University of Calgary in Canada. They rightly emphasize that subjectivity is relational and contextual as experienced and lived by the individual, and they propose that respecting patient vulnerability should share those features (Bell et al. 2014; see also Ford 2009). Fins has warned against a neuroethics that would be as static and flat as the fMRI scans driving so much of neuroscience toward depersonalized dogmas about brain functioning. His 2008 declaration still rings true: "Neuroimages are not in fact transparent, they are reflective" (Fins 2008: 50). How brain scans are interpreted says far more about what those who interpret scans are prepared to see, than what subjects' brains are actually doing. Fins's cautionary words continue:

> "I for one want to stay practical—as a physician and medical ethicist—and work in the service of patients who have been historically marginalized. I will leave the speculative work to the non-pragmatists. If we confine ourselves to the instrumentality of research or clinical applications of our work—avoiding the more speculative domains—I believe we will be better able to manage developments in our field and not be distracted by rhetoric and ideology, which can become powerful barriers to meeting a dire clinical need (Fins 2008: 50).

If neurological and psychiatric patients are to be regarded as subjects in the double sense as being subject to research and treatment agendas, while trying to live their subjective lives with meaning, then neuroethical analyses and guidance of this reality carry special ethical responsibility indeed. That responsibility is surely distributed everywhere that neurological research is fostered and applied. Herein, we suggest an initial division of labor into "research neuroethics" and "interrogative neuroethics." A third mode of neuroethics, "interoperable neuroethics," is then tasked in a later section with communicating with other areas of scientific and social knowledge and application.

16.4 Research Neuroethics

To analyze what is happening in exploratory and experimental research, where hopes for neuroscientifically based treatments are translated into therapeutic and rehabilitative regimens, it becomes important to understand the motivations and priorities of the people involved. Research neuroethics takes the research context and the methodical work of researchers as its field for targeted inquiries. For some novel application of neuroscience and its technologies (i.e., neuroS/T), the development and availability of neuroS/T research can be queried and assessed at many levels (Giordano 2015, 2016). These levels include a range of questions about research ethics:

How do people experience the neuroS/T research regimen as its subjects, from their personal perspective?

How do people assess this neuroS/T research for themselves, in terms of their medical needs and health priorities?

How do people express their disagreements among each other about prioritizing this, and similar types, of neuroS/T research, assemble into health-related interest groups, and justify their respective standpoints?

How do allied disciplines (e.g., medicine, psychiatry) assess the potential of this neuroS/T research for meeting therapeutic needs, with due regard for priorities expressed by interest groups, ethno-cultural populations, and (if any) broader social consensus?

How do other professions (e.g., law) and institutions (e.g., churches, government agencies) assess the conduct and implications of neuroS/T research in terms of the values and norms they presume and uphold?

How has the development of this neuroS/T research, and the brain science that undergirds it, been evaluated by standards of research risk assessment, responsible innovation, genuine social need, and justice for members of society?

How is society in general assessing the progress and impact of this neuroS/T research on people's lives, according to its customary values, cherished ideals, and established laws?

Research neuroethics must be entangled with intensely personal health needs in the context of individual life plans, on the one side, and with broadly social capacities for the public understanding and assessment of medical information on the other. It will share a great deal with the settled standard of research ethics in general and will encourage professional conduct in all research settings. Research neuroethics will also work closely with behavioral ethics on the more individual level and with social ethics on broad(er) levels. A second mode of neuroethics, interrogative neuroethics, more closely investigates what individuals presently think about new neuroS/T and the ways that such innovations can be clinically implemented.

16.5 Interrogative Neuroethics

To better anticipate what is happening in research laboratories and clinical settings as the need for neuroscientifically based treatments is translated into therapeutics, the experiences, motivations, values, and priorities of the people involved must be understood. Interrogative neuroethics takes the social context and peoples' settled views as its field for targeted inquiries. Here, for any novel neuroS/T development and availability, key questions include:

How do people experience the application of this neuroS/T, in the ongoing course of their lives?

How do people assess this neuroS/T's value for themselves, in terms of their health-care needs and their stable self-conceptions and values?

How do people express their disagreements among each other about this neuroS/T's value, form health-related interest groups, and justify their respective standpoints?

How do related disciplines (e.g., medicine, psychiatry) developing and utilizing this neuroS/T assess its capacity for respecting and supporting individuals as persons, with due regard for their membership in interest groups and/or social identity with an ethno-cultural population?

How do other professions (e.g., law) and institutions (e.g., churches, governments) assess neuroS/T's use in terms of the values and norms that those professions presume and uphold?

How has the application of this neuroS/T, and the brain science upon which it is based and derived, been evaluated by scientific and clinical participants for responsible innovation, genuine social need, and justice for members of society?

How is society, in general, assessing neuroS/T's impacts on people's lives, according to customary values, cherished ideals, and established laws?

Answers to these questions possess great significance and serve as a vital resource for more expansive deliberations about the potential and actual impact of neuroS/T on the needs and priorities of patients and their families, clinical care providers, health-care institutions and systems, public interest groups, and government agencies. In turn, well-informed and comprehensive policy formulations and implementations that guide and sustain support for research and use of neuroS/T should take these deliberations into careful consideration.

When ethical concerns need to be raised for consideration, interrogative neuroethical methods do not impose normative principles, for interrogative neuroethics wields none. Rather, familiarity with the moral values held by this or that sector of society, or society as a whole, allows an interrogative neuroethics to relay that further (and deeper) ethical dialogue is needed. Surveying attitudes and expectations about brain modifications can disclose both aversions and approvals toward neurotechnological advances.

Peter Reiner of the National Core for Neuroethics, at the University of British Columbia, and colleagues have sampled views among physicians and the public about brain stimulation, cognitive enhancement, and related neuroS/T (Fitz et al. 2014; Cabrera et al. 2015; see also Franke et al. 2014). Consumer demand for pharmaceuticals that supposedly enhance mental acuity and stamina is also worthy of close monitoring. Studies of college students around the world—typically conducted by teams comprised of experts in mental health, public health, pharmacology, and bioethics—are a notable category (Sattler et al. 2013; Singh et al. 2014; Maier et al. 2015; Schelle et al. 2015; Jensen et al. 2016). Elisabeth Hildt (then at the University of Mainz) worked with Andreas Günter Franke, of Mainz University Medical Center, on a qualitative study discerning why and how students seek academic advantage within the complex context of life's many pressures and challenges (Hildt et al. 2014). Interrogative neuroethics may even select an entire nation for study, as neuroethicist Arlene Salles undertook for Argentina (Salles 2014). How constituencies understand what neuroS/T can do, and how they evaluate any types of use by their own priorities, are the primary areas of interest for interrogative neuroethics.

A survey led by Jens Clausen (Institute for Ethics and History of Medicine, University of Tübingen) and Femke Nijboer (Human Media Interaction, University of Twente) further illustrates the interrogative approach. Clausen and Nijboer assessed answers from 145 respondents at an international Brain-Computer-Interface (BCI) conference, concerning terminology and definitions for BCI, marketability of BCI technologies, ethical aspects of BCI research and application, and perceived urgent matters about BCI (Nijboer et al. 2013).

Interrogative neuroethics easily integrates with issues of research neuroethics. Experimental diagnostic techniques may yield ambiguous gauges of neurological impairment, so estimates on degrees of cognitive functioning must be paired with clinical and family judgments about patient decision-making and quality of life. For example, a group of medical ethicists and physicians in Canada are pursuing projects to query the role of neuroimaging in determining awareness and sentience after severe brain injury, providing for patient participation in decisions, and explaining neurological prognostications to families (Weijer et al. 2014).

Perhaps the disciplinary extent of neuroethics could halt here, where research neuroethics and interrogative neuroethics are entwined. However, we envision additional territory to cover. That territory has been described by members of the Presidential Commission for the Study of Bioethical Issues, who in their 2014/15 *Gray Matters* reports, posited a greater ethical burden—and responsibility—evoked by the brain sciences. Summarizing those reports, neurologist and Commission member Stephen Hauser published a paper entitled *What ethics integration looks like in neuroscience research*. As might be anticipated given the growth of both neuroscience and neuroethics, Hauser's expectations for interdisciplinary collaboration on far-ranging neuroethical issues are even more comprehensive than the territory covered by the 2002 special issue of *Brain and Cognition* on neuroethics. Hauser states:

> Ethics integration entails collaboration between researchers and ethics professionals to acknowledge and understand the societal and ethical issues and implications of their work. Integration should equip scientists to recognize and address ethical issues as they arise, and ethicists to understand the science and technology with which they engage (Hauser 2014: 623).

These expectations are reasonable, in our view, but how should those more extensive collaborations proceed? The next section surveys some neuroethical inquiries offering answers to that question.

16.6 Neuroethics Operating Across Discourses

The targeted focus that is characteristic of both research neuroethics and interrogative neuroethics permits tremendous utility for near-term deliberations and planning. In the longer run, however, minds are gradually changed by intriguing news about the brain. Neither research neuroethics nor interrogative neuroethics are designed for anticipating or registering what happens when these factors become highly salient, such as when:

Fresh neuroscience findings call for altered conceptions of psychological matters.

People change how they regard psychological matters in light of news about neuroscientific findings.

Social subgroups and ethno-cultural populations compare their views on psychological matters with neuroscientific information.

Various related scientific and humanities fields assess and critique neuroscientific claims about the validity of findings and the relevance of clinical applications.

Different professions disagree about how to deal with drifting conceptions and evolving models of mental/psychological/behavioral matters.

Society's many reactions to emerging ways that conceptions of selves and persons are called into question.

How shall neuroscientists deliberate with ethicists, and how would both of these groups communicate with a wider public? That answer must start from the way that the new and unfamiliar must somehow be meaningfully linked with the comfortable and familiar. Neuroethical discourse must work with information provided by brain sciences, even as it addresses human matters in terms set by society. Martha Farah of the Center for Neuroscience and Society at the University of Pennsylvania has written

> [N]euroscience can now be brought to bear in many different spheres of human life, beyond the traditional application area for biological science, medicine. Any endeavor that depends on being able to understand, assess, predict, control, or improve human behavior is, in principle, a potential application area for neuroscience. This includes diverse sectors of society, for example, education, business, politics, law, entertainment, and warfare (Farah 2012: 573).

These far-reaching prospects for neuroscience all have one thing in common: they are premised on the ability of those social sectors to appreciate and utilize neuroscience's discoveries for their own specialized work. Explaining the intricate implications of neuroscience could be avoided of course, by taking the easy road of depicting science as the supreme truth and chief debunker. Formulaic headlines in this tenor seem so familiar by now. Statements such as "Brain science says X, so widely-held views about Y are wrong" presume that a conception of Y (e.g., feeling love or some moral judgment) must be vulnerable to a hypothesis about X (e.g., oxytocin levels or prefrontal cortical activity). Assertions like "Neuroscience cannot find X in the brain, so X is a myth" presume that the only reality X could have (e.g., deeply loving or freely choosing) is a neurological reality. Neuroethics need not be reduced to crude myth-busting, if it first investigates and revises any presumptions.

Domains of knowledge are ontologies in their identification and categorization of known matters, and their interrelationships, in turn permitting information storage, efficient retrieval, and responsive querying. Any area of human knowledge displays an ontology, usually in a poorly organized and less-than-logical way. Disciplinary fields attempt to improve their ontologies for better methodological utility. Life sciences and health sciences put vast efforts into regimenting and coordinating their ontologies—medical informatics is a prominent example among many kinds of biomedical ontologies.

But no matter how regimented and logical a domain's ontology may be, a neighboring discipline probably has a somewhat different ontology—because different things matter in differing ways to different groups. Even if two related subfields both use the same term for an entity or process, those subfields may not mean precisely the same thing, even in those situations in which all may agree that only one subject matter is involved. Informatics ontology can advise fields on improved precision for theoretical terms and their relations and operational definitions of as many significant concepts as possible. That precision and operationality in turn permits better mutual comprehension between neighboring fields. Ideally, even where two fields do not have closely similar ontologies, they would be able to understand and use scientific information from each other, thanks to a degree of terminological and ontological interoperability. Neuroscience's accounts about neural systems and functions will be fairly closely aligned with behavioral and cognitive psychology. At their intersection, the emerging field of cognitive neuroinformatics is an example of interoperable ontology at work (Poldrack and Yarkoni 2016). Neuroethics must itself become more interoperable, so that it may meaningfully engage with many other fields.

This interoperability can extend to any aspect of psychology and mentality, if (and perhaps, only if) neuroethics does not deviate from the interrogative mode. An example involving neurological adjustments to reduce compulsive behaviors will serve. Grant Gillett of the University of Otago, New Zealand, and Berthold Langguth of the University of Regensburg, Germany, collaborated with medical and neuroscience colleagues to explore the psychological ramifications of neuromodulating cortical regions associated with predicting, considering, and choosing among optional actions (De Ridder et al. 2016). Because patients undergoing successful treatment typically report greater freedom to choose actions and deny compulsions, these individuals can be said to be acquiring "freer wills." On the ethical level, neuroethicists are invited to conclude that such a psychological outcome is rightly regarded as a morally positive result.

Law, political science, and economics are further examples of fields of knowledge, and beyond all disciplinary ontologies lies the vaster realm of socially accepted ontologies (in a loose sense) that encompass what people typically take to be real. These discordant and even conflicting ontologies, whether disciplinary or popular, provoke deep puzzlement and philosophical inquiry. Knowledge about brain structures and processes cannot dictate how people must understand their ordinary, and extraordinary, behaviors and projects of daily life. However, when the brain sciences are asked to craft questions about causes of behavior in order to apply social categorizations to a person's conduct (as willful, deliberate, autonomous, negligent, compulsive, and so on), the scientific answers are under no obligation to conform to those categorizations. Nicole Vincent rightfully notes that neuroscientific evidence about a specific type of poor brain function cannot, by itself, indicate whether a person suffers from a blameless incapacity or exhibits an immoral character. Vincent calls into question that false dichotomy:

> [I]n my view the capacity-character problem … reflects a need for theories of moral and legal responsibility to recognize a broader and more finer-grained taxonomy of the different kinds of moral appraisal, and to take these different kinds of moral appraisal into account in the right way (Vincent 2015: 494).

Neuroethics can become fully interoperable, helping to sustain mutual comprehension between advances in brain research, multiple disciplinary fields, and sectors of the public trying to keep pace with such developments. It must prepare for conversing with the polysemous discourses already heard today and with the emerging discourse(s) of tomorrow.

16.7 Interoperable Neuroethics

Communication about science must understand society just as carefully as society should try to understand science. Brain science is no exception. Neuroethicists often pronounce that some domain of social discourse is predicated on a certain conception of mental events and then warn that nothing in the brain indicates that such mental events really occur. Yet that discourse may be misleading. What we like to say about ourselves need not match our actual expectations, and what we like to say about our social institutions need not be accurate about how they actually work. Stephen Morse, citing extensive legal scholarship, points to the realm of legal proceedings:

> Contrary to what many people believe, and what judges and others sometimes say, free will is not a legal criterion that is part of any doctrine, and it is not even foundational for criminal responsibility. … when adopting a compatibilist metaphysics about responsibility, criminal law doctrines are fully consistent with the truth of determinism or universal causation that allegedly undermines the foundations of responsibility (Morse 2015: 54).

According to Morse, and we agree, jurisprudence and legal philosophy should be closely consulted before any neuroethical verdict is summarily rendered.

The intersection of legal justice, social values, and personal ethics is particularly fraught with confusions and tensions across discourses. The area of forensic psychology is displaying the transformative effects of neuroscience in the courtroom, which in turn invites unexpected social intrusions into the realm of law. Georgia Gkotsi of the Institute of Legal Psychiatry, Switzerland, has surveyed a variety of civic priorities, such as public safety and the medicalization of violence, pondering how that consequentialism may diminish offenders' autonomy and rights (Gkotsi and Gasser 2016). Farah Focquaert of Ghent University, Belgium, assessed a similar civic overlap, where interests of public welfare may shift the purpose of punishment over to a lifetime of mandatory neurological "correction" (Focquaert 2014).

Jennifer Chandler of the University of Ottawa Law Faculty, working together with Eric Racine and colleagues provided another illustration by considering the legal implications of a victim of violent crime who undergoes therapy to reduce memory trauma prior to the criminal trial. Are legal proceedings that rely heavily on evidence of reliable memory and emotional distress going to be compromised (Chandler et al. 2013)? As legal systems begin to grapple with such novel questions, neuroethics can monitor discourses from many countries around the world, observing how novel neuroS/T impact the practice of law and how legal climates are shaping neuroscientific research (Farahany 2009; Spranger 2012; Pardo and Patterson 2013). Neuroethics can also monitor international concerns about mental health and human rights (Stein and Giordano 2015).

A highly conversant neuroethics will be an interoperable neuroethics, grounded in perceptive attention to nuanced discourses occurring in diverse domains. In this way, we could pose the following questions about novel neuroS/T, building from simpler questions posed in prior sections:

How might people understand this neuroS/T, as applied to their own mental lives, and think about potentially altering their notions of mental/psychological matters or even modifying their entire self-conceptions?

How might people express their disagreements with others over this neuroS/T's seeming revision of mental/psychological matters, align together in interest groups, and justify their disagreements about each other's self-conceptions?

How might brain-related disciplines agree or disagree about the potential for this neuroS/T to require revisions to scientific conceptions of mental/psychological matters?

How may philosophical analyses and assessments of disciplinary views on this neuroS/T expose fallacious, confused, or biased thinking and moderate excessive claims about drastic revisions to conceptions of mental/psychological matters?

How may brain-related disciplines disseminating neuroS/T's revision of certain mental/psychological matters assess implications for respecting and supporting individuals, as those individuals conceive of themselves, with due regard for their membership in interest groups and/or social identity with an ethno-cultural population?

How might other professions and institutions assess disciplinary claims made about this neuroS/T's revision of mental/psychological matters and judge resulting effects on individuals' self-conceptions, in terms of the values and norms that those professions exemplify and uphold?

How may society in general assess neuroS/T's revision of mental/psychological matters and effects on individuals' self-conceptions, according to its customary values, cherished ideals, and established laws?

Gaining answers to these questions would allow neuroethical inquiries to diagnose and help alleviate conflicting views over the import and worth of neuroS/T innovation and advances. What appear to be sharply dissenting viewpoints may only require some careful translation to help parties understand that and how their differences are not so great. On the other hand, parties using the same terms may be camouflaging quite discrepant views, which need to be elucidated and made plain for critique.

Translation can be done both well and poorly, but it is surely unavoidable regardless. Neither neuroscience nor neuroethics should imagine that scientific knowledge or technological prowess is straightforwardly conveyed to the public. Racine, Illes, and DuRousseau investigated the media coverage of a widely-reported scientific achievement in 2002: "Rat navigation guided by remote control" as published in *Nature*. They noted how value-laden phrases such as "remote-controlled rats," "mind control," and "threat to autonomy" proliferated throughout the ensuing reactions in the press.

> Our study supported a broadened, multidirectional approach to the understanding and practice of science communication … We found that when research moved from the bench to headline, findings were not simply transmitted but instead were *translated*. Language literally changed, and with it changed the meanings ascribed to both the study and the fundamental goals of science communication (Racine et al. 2012: 186).

Neuroethics cannot forestall this kind of fast-paced process of dissemination and digestion. However, versatile and fluent neuroethicists could lend a conversant academic voice to ensuing public debates.

Interoperable neuroethics has a much-needed role at the intersection of medicine, government regulation, and social welfare. One example is the question whether professionals such as pilots or doctors should be permitted—or even encouraged—to seek off-label use of certain drugs for their alleged ability to promote alertness, concentration, memory, sharpness, and so on (Franke et al. 2013).

Another opportunity for interoperable neuroethics is provided by the growing availability of cognitive enhancement devices (CEDs). Thomas Douglas and Julian Savulescu of the Uehiro Centre for Practical Ethics at the University of Oxford, along with Oxford colleagues and Neil Levy of the University of Melbourne, argued that CEDs should be classified with medical devices and regulated accordingly, per the EU's Medical Devices Directive (MDD). They forcefully claim:

> CEDs are not categorically different from medical devices; in fact, the very same device may be used both for therapeutic and enhancement purposes, in some cases using similar parameters. CEDs, as devices that modify brain function to improve cognitive performance are, in important respects, the same *sorts* of devices that the MDD covers: they intervene to modify physiological processes and present varying degrees of physiological risks and side effects (Maslen et al. 2014: 79).

They go on to point out the heavy (and unjustifiable, in their view) burdens upon device manufacturers and the public wanting those devices if new regulations are imposed. Arguing on behalf of medicine's contribution to social welfare rightly carries great weight in these sorts of debates.

However, an interoperable neuroethics can also ponder how medicine and the public may be poorly informed about the real-world effects of such devices. CEDs do objectively modulate neurophysiological functions, but the subjective results will surely vary from individual to individual, in various contexts, and across cultures. What does "cognitive" normality or improvement really mean? The contrary position against lightly regulating CEDs has been taken by James Giordano of Georgetown University and John Shook of the University at Buffalo.

> Neuroethical analyses and explorations into cognitive enhancement must keep abreast of relevant findings from many fields, such as personal genomics, developmental psychology, social neuroscience, cultural neuroscience, cross-cultural psychology, and cultural anthropology. As any of these fields can indicate, there will always be debate as to what constitutes the "cognitively normal" human brain, and rightly so. What exactly counts as constituting a cognitive deficit, disorder, distortion, or bias will not converge across cultures or even within societies. It is naïve to suppose that a compensatory adjustment, much less an enhancing adjustment, could be generically assigned any validity across all of humanity (Shook and Giordano 2016: 84).

Interoperable neuroethics, as this debate exemplifies, points the way toward deep philosophical issues, while displaying interoperability's inherent limitations.

16.8 Disciplined Neuroethics

Sensitivity to the diverse interpretations of neuroS/T that people and institutions will form for themselves, and their divergent views of mental and personal matters that result, establishes that interoperable neuroethics should remain acutely aware of tectonic shifts in conceptions of mentality, agency, the self, the person, and so on. But interoperable neuroethics is powerless, by itself, to do anything about those shifts. Adjudicating matters concerning the right or best views on those core psychological and ethical matters falls outside its competence.

In this regard, interoperable neuroethics is *not* designed for resolving issues such as:

- Neuropsychological revisions to folk psychological conceptions that in turn alter people's sense of their values, their moral capacities, and their self-identities
- Revisions to people's values, morals, and self-valuations happening in different ways across populations, producing sharp disagreements on fundamental matters
- Social subgroups and ethno-cultural populations that adopt highly variable stances toward putative revisions by neuroscience to communal values, morals, and identities
- Professions committed to visions of the self, the bases of self-worth, and the dignity of personhood that cease assuming that these should be univocal matters
- Societies relying on stable institutions and resolute professions (e.g., education, law) that may find those foundations eroding and fragmenting

Neuroethics becomes deeply philosophical as these kinds of issues arise. However, its interdisciplinary and culturally fluent inquiries from the foundational stages of research, interrogative, and interoperable neuroethics will serve it well. Two more examples from recent neuroethical literature serve to justify this hope.

A team of Japanese researchers, led by Eisuke Nakazawa of the University of Tokyo and Keiichiro Yamamoto of the National Center for Global Health and Medicine, is acquiring impressive expertise with real-time fMRI-based neurofeedback techniques for modifying both behaviors and mental judgments (Nakazawa et al. 2016). They are also exploring considerable neuroethical territory, as they ponder the workability and ethicality of modifying moral judgment through the use of this approach. Beyond questions of research neuroethics (e.g., about the safety, efficacy, reversibility of the techniques, etc.) and interrogative neuroethics (e.g., about subjects' compliance, goals, and self-transformation), interoperable neuroethics poses additional questions, such as:

How authentic could these moral modifications become?

Who will approve this technique as therapeutic or possibly as enhancing?

By what standard could neuroscience affirm that moral improvement is achievable?

Who will receive society's encouragement (or coercion) to undergo this moral modification?

John Trimper, Paul Root Wolpe, and Karen Rommelfanger of Emory University are addressing "brain-to-brain interfacing" technologies (Trimper et al. 2014). Again, concerns at the stages of research and interrogative neuroethics are accompanied by ethical quandaries calling for interoperable neuroethics. Here, questions include:

In what sense could two brains be mentally connected?

How would that mental connection constitute a psychological overlap or continuity?

Does that relationship amount to a diminishment of subjectivity or an enlargement?

Would a subject's autonomy be diminished or possibly enhanced?

Where does any boundary to the "self" have to be re-situated, or is it blurred entirely?

These matters in turn inspire more difficult questions, indeed, fully philosophical questions, about the opportunities and risks for individuality and personhood that are incurred when neuroS/T is employed to manipulate psychological capacities.

We have stated before, and re-iterate here, that neuroethics, as a discipline and a set of practices, has come of age and must now face the future of rapidly developing brain science, ever more diverse applications and effects of the knowledge and capabilities that brain science affords, and last, but certainly not least, of its own viability, engagement, and value, to the foci and scope of brain science and its translation, and to society, writ-large (Giordano 2010, 2014). While coming of age, it is our view that neuroethics will mature as a discipline, developing a robust portfolio of tools and principles having proven value in global contexts (Shook and Giordano 2014; Lanzilao et al. 2013), and, in these ways, fully contribute to philosophical debates in the modes of research, interrogative, and interoperable neuroethics. That maturation is evidently well underway, and its impressive growth is measured by the ample neuroethical collaborations having both multidisciplinary and multinational dimensions.

References

Al-Delaimy WK (2012) Ethical concepts and future challenges of neuroimaging: an Islamic perspective. Sci Eng Ethics 18(3):509–518

Anderson J, Mizgalewicz A, Illes J (2012) Reviews of functional MRI: the ethical dimensions of methodological critique. PLoS One 7(8):e42836

Becker K, Shook JR, Darragh M, Giordano J (2017) A four part working bibliography of neuroethics: part 4 – "Neuroethics of clinical and social applications of brain science". Phil Ethics Humanit Med 12(2)

Bell E, Racine E, Chiasson P, Dufourcq-Brana M, Dunn LB, Fins JJ et al (2014) Beyond consent in research: revisiting vulnerability in deep brain stimulation for psychiatric disorders. Camb Q Healthc Ethics 23(3):361–368

Booth T, Boyd-Ellison JM (2015) The current impact of incidental findings found during neuroimaging on neurologists' workloads. PLoS One 10(2):e0118155

Borgelt E, Anderson JA, Illes J (2013) Managing incidental findings: lessons from neuroimaging. Am J Bioeth 13(2):46–47

Bracken-Roche D, Bell E, Racine E (2016) The "vulnerability" of psychiatric research participants: why this research ethics concept needs to be revisited. Can J Psychiatr 61(6):335–339

Brosnan C, Cribb A, Wainwright SP, Williams C (2013) Neuroscientists' everyday experiences of ethics: the interplay of regulatory, professional, personal and tangible ethical spheres. Sociol Health Illn 35(8):1133–1148

Buniak L, Darragh M, Giordano J (2014) A four part working bibliography of neuroethics: part 1: overviews and reviews—defining and describing the field and its practices. Phil Ethics Humanit Med 9(9)

Cabrera LY, Fitz NS, Reiner PB (2015) Empirical support for the moral salience of the therapy-enhancement distinction in the debate over cognitive, affective and social enhancement. Neuroethics 8(3):243–256

Chandler JA, Mogyoros A, Rubio TM, Racine E (2013) Another look at the legal and ethical consequences of pharmacological memory dampening: the case of sexual assault. J Law Med Ethics 41(4):859–871

Darragh M, Buniak L, Giordano J (2015) A four part working bibliography of neuroethics: part 2—neuroscientific studies of morality and ethics. Phil Ethics Humanit Med 10(1)

De Ridder D, Vanneste S, Gillett G, Manning P, Glue P, Langguth B (2016) Psychosurgery reduces uncertainty and increases free will? A review. Neuromodulation 19(3):239–248

Deslauriers C, Bell E, Palmour N, Pike B, Doyon J, Racine E (2010) Perspectives of Canadian researchers on ethics review of neuroimaging research. J Empir Res Hum Res Ethics 5(1):49–66

Farah MJ (2012) Neuroethics: the ethical, legal, and societal impact of neuroscience. Annu Rev Psychol 63:571–591

Farahany N (ed) (2009) The impact of behavioral sciences on criminal law. Oxford University Press, Oxford

Fine C (2013) Neurosexism in functional neuroimaging: from scanner to pseudo-science to psyche. In: Ryan M, Branscombe N (eds) The Sage handbook of gender and psychology. Sage, London, pp 45–60

Fins JJ (2008) Neuroethics and neuroimaging: moving toward transparency. Am J Bioeth 8(9):46–52
Fitz NS, Nadler R, Manogaran P, Chong EW, Reiner PB (2014) Public attitudes toward cognitive enhancement. Neuroethics 7(2):173–188
Focquaert F (2014) Mandatory neurotechnological treatment: ethical issues. Theor Med Bioeth 35(1):59–72
Ford PJ (2009) Vulnerable brains: research ethics and neurosurgical patients. J Law Med Ethics 37(1):73–82
Franke AG, Bagusat C, Dietz P, Hoffmann I, Simon P, Ulrich R et al (2013) Use of illicit and prescription drugs for cognitive or mood enhancement among surgeons. BMC Med 11:102
Franke AG, Papenburg C, Schotten E, Reiner PB, Lieb K (2014) Attitudes towards prescribing cognitive enhancers among primary care physicians in Germany. BMC Fam Pract 15:3
Fujita M, Hayashi Y, Tashiro S, Takashima K, Nakazawa E, Akabayashi A (2014) Handling incidental findings in neuroimaging research in Japan: current state of research facilities and attitudes of investigators and the general population. Health Res Policy Syst 12:58
Giordano J (2010) Neuroethics—coming of age and facing the future. In: Giordano J, Gordijn B (eds) Scientific and philosophical perspectives in neuroethics. Cambridge University Press, Cambridge, pp xxv–xxix
Giordano J (2013) Ethical considerations in the globalization of medicine. An interview with James Giordano. BMC Med 11(1):article 69
Giordano J (2014) The human prospect(s) of neuroscience and neurotechnology: domains of influence and the necessity—and questions—of neuroethics. Hum Prospect 4(1):1–11
Giordano J (2015) A preparatory neuroethical approach to assessing developments in neurotechnology. AMA J Ethics 17(1):56–61
Giordano J (2016) Toward an operational neuroethical risk analysis and mitigation paradigm for emerging neuroscience and technology (neuroS/T). Exp Neurol
Gkotsi GM, Gasser J (2016) Neuroscience in forensic psychiatry: from responsibility to dangerousness. Int J Law Psychiatry 46:58–67
Hauser SL (2014) What ethics integration looks like in neuroscience research. Ann Neurol 75(5):623–624
Heinrichs JH (2012) The sensitivity of neuroimaging data. Neuroethics 5(2):185–195
Hildt E, Lieb K, Franke AG (2014) Life context of pharmacological academic performance enhancement among university students—a qualitative approach. BMC Med Ethics 15:23
Horvath J, Perez J, Forrow L, Fregni F, Pascual-Leone A (2013) Transcranial magnetic stimulation: future prospects and ethical concerns in treatment and research. In: Chatterjee A, Farah MJ (eds) Neuroethics in practice. Oxford University Press, Oxford, pp 209–234
Illes J, Raffin T (2002) Neuroethics: an emerging new discipline in the study of brain and cognition. Brain Cogn 50(3):341–344
Jensen C, Forlini C, Partridge B, Hall W (2016) Australian university students' coping strategies and use of pharmaceutical stimulants as cognitive enhancers. Front Psychol 7:article 277
Kehagia A, Tairyan K, Federico C, Glover GH, Illes J (2012) More education, less administration: reflections of neuroimagers' attitudes to ethics through the qualitative looking glass. Sci Eng Ethics 18(4):775–788
Lanzilao E, Shook J, Benedikter R, Giordano J (2013) Advancing neuroscience on the 21st century world stage: the need for—and proposed structure of—an internationally relevant neuroethics. Ethics Biol Eng Med 4(3):211–229
Lipsman NI, Glannon W (2013) Brain, mind and machine: what are the implications of deep brain stimulation for perceptions of personal identity, agency and free will? Bioethics 27(9):465–470
Maier LJ, Liakoni E, Schildmann J, Schaub MP, Liechti ME (2015) Swiss university students' attitudes toward pharmacological cognitive enhancement. PLoS One 10(12):e0144402
Martin A, Becker K, Darragh M, Giordano J (2016) A four part working bibliography of neuroethics: part 3— "The ethics of neuroscience". Phil Ethics Humanit Med 11(2):7

Maslen H, Douglas T, Kadosh RC, Levy N, Savulescu J (2014) The regulation of cognitive enhancement devices: extending the medical model. J Law Biosci 1(1):68–93

Morse S (2015) Criminal law and common sense: an essay on the perils and promise of neuroscience. Marquette Law Rev 99(1):39–72

Nakazawa E, Yamamoto K, Tachibana K, Toda S, Takimoto Y, Akabayashi A (2016) Ethics of decoded neurofeedback in clinical research, treatment, and moral enhancement. AJOB Neurosci 7(2):110–117

Nijboer F, Clausen J, Allison BZ, Haselager P (2013) The Asilomar survey: stakeholders' opinions on ethical issues related to brain-computer interfacing. Neuroethics 6(3):541–578

Palmour N, Affleck W, Bell E, Deslauriers C, Pike B, Doyon J et al (2011) Informed consent for MRI and fMRI research: analysis of a sample of Canadian consent documents. BMC Med Ethics 12(1):article 1

Pardo M, Patterson D (eds) (2013) Minds, brains, and law: the conceptual foundations of law and neuroscience. Oxford University Press, Oxford

Poldrack R, Yarkoni T (2016) From brain maps to cognitive ontologies: informatics and the search for mental structure. Annu Rev Psychol 67:587–612

Racine E, DuRousseau D, Illes J (2012) Ethical issues in performance-enhancing technologies: from bench to headline. In: Giordano J (ed) Neurotechnology: premises, potential, problems. CRC Press, Boca Raton, pp 175–190

Rippon G, Jordan-Young R, Kaiser A, Fine C (2014) Recommendations for sex/gender neuroimaging research: key principles and implications for research design, analysis, and interpretation. Front Hum Neurosci 8:650

Rossi S, Hallett M, Rossini PM, Pascual-Leone A (2009) Safety, ethical considerations, and application guidelines for the use of transcranial magnetic stimulation in clinical practice and research. Clin Neurophysiol 120(12):2008–2039

Salles A (2014) Neuroethics in a "Psy" world: the case of Argentina. Camb Q Healthc Ethics 23(3):297–307

Sattler S, Sauer C, Mehlkop G, Graeff P (2013) The rationale for consuming cognitive enhancement drugs in university students and teachers. PLoS One 8(7):e68821

Schelle KJ, Olthof BM, Reintjes W, Bundt C, Gusman-Vermeer J, van Mil AC (2015) A survey of substance use for cognitive enhancement by university students in the Netherlands. Front Syst Neurosci 9:10

Semrud-Clikeman M, Bledsoe J (2014) Understanding the neuroscience of clients with Asian heritage. In: Mark Davis J, D'Amato RC (eds) Neuropsychology of Asians and Asian-Americans. Springer, New York, pp 117–133

Shechner T, Wakschlag N, Britton JC, Jarcho J, Ernst M, Pine DS (2013) Empirical examination of the potential adverse psychological effects associated with pediatric fMRI scanning. J Child Adolesc Psychopharmacol 23(5):357–362

Shook JR, Giordano J (2014) A principled, cosmopolitan neuroethics: considerations for international relevance. Philos Ethics Humanit Med 9(1):article 1

Shook JR, Giordano J (2016) Defining contexts of cognitive (performance) enhancements: neuroethical considerations, and implications for policy. In: Jotterand F, Dubljevic V (eds) Cognitive enhancement: ethical and policy implications in international perspectives. Oxford University Press, Oxford, pp 76–98

Singh I (2013) Brain talk: power and negotiation in children's discourse about self, brain and behaviour. Sociol Health Illn 35(6):813–827

Singh I, Bard I, Jackson J (2014) Robust resilience and substantial interest: a survey of pharmacological cognitive enhancement among university students in the UK and Ireland. PLoS One 9(10):e105969

Spranger TM (ed) (2012) International neurolaw: a comparative analysis. Springer Science, Berlin

Stein DJ, Giordano J (2015) Global mental health and neuroethics. BMC Med 13:44

Trimper JB, Root Wolpe P, Rommelfanger KS (2014) When "I" becomes "We": ethical implications of emerging brain-to-brain interfacing technologies. Front Neuroeng 7:article 4

Ulmer S, Booth TC, Widdershoven G, Jansen O, Fesl G, von Kummer R et al (2013) Incidental findings in neuroimaging research: ethical considerations. In: Ulmer S, Jansen O (eds) fMRI: basics and clinical applications. Springer, Berlin, pp 311–318
Vincent NA (2015) A compatibilist theory of legal responsibility. Crim Law Philos 9(3):477–498
Wardlaw JM, Davies H, Booth TC, Laurie G, Compston A, Freeman C et al (2015) Acting on incidental findings in research imaging. BMJ 351:h5190
Weijer C, Peterson A, Webster F, Graham M, Cruse D, Fernández-Espejo D et al (2014) Ethics of neuroimaging after serious brain injury. BMC Med Ethics 15:41

The Biopolitics of Neuroethics

17

Fabrice Jotterand and Marcello Ienca

Abstract

Neuroethics is an interdisciplinary endeavor committed to promoting responsible innovation and ethical understanding in neuroscience and neurotechnology. As the field develops, closer examination will be needed concerning how neuroethics discourse could, either implicitly or explicitly, shape our brains and our minds, and ultimately, what it means to be human. Borrowing the notion of biopolitics from Michel Foucault, this chapter explores and critically examines the biopolitical dimension of neuroethics. In the first section, the concept of biopolitics is examined through the lens of Michel Foucault's analysis with a particular focus on the politicization of science and technology. The second section outlines the key features of neuroethics and the main challenges of attempting to pinpoint the identity of the neuroethics community. The third section delineates some positive developments in establishing the field and in building a community of scholars and researchers interested in the ethical and social implications of advances in neuroscience. The fourth section provides a rationale as to why the neuroethics community has the moral obligation to scrutinize the socioeconomic motivations and political drives behind any research project to ensure both as much transparency as possible and state neutrality. The final section of the chapter discusses the challenges of establishing a dominant discourse in light of political and moral pluralism and suggests a potential approach (i.e., procedural ethics grounded on a deliberative democracy approach) to shape such discourse.

F. Jotterand, Ph.D., M.A. (✉)
Center for Bioethics and Medical Humanities, Medical College of Wisconsin, Milwaukee, WI, USA

Institute for Biomedical Ethics, University of Basel, Basel, Switzerland
e-mail: fjotterand@mcw.edu

M. Ienca, M.Sc., M.A.
Institute for Biomedical Ethics, University of Basel, Basel, Switzerland
e-mail: marcello.ienca@unibas.ch

E. Racine, J. Aspler (eds.), *Debates About Neuroethics*, Advances in Neuroethics,
DOI 10.1007/978-3-319-54651-3_17

17.1 Introduction

Progress in neuroscience and neurotechnology is often portrayed as the next step toward addressing intractable issues in research and medicine, as well as in society more broadly, such as criminal behavior, treatment resistant psychiatric disorders, and chronic neurological conditions (e.g., cerebral palsy) (Chandler and Dodek 2016; Jotterand and Giordano 2015; Persson and Savulescu 2008, 2012, 2013; Jotterand forthcoming). Novel applications (e.g., cognitive neuroenhancers, brain-computer interfaces, neuroprosthetics) raise ethical, legal, and social concerns that neuroethics claims to carefully address. As it develops, however, a closer consideration should take place concerning how neuroethics discourse could, either implicitly or explicitly, shape our brains and our minds and, ultimately, what it means to be human. Borrowing the notion of biopolitics from Michel Foucault, this chapter explores and critically examines the biopolitical dimension of neuroethics (1) as a second-order discipline and (2) with regard to the identity of the neuroethics community. To this end, the chapter discusses the implications of such biopolitics on the process of discourse building and on the creation of the neuroethics community. In the first section, the concept of biopolitics is examined through the lens of Michel Foucault's analysis with a particular focus on the politicization of science and technology. The second section outlines the key features of neuroethics and the main challenges of attempting to pinpoint the identity of the neuroethics community. The third section delineates some positive developments in establishing the field and in building a community of scholars and researchers interested in the ethical and social implications of advances in neuroscience. The fourth section provides a rationale as to why the neuroethics community has the moral obligation to scrutinize the socio-economic motivations and political drives behind any research project to ensure both as much transparency as possible and state neutrality. The final section of the chapter discusses the challenges of establishing a dominant discourse in light of political and moral pluralism and suggests a potential approach (i.e., procedural ethics grounded on a deliberative democracy approach) to shape such discourse.

17.2 Biopolitics

The term biopolitics was popularized by the French philosopher Michel Foucault in 1974 at a conference on social medicine (Foucault 2000). However, he only provided a systematic analysis of the concept two years later, in his lectures delivered at the Collège de France and in his book *The History of Sexuality, Vol. 1* (Foucault 1980). In his work, Foucault examines various forms of exercising power from a historical and analytical perspective, particularly its modern articulation with regard to life (i.e., bio-power). Biopolitics, as Foucault understood it, designates a shift in how power, or sovereign power in its political sense, is managed and displayed. The juridical form of sovereign power describes the right of a ruler or the state "to seize things, time, bodies, ultimately the life of subjects" (Rose and Rabinow 2003); however, this model of power underwent a transformation in the eighteenth century given what Foucault calls biopower, which "designate[s] what brought life and its mechanisms into the realm of explicit calculations and made knowledge-power an agent of transformation of human life" (Foucault 1980: 143). In other words, a transformation occurred in political thinking and practice during the eighteenth century which allowed, on the one hand, an understanding of biological existence as political existence and, on the other hand, a shift from mastery over the individual body to the regulation of whole populations through social interventions. As Foucault puts it in his *History of Sexuality*:

> "...[f]or the first time in history ... biological existence was reflected in political existence ... Power would no longer be dealing simply with legal subjects over whom the ultimate dominion was death, but with living beings ... [W]hat might be called a society's 'threshold of modernity' has been reached when the life of the species is wagered on its own political strategies. For millennia, man remained what he was for Aristotle: a living animal with the additional capacity for a political existence; modern man is an animal whose politics places his existence as a living being in question" (142–143).

This succinct analysis demonstrates not only how the role of state power over life becomes a way for control of the masses but also a mechanism for establishing specific goals for living beings tied to political agendas. The placement of life at the core of the political order constitutes an attempt to address issues about the survival of the human species, which has taken various approaches, such as ecological biopolitics and more recently technocentric biopolitics. According to German political scientist Dietrich Gunst, ecological biopolitics refers to health policy and the regulation of the population tied together with environmental protection and issues regarding the future of the human species (e.g., debate over genetically modified organisms [GMOs]) (Gunst 1978, cited in Lemke 2011: 24). In the same vein, but more relevant to the focus of this chapter on the biopolitics of neuroethics, German sociologist Wolfgang van den Daele articulates a different strand of biopolitics, i.e., technocentric biopolitics (Van Den Daele 2005). With the emergence of novel technologies, the protection of life is not only addressed through environmental considerations but also with regard to reflections on the social acceptability of technological advances. This version of biopolitics refers to:

> "the approximately twenty-year societal thematization and regulation of the application of modern technologies and natural science to human life. Within the purview of these policies stand, above all, reproductive medicine and human genetics. Increasingly, however, one finds brain research, as well as the scientifically and technically rather uninteresting field of cosmetic surgery. Biopolitics responds to the transgression of boundaries. It reacts to the fact that the boundary conditions of human life, which until now were unquestioned because they lay beyond the reach of our technical capabilities, are becoming accessible to us ... The results of such transgressions are oral controversies and debates about regulation that come down to the old question: Just because we can, should we?" (Van Den Daele 2005: 8 cited in Lemke 2011: 26–27; italics ours).

It is important to note that within the context of technocentric biopolitics, the issue of power over life is no more a question of exerting power over life but rather, setting boundaries for transgressions of human biological limitations. In addition, the application of modern technologies is not limited to traditional questions of genetics and reproductive technologies but increasingly encompasses brain research and its regulatory and public policy implications. However, the relationship between neuroscientific discoveries and their implications for our understanding of issues such as the mind-brain problem, the use of neuroscientific evidence in court, neurodeterminism, moral agency, etc. remains constrained by sociopolitical factors and ideologies, as well as potential competing interests. In short, neuroscience is not immune to political control. As Henry and Plemmons point out "science is set in a socio-political context … the politics of conflicting and ideologically shaped interests mediate the outcome of public policies based on scientific research, often with negative impacts on society's most vulnerable groups. Neuroscience and, therefore, neuroethics, are inherently political" (Henry and Plemmons 2012). What follows is that biopolitics is integral to neuroethics, and therefore the moral assessment of advances in neuroscience research and their applications in medicine and society at large must always recognize the potential risk of imposing ideologies for the control of populations.

This point becomes even more salient when we consider various stages of technological progress throughout history, particularly the current one. According to Marx W. Wartofsky, technological progress can be summarized in four revolutions: (1) the "revolution of the hand tool," which refers to the very creation of technology and the use of tools; (2) the "industrial revolution," which denotes the transition from hand tools to machine; (3) the use of "calculating or computing machines in conjuncture with other machines," which led to the automatization of production; and (4) "the politicization of technology," which refers to the convergence of science, technology, politics, economics, and values (1992).

In the last of these revolutions, high technology—sophisticated technology that requires specialized devices or systems—plays a central role (Wartofsky 1992). The use of techno-science as a source of economic and political power, and by extension, military power, has become an imperative in the political and economic life of a nation (Wartofsky 1992). John Ziman rightly observes that contemporary "science is being pressed into the service of the nation as the driving force in a national [research and development (R&D)] system, a wealth-creating techno-scientific motor for the whole economy" (Ziman 2002). Neuroscience and neurotechnology are no exception, and there are many incentives, other than altruism, for a nation to invest in neuroscience R&D. The challenge is to establish a moral and regulatory framework enabling neuroscience research to move from the lab to the real world in a prudent manner and to find public support for further research. If neuroethics is to play a role in shaping ethical and regulatory discourse, the following critical question ought to be raised: considering that neuroethics is still a young field, to what extent is it suited to build networks and capacity to address ethical issues arising in neuroscience? Before we can answer this question, we need to take a closer look at the identity of the current neuroethics community.

17.3 Identity of the Neuroethics Community

Demarcating the identity of the neuroethics community is difficult for a variety of reasons. First, due to the interdisciplinary character of the neuroethics enterprise, the neuroethics community encompasses scholars and researchers from various fields of research, including, but not restricted to, neuroscience, psychology, cognitive science, philosophy, bioethics, law, psychiatry, clinical neuroscience and neurology, computer science, engineering, science and technology studies, and the social sciences. While one of the major goals of neuroethics is precisely to bridge the gaps between these disciplines and develop a comprehensive theoretical framework and a common language, interdisciplinary and intradisciplinary differences (as in the case of the analytic-continental divide in philosophy) have not been entirely integrated and might never be. Consequently, discipline-specific language barriers are likely to arise.

Second, the interdisciplinarity of neuroethics and the lack of consensus about basic universal norms and values challenge the ability of the field to provide indisputable moral guidance[1]. Neuroethics in its current stage of development lacks moral and conceptual consensus, and so it struggles to provide normative guidance at the policy level. Conflicting ethical frameworks and philosophical paradigms compete, so neuroethical issues are analyzed within different social contexts and according to particular assumptions about what constitutes the right action, appropriate decision, or proper interpretation of neuroscientific data. This lack of normativity requires a careful examination of the significance of universal moral norms with regard to moral content. Neuroethics as a form of biopolitics inherently conveys political (and moral) understandings concerning how neuroscientific research should be implemented, the meaning of neuroscientific discovery, and ultimately how such research determines human ends and goals.

In addition, while there might be a consensus among neuroethics researchers that some cross-disciplinary topics eminently belong to the neuroethics domain (e.g., neuroenhancement, free will, the use of brain imaging in criminal justice, the ethics of brain interventions, emerging neurotechnology), other relevant topics are not always immediately linked to neuroethics—and only have a marginal place in either the neuroethics literature or in neuroethics conference programs. These topics include, among many others, the ethics of biologically inspired computing and artificial intelligence, big data issues arising in neuroscience research and medical data bases, philosophical issues in moral psychology and philosophy of mind (e.g., moral sensitivity, internalism vs. externalism, agency), the politics of neuromarketing, the ethical use of assistive robotics in neurological care, healthy aging and neurogeriatric care in the aging world, the problem of animal cognition and animal rights, action theory, critical issues in evolutionary psychology, etc. The relevance of these topics for neuroethical discourse is undeniable, yet their "membership" in neuroethics as an area of inquiry is unclear, given their low prevalence in the literature and the lack of unambiguous demarcation of neuroethics as a discipline.

[1] Of course, this challenge is not unique to neuroethics but applies to the entire applied ethics domain. For example, similar problems have been addressed by scholars within the fields of bioethics and medical ethics (Hellsten 2008; Tangwa 2004). In addition, the very idea of moral universalism is often disputed in moral philosophy by proponents of moral relativism. While taking a stance in the debate on moral universalism vs. moral relativism is beyond the scope of this chapter, it suffices to consider here that if universal moral norms were true, the lack of consensus about them within neuroethics would undermine its ability to provide indisputable moral guidance.

Leefmann et al. conducted a systematic literature review to investigate the quantitative development of scientific publications in neuroethics from 1995 to 2012 (Leefmann et al. 2016). Their results showed that "the academic backgrounds and affiliations of many neuroethics researchers speak for a very close entanglement of neuroscience and neuroethics," confirming the narrow focus of neuroethics discourse. Most importantly, the authors observe that "after more than one decade there still is no dominant agenda for the future of neuroethics research," the reason for which they call for more reflection on "the theoretical underpinnings and prospects to establish neuroethics as a marked-off research field distinct from neuroscience and the diverse branches of bioethics" (Leefmann et al. 2016).[2] This uncertainty about whether gray zones in the neuroscience-ethics entanglement should be legitimately considered components of neuroethics might negatively affect the identity of researchers working in these areas as members of the neuroethics community.

The fourth difficulty arising when demarcating the neuroethics community and its identity is the problem of primary and secondary identity. It is uncertain, in fact, whether neuroethics researchers perceive the neuroethics community as their primary academic identity or whether they primarily consider themselves to be neuroscientists, philosophers, lawyers, bioethicists, psychiatrists, psychologists, engineers, social scientists, etc.—who have neuroethics among their secondary academic interests and areas of expertise. This academic "identity problem" is relatively common in the early stages of development of new disciplines. By analogy, it might be observed that in the early years of other multidisciplinary endeavors such as bioethics and cognitive science, researchers working in these fields were more inclined to describe themselves according to their discipline-specific academic background rather than as, respectively, "bioethicists" and "cognitive scientists."

Simply stated, neuroethics is a new field of investigation that examines the implications of neuroscience research for ethics and policy. Neuroethics is distinct from bioethics in that it focuses on the knowledge gained from advances in neuroscience to address problems associated with neurological disorders and brain research. Nevertheless, neuroethics and bioethics have a common feature: they are both "second-order" disciplines. Albert Jonsen, commenting on whether bioethics is a discipline, writes:

> A discipline is a coherent body of principles and methods appropriate to the analysis of some particular subject matter. Bioethics has no methodology, no master theory … bioethics might be called a "demi-discipline". Only half of bioethics counts as an ordinary academic discipline… the other half of bioethics is the public discourse … (Jonsen 1998: 345–346).

The dual dimension of the identity of bioethics also characterizes neuroethics. Neuroethics has the features of an ordinary academic discipline but also engages in public discourse on a variety of topics that include issues such as the use of

[2] The point here is not to discuss the desirability of a dominant neuroethics agenda. For further clarification on this issue, see sections 17.5 and 17.6 where we suggest a pluralistic approach: a procedural ethics grounded on a deliberative democracy framework.

neuroscientific evidence in the court, psychiatric disorders and social violence, and the nature of moral decisions. In addition, a second-order discipline attracts people who can address problems collaboratively, encouraging the inclusion of perspectives from various professions, disciplines, and academic fields (Kopelman 2009). This is particularly true in neuroethics. The type of ethical and policy issues arising from neuroscience research require a broad range of expertise that no one can display in its entirety. Finally, members of a second-order discipline set standards of competency and pedagogical criteria specific to their discipline of origin. This means that no one can claim to be an expert in neuroethical issues but rather, someone with a particular expertise who participates in a broader discussion beyond one's own discipline. That said, it is not infrequent to see scholars referring to themselves as or being called bioethicists or cognitive scientists. Consequently, it is reasonable to predict that neuroethics may follow a similar trajectory: in the near future, neuroethics researchers might primarily consider themselves and be considered by others to be "neuroethicists," perceiving the neuroethics community as their primary academic identity.

Fifth, the neuroethics community faces the problem of determining what type of regime of authority it is capable of and entitled to exert over the knowledge and technology produced by neuroscience. In fact, most neuroethicists would agree that neuroethics is committed to responsible development in neuroscience and neurotechnology, and it must have an important advisory function on guaranteeing the ethical sustainability of neuroscience research and neurotechnology development and use.[3] However, no unanimous consensus seems to have been reached in the community, nor among legislators and institutional bodies, about how the neuroethics community should enact its role as a guarantor.

Finally, in even more subtle biopolitical fashion, the neuroethics community is confronted with the problem of recognizing that the knowledge and technologies produced by advancing neuroscience and neurotechnology might be interpreted as "biopolitical technologies of governance," which could potentially be used to exert control and power over neural processes and ultimately human behavior. Since neural processes are essential constituents of what Foucault calls "the processes of life" (Lemm and Vatter 2014), neuroethics has the potential to both represent a critical instance of biopower and be one of the guarantors and guardians of neural processes from the expanding political control and power over them.

Two examples may be useful to better appreciate this problem. First, as several off-the-counter neurotechnologies enable the collection, aggregation, storage, mining, and reuse of users' data, they could be used to monitor and exert control over human social and biological processes. Even more strikingly, several emerging neurotechnologies, such as brain-computer interfaces and neurostimulators, have dual-use potential and are currently being used for both civil (e.g., clinical or commercial)

[3] For example, the International Neuroethics Society (INS) describes its mission as "to promote the development and responsible application of neuroscience through interdisciplinary and international research, education, outreach and public engagement for the benefit of people of all nations, ethnicities, and cultures." See: http://www.neuroethicssociety.org/history.

and military applications (Miranda et al. 2015). However, neuroethics research shows that awareness of such dual-use phenomena is reportedly low among researchers and developers (Tennison and Moreno 2012).

Since military applications of neurotechnology are paradigmatic examples of biopolitical technologies of governance that can be used to oversee human social and biological processes, the neuroethics community faces the critical problem of developing viable strategies for acting as a guardian and mediator of such transferal of sovereign power into the neural (and mental) dimension in democratic societies.[4]

17.4 Building the Global Neuroethics Community

Despite the challenges outlined above in both building the neuroethics community and developing the field into a coherent discipline, many positive developments have taken place in the last decade. An important step in the building of the neuroethics community was the establishment of the International Neuroethics Society (INS), a professional international organization gathering over 370 international researchers and students working on the social, legal, ethical, and policy implications of advances in neuroscience. Since its foundation in 2006, the INS has been a chief aggregator of neuroethics researchers and a catalyzer of intellectual exchange in neuroethics scholarship. In addition, the organization promotes the public understanding of the issues raised by neuroscience research and the powerful new tools being developed in the field, such as advanced brain imaging and neurostimulation. In parallel with the INS, a number of national neuroethics organizations have been established, such as the Società Italiana di Neuroetica (SINe) in Italy, the Asociación Mexicana de Nueroética (AMNE) in Mexico, and the Swiss Network for Neuroscience, Ethics and Law (SNNEL).

While the integration and harmonization of national or local neuroethics organizations into an international one would be needed for the creation of a global neuroethics community, it is still uncertain how such different perspectives might be integrated into a more comprehensive framework. Different world regions and different countries have specific public health needs, medical infrastructures, research management systems, research funding priorities, science policies, and even different cultural, philosophical, and socioeconomic traditions. Finding unity in such diversity is a major challenge, not only in theory but also in the practical development of a potential global agenda for the future of neuroethics research. In fact, while some critical issues in contemporary neuroethics might have high significance in certain world regions, they might not thereby meet real socioeconomic challenges in other areas of the globe.

[4]The metaphor of the "guardian and mediator" should not be interpreted as a constituted body or force, like an Orwellian "ethics police," but as a system of knowledge and expertise with a critical advisory function.

For example, while the commercialization of direct-to-consumer neurodevices is rapidly emerging as a critical neuroethical issue in Western countries—where such devices are increasingly available as commercialized products—such debate is of much more marginal relevance in developing countries plagued by widespread poverty and communicable disease morbidity. Consequently, different areas of the world will require different neuroethics agendas, as well as political and moral frameworks based on the contextual priorities faced by the populations living in such areas. In addition, some regions of the world might experience unique problems as a consequence of geographically confined mental health emergencies, policies, or sociotechnological trends. For example, the need for policy interventions and ethical responses to the microcephaly-causing Zika virus (ZIKV) is extremely urgent in countries and territories with active ZIKV transmission (e.g., Central and South America) (Johansson et al. 2016). Even more narrowly, neuroethical issues associated with increased rates of chronic traumatic encephalopathy (CTE) and brain or cervical spinal injuries among former American football (Bachynski and Goldberg 2014; McNamee 2014; Rihn et al. 2009) or ice hockey players (Bachynski and Goldberg 2014; Caron and Bloom 2015) are of great relevance in the countries where such sports are professionally played, but are less important topics of debate in those countries where such sports are absent. Therefore, in order to build a truly international and global neuroethics community, there is an unavoidable need for pluralistic and more indigenous approaches that are capable of generating unity among researchers without thereby obliterating regional diversity. This form of pluralism will be further elaborated in the concluding section.

17.5 Creating Spaces of Neuro- and Biopolitical Scrutiny

The coexistence of different and sometimes partly conflicting agendas in international neuroethics research is further challenged by political decisions and governmental involvement in the research enterprise. Public private initiatives in neuroscience and related areas of research are often elicited, promoted, and partly financed by political institutions. For example, back in 1982, Japan's Ministry of International Trade and Industry had launched the *Fifth Generation Computer Systems* (FGCS), an initiative aimed at creating a computer using massively parallel computing, which provided—in spite of many limitations—a critical platform for future developments in artificial intelligence. Five years later, in 1987, the Reagan Administration submitted to Congress the budget of the Human Genome Project (HGP), which was later formally founded by the US Department of Energy and the National Institutes of Health and which resulted in unprecedented advancements in neurogenetics. More recently, the BRAIN Initiative (*Brain Research through Advancing Innovative Neurotechnologies*) was launched and promoted by the US White House during the second Obama administration on April 2, 2013. Around the same time, the Human Brain Project (HBP) started on October 1, 2013 under the European Commission Future and Emerging Technologies Flagship and was promoted and funded by the European Union.

While such politically supported public-private initiatives are vital for generating disruptive and long-lasting innovation in neuroscience and related fields, it is undeniable that they represent realizations of the specific agendas of their institutional promoters and funders. As such, they represent, in Foucauldian terms, various manifestations of how political power is exercised on whole populations in the specific aspects of human life associated with and investigated by brain research (Foucault 2007). These manifestations might not necessarily be morally problematic per se, but it is critical for a mature neuroethics that such biopolitical aspects are not obliterated but extensively assessed and reflected upon. As big neuroscience projects grow in number and variety, neuroethics researchers have an additional duty toward scrutinizing the socioeconomic motivations and political goals underlying the creation and promotion of such programs at the governmental level to defend transparency and, as much as possible, the respect for state neutrality—unless national security is at stake.

This biopolitical dimension becomes even more explicit in relation to neuroscience research for military purposes. National security organizations in the United States, China, Russia, and the European Union, including the armed services and the intelligence community, have developed a strong interest and a close relationship with neuroscience and neurotechnology. As reported by Tennison and Moreno, "current state-of-the-art neuroscience, including new forms of brain scanning, brain–computer interfaces (BCIs), and neuromodulation, is being tapped for warfighter enhancement, deception detection, and other cutting-edge military applications to serve national security interests" (Tennison and Moreno 2012). In particular, the US Defense Advanced Research Projects Agency (DARPA) is currently funding research and technology development in the field of BCI (Miranda et al. 2015).

While holding a great potential for sparking technological innovation, such military applications of neuroscience are not value-neutral but instrumental to a specific biopolitical and military agenda. This agenda involves not only the "weaponization of neurotechnology" (Noll 2014) but also the weaponization of the brain itself as this becomes the next battlespace of international conflict. Such militarization falls into a chasm in neuroethics and international law and adds an additional level of control of governments and national security bodies onto individual citizens. However, awareness of this chasm among the neuroethics community appears still relatively low. For instance, the 200-page report on the ethical implications of the BRAIN Initiative, released in full in March 2015, did not include any mention of the terms "dual use" or "weaponization" nor explicit references to neurosecurity issues, hence omitting the entire biopolitical problem of brain control in the light of national security from its analysis.

17.6 Creating Spaces of Neuro- and Biopolitical Diversity

The creation and expansion of a truly international neuroethics community is one of the major structural and organizational challenges ahead for researchers working on the ethical, legal, and social implications of neuroscience. While the proliferation of national neuroethics societies and, most importantly, the birth of the INS represent milestones in the academic establishment of this discipline, it remains an open challenge to determine what type of discourse should shape the community. As we have argued in the previous section, the neuroethics community is still facing the problem of bridging interdisciplinary and intradisciplinary differences between the various academic backgrounds of neuroethics researchers, as well as the problem of determining what type of regime of authority should be exerted over the knowledge it produces. In addition, different countries or global areas may face different public health priorities related to neuroscience and mental health, different political agendas of funding institutions, different religious or sociocultural influences, etc. In response to this context, we argue that it is critical for the neuroethics community to recognize and consequently assess the biopolitical significance of the knowledge and technology produced in neuroscience research by creating spaces of neuro- and biopolitical scrutiny. As neuroscience and, therefore, neuroethics, "are inherently political" (Henry and Plemmons 2012, p. 574), biopolitics is integral to neuroethics. This implies that the ethical evaluation of advances in neuroscience research and their applications in medicine and society at large must always recognize their biopolitical significance and anticipate potential impositions of political and religious ideologies or surveillance and national security strategies for the control of populations.

Concomitantly, in light of such geographical, organizational, political, and sociocultural diversity, there is a need for the creation of spaces of neuro- and biopolitical diversity. The creation and protection of such open spaces is crucial to prevent the establishment of a monolithic narrative about how to address neuroethical issues which may be instrumental to a specific political, sociocultural, or religious agenda. Requirements for the creation of such spaces are (1) an evidence-based approach to the discovery of the facts that are relevant and that inform the ethical analysis—that is, based on the underlying neuroscience and neuroengineering research findings—and (2) the acceptance of some form of political pluralism within the neuroethics community. In fact, a solid foundation based on evidence is a critical protective measure to prevent that neuroethics discourse might, in the long run or in certain countries, be hijacked by contingent political, military, or religious agendas. As Charles Darwin famously observed: "false facts are highly injurious to the progress of science, for they often long endure; but false views, if supported by some evidence, do little harm, as everyone takes a salutary pleasure in proving their falseness; and when this is done, one path towards error is closed and the road to truth is often at the same time opened" (Darwin 1888).

Concurrently, at the normative level, there is a need for protecting political pluralism in the emerging international neuroethics community. This should start with the observation that there are different value systems in the world as a consequence of different contingent facts and local political agendas, and there are various biopolitical positions that arise out of that observation. Therefore, the protection of political pluralism in the neuroethics community is crucial to protect people's freedom from restrictions that governments, churches, and national security agencies can put on them based on their value systems and political agendas. It is worth noting that this form of liberal pluralism does not logically imply a radically relativist version of political pluralism according to which all value systems are equally true or morally acceptable. Rather, it implies that different value systems ought to be tolerated. The reason for that stems from the fact that a pluralism at the level of value systems or viewpoints does not necessarily imply an equal pluralism at the level of values themselves.

A possible approach in the light of political pluralism (which we only outline briefly here due to the limited scope of our chapter) for the assessment of emerging neurotechnologies and guidance of policy analysis and public engagement has been developed by Gutmann and Thompson in their work on deliberative democracy. According to deliberative democracy theorists, our ability to reach consensus on a foundational level is limited by political, ethical, and cultural pluralism (Gutmann and Thompson 2000). Therefore, to adjudicate disagreement concerning the implications of (neuro)science and (neuro)technology advances, political and moral discourse in the public square should take the form of a proceduralist ethics—an ethics grounded on the individual as the source of authority and permission as the safeguard for the respect of such individual authority (Engelhardt 1996). This approach acknowledges the confines of social collaboration without necessarily ruling out the possibility, through a political and discursive process, to reach a managed agreement—i.e., conditional proceduralism (Gutmann and Thompson 2000). Furthermore, it considers the moral, political and economic conjectures that constrain and shape the goals, methods, and hypotheses of neuroscience and neurotechnology. Since these assumptions determine the justifiability of neuroscientific and neurotechnological progress, they need to be integrated proactively in the overall assessment of potential scientific controversies associated with neuroscience (Estlund 1997).

Concluding Remarks

As we have outlined in this chapter, the identity of the neuroethics community is conditional upon the resolution of five problems inherent in demarcating the field of neuroethics as an academic discipline (see Table 17.1). First, the identity of the neuroethics community is conditional upon the possibility of bridging the interdisciplinary and intradisciplinary differences between the various academic backgrounds of neuroethics researchers and breaking down language barriers among them. Second, such an identity can be strengthened by demarcating the field of neuroethics in an inclusive manner, without repudiating researchers who are working in critical areas of the neuroscience-ethics entanglement on topics and issues that are not mainstream in the neuroethics literature. Third, the transient problem of primary and secondary identities within the neuroethics community is likely to be overcome in the near future as a consequence of the growing maturity of the neuroethics community and the progressive establishment of neuroethics as a consistent academic discipline. This process can be accelerated by developing a straightforward agenda for the future of neuroethics research and through the creation of associative platforms for the coordination of neuroethics research and the building of the neuroethics community. Ultimately, the neuroethics community faces the intimately related problems of determining what type of regime of authority it could and should exert over the knowledge

Table 17.1 Challenges in the Identity of Neuroethics

Problems	Obstacles
Identity of the neuroethics community	Interdisciplinary and intradisciplinary differences between the various academic backgrounds
Demarcating the field of neuroethics	Non-inclusive approaches to thematic choices and uncertain relations to other fields of investigation
Primary and secondary identity	Identity conflict with preexisting academic affiliations
Type of regime of authority	Different normative and prescriptive stances
	Different regulatory approaches
Recognizing the biopolitical significance	Lack of self-awareness among researchers

and technology produced by neuroscience and of recognizing the biopolitical significance of such knowledge and technology.

Despite these challenges, in our analysis, we also highlighted many positive developments that occurred in the last decade in building the neuroethics community and growing the field into a coherent discipline, the most important of which being the establishment of the INS in 2006. However, a truly international perspective necessitates the establishment of a pluralistic approach that is capable of generating unity among researchers without thereby obliterating regional, cultural, ethical, and political diversity. To this end, we have suggested that a deliberative democracy framework is a fruitful approach to address the challenge of political pluralism. By this notion, we describe a deliberative model in which pluralistic views on the ethical dimensions of neuroscience and neurotechnology are coordinated within a framework of proceduralist ethics.

In addition, we have argued that it is critical for the neuroethics community to recognize and consequently assess the biopolitical significance of the brain-society entanglement. Creating spaces of neuro- and biopolitical scrutiny allows for the ethical evaluation of advances in neuroscience research and their applications in medicine and society and incorporates a process of critical anticipation of potential impositions of political ideologies or surveillance and national security strategies to control populations. Such spaces are critical venues for scrutiny in the emerging future of neuropolitics and neurowarfare. As neuroscience and neurotechnology rapidly become of primary significance for governments, national security agencies, and armed forces, neuroethicists have an additional duty toward the investigation of the socioeconomic motivations, political goals, and dual uses underlying the creation and promotion of such programs to defend transparency and, as much as possible, the respect for state neutrality—unless national security is at stake.

Recognizing the inherent biopolitical dimension of neuroethics is an important step toward the responsible conduct of research in neuroscience and its applications in medicine and society for the reasons outlined in this chapter. That said, in order to ensure that the field establishes itself as a resource for ethical analysis, the neuroethics community has a responsibility to reject any monolithic discourse that does not promote a plurality of perspectives and does not encourage critical assessment.

References

Bachynski KE, Goldberg DS (2014) Youth sports & public health: framing risks of mild traumatic brain injury in american football and ice hockey. J Law Med Ethics 42(3):323–333

Caron JG, Bloom GA (2015) Ethical issues surrounding concussions and player safety in professional ice hockey. Neuroethics 8(1):5–13

Chandler JA, Dodek AM (2016) Cognitive enhancement in the courtroom. Cognitive enhancement: ethical and policy implications in international perspectives, 329

Darwin C (1888) The descent of man, and selection in relation to sex, vol 1. Murray, London

Engelhardt HT (1996) The foundations of bioethics. Oxford University Press, New York

Estlund D (1997) Beyond fairness and deliberation: the epistemic dimension of democratic authority. Deliberative democracy: essays on reason and politics, 173

Foucault M (1980) An introduction: the history of sexuality, vol 1. Hurley R (trans). Vintage, New York

Foucault M (2000) The birth of social medicine. Power 3:1954–1984

Foucault M (2007) Security, territory, population. Springer, New York

Gunst D (1978) Biopolitik zwischen Macht und Recht. Mainz: v. Hase und Köhler Verlag.

Gutmann A, Thompson D (2000) Why deliberative democracy is different. Social Philos Policy 17(01):161–180

Hellsten SK (2008) Global bioethics: utopia or reality? Dev World Bioeth 8(2):70–81

Henry S, Plemmons D (2012) Neuroscience, neuropolitics and neuroethics: the complex case of crime, deception and fMRI. Sci Eng Ethics 18(3):573–591

Johansson MA, Mier-y-Teran-Romero L, Reefhuis J, Gilboa SM, Hills SL (2016) Zika and the risk of microcephaly. N Engl J Med 375(1):1–4

Jonsen AR (1998) The birth of bioethics. Oxford University Press, New York

Jotterand F (forthcoming) Cognitive enhancement of today may be the normal of tomorrow. In: Illes J, Hossain S (eds) Neuroethics: defining the issues in theory, practice and policy, 2nd edn. Oxford University Press, Oxford

Jotterand F, Giordano J (2015) Real-time functional magnetic resonance imaging–brain-computer interfacing in the assessment and treatment of psychopathy: potential and challenges. In: Handbook of neuroethics. Springer, pp 763–781

Kopelman LM (2009) Bioethics as public discourse and second-order discipline. J Med Philos 4:261–273

Leefmann J, Levallois C, Hildt E (2016) Neuroethics 1995–2012. A bibliometric analysis of the guiding themes of an emerging research field. Front Hum Neurosci 10:336

Lemke T (2011) Biopolitics: an advanced introduction. New York University Press, New York

Lemm V, Vatter M (2014) The government of life: Foucault, biopolitics, and neoliberalism. Oxford University Press, Oxford

McNamee M (2014) Professional football, concussion, and the obligation to protect head injured players. Sport Ethics Philos 8(2):113–115

Miranda RA, Casebeer WD, Hein AM, Judy JW, Krotkov EP, Laabs TL et al (2015) DARPA-funded efforts in the development of novel brain–computer interface technologies. J Neurosci Methods 244:52–67

Noll G (2014) Weaponising neurotechnology: international humanitarian law and the loss of language. Lond Rev Int Law 2(2):201–231

Persson I, Savulescu J (2008) The perils of cognitive enhancement and the urgent imperative to enhance the moral character of humanity. J Appl Philos 25(3):162–177

Persson I, Savulescu J (2012) Unfit for the future: The need for moral enhancement. OUP, Oxford

Persson I, Savulescu J (2013) Getting moral enhancement right: the desirability of moral bioenhancement. Bioethics 27(3):124–131

Rihn JA, Anderson DT, Lamb K, Deluca PF, Bata A, Marchetto PA et al (2009) Cervical spine injuries in American football. Sports Med 39(9):697–708

Rose N, Rabinow P (2003) Thoughts on the concept of biopower today. http://www.molsci.org/research/publications_pdf/Rose_Rabinow_Biopower _Today. Pdf. Accessed 1 June 2007

Tangwa GB (2004) Between universalism and relativism: a conceptual exploration of problems in formulating and applying international biomedical ethical guidelines. J Med Ethics 30(1):63–67

Tennison MN, Moreno JD (2012) Neuroscience, ethics, and national security: the state of the art. PLoS Biol 10(3):e1001289

Van Den Daele W (2005) Einleitung: Soziologische Aufklärung zur Biopolitik *Biopolitik*. Springer, pp 7–41

Wartofsky MW (1992) Technology, power, and truth: political and epistemological reflections on the fourth revolution. In: Democracy in a technological society. Springer, pp 15–34

Ziman J (2002) Real science: what it is and what it means. Cambridge University Press, Cambridge

Conclusion

We hope that readers have enjoyed the contributions featured in this book. As editors, we are acutely aware of the need to pursue public discussions and scholarship beyond this volume about the ethical challenges generated by developments in neuroscience, as well as its more philosophical and theoretical implications. Indeed, a number of scientific advances, technological achievements, and important challenges associated with neurological and psychiatric care motivate such an effort. That said, it is important to reflect not only on the responses we offer to such challenges but also on their historical underpinnings, the theories and methodologies that support them, and the training and research programs needed to prepare future generations of scholars to meaningfully tackle them. As we saw the history and development of neuroethics is not monolithic, bringing forth different visions about neuroethics and its identity. It will be crucial for scholars involved in neuroethics to reflect further on what might make neuroethics distinct or else what makes it a specific endeavor in the vast constellation of existing fields and disciplines. Furthermore, the actual value and drawbacks of emphasizing this specificity need to be critically and thoughtfully considered. In terms of theories and methods, neuroethics has much work to do with regard to defining whether it will offer new theories and methods and/or borrow and refine existing ones. Moreover, there is a need to initiate dialogues about what its research objects (foci) should be. Finally, the field has grown significantly, as attested to by both the continued creation of new programs and the sustainability of older ones (notably in North America); however, there likely needs to be more attention paid to how such models can inspire (or fail to inspire) the development of different research and training programs in the context of different national priorities.

Debates about neuroethics is the first volume to explicitly feature debates about the development, focus, and future of neuroethics. It is also one of the first volumes to provide reflections on the strengths and weaknesses of various aspects of neuroethics research and discourse (e.g., methodology, theories, training programs, networks). We hope that the conversation about these matters continues, that it gains in depth and diversity, and that the kind of self-reflection that neuroethics encourages in neuroscience research, clinical care, and public policy also becomes a central feature of neuroethics itself. Any gaps in the content of this volume should be understood as an invitation and an opportunity for readers to usher the field forward.

E. Racine, J. Aspler (eds.), *Debates About Neuroethics*, Advances in Neuroethics,
DOI 10.1007/978-3-319-54651-3

MIX
Papier aus verantwortungsvollen Quellen
Paper from responsible sources
FSC® C105338

If you have any concerns about our products,
you can contact us on
ProductSafety@springernature.com

In case Publisher is established outside the EU,
the EU authorized representative is:
Springer Nature Customer Service Center GmbH
Europaplatz 3, 69115 Heidelberg, Germany

Printed by Libri Plureos GmbH
in Hamburg, Germany